History of Science in the United States

GARLAND REFERENCE LIBRARY OF THE HUMANITIES
VOLUME 1711

History of Science in the United States
A Chronology and Research Guide

Clark A. Elliott

Garland Publishing, Inc.
New York and London
1996

Copyright © 1996 by Clark A. Elliott
All rights reserved

Library of Congress Cataloging-in-Publication Data

Elliott, Clark A.
  History of science in the United States : a chronology and research guide / Clark A. Elliott.
      p.   cm. — (Garland reference library of the humanities ; v. 1711)
  Includes bibliographical references and index.
  ISBN 0-8153-1309-8 (alk. paper)
    1. Science—United States—History—Chronology. 2. Science—United States—History—Handbooks, manuals, etc.   I. Title. II. Series.
Q125.E45   1996
509.73—dc20   95-43891
                                                                            CIP

Printed on acid-free, 250-year-life paper
Manufactured in the United States of America

Contents

Preface	vii
1. Chronology	3
2. Chronology: Scientist Cohorts by Decade Reaching Age 25	351
3. Research Guide	445
4. Research Guide: Bibliography	483
Index: Chronology and Scientist Cohorts	515

Preface

Purpose

The intention of this volume is two-fold: first, to give a chronologically arranged overview of selected data on the history of science in the United States, and second, to orient the reader to the substantial reference literature and research sources as guidance to further study of the topic. The subject areas that are covered include astronomy, biology, chemistry, geology, mathematics, physics, and their related disciplines; areas such as anthropology and psychology are covered to a lesser extent. Medicine, engineering, and technology are included, for the most part, only as they relate to, or shade into, the history of science. Science is the central focus, but the content of the work recognizes that the boundaries between subjects or activities are not absolute and certainly not when coverage spans several centuries.

Chronology

The major part of this volume is the chronology, given in two sections, first the primary array of general data, and then a supplementary listing of scientist cohorts (by the decade in which they reached age twenty-five). The history of science has many aspects, and reference works can be constructed in different ways, as the research guide that follows the chronology vividly demonstrates. Survey histories will follow the unfolding of history over time in a connected and interpretive way. A chronology such as this one covers the same temporal terrain through the vehicle of individual events. It is, in the first instance, a means of arranging and accessing information. But it also can be consulted for extended historical periods and even read from beginning to end.

The chronology covers the period from the earliest colonial times through 1990. It is limited to the history of science as delineated above and presupposes that users either have general knowledge of American history or will consult other reference or secondary works for background information on the political, social, and intellectual context within which the history of science in the United States has developed.

Retrospective chronology (as opposed to one that might be maintained while the events themselves unfold, as in a diary) is bound to be derivative. This one is no exception, as the note on sources that accompanies the chronology indicates. But *the* chronology of the history of science in the United States is an unlikely product, given the inevitable selectivity that must be done in the execution. This one tries to incorporate a very broad definition of history of science, encompassing a number of aspects of the subject while remaining manageable in size. There is, of course, no end to the detail that a chronology might include and by no means are all of the entries here of equal significance. Both by their singularity, and by their representativeness, they form a historical structure and suggest a multitude of stories. Individual lives are a crucial part of the history of science and that facet is reflected especially by drawing on an analysis of American topics in the volumes of the *Dictionary of Scientific Biography*. Science as viewed in its national context is more than lives, however, and the chronology has tapped secondary works and other repositories of information to give a varied content. When publications are listed, reasonably complete information ordinarily is given, so that the chronology also serves as a selective historical bibliography.

Each entry in the chronology gives the date (with varying degrees of specificity), a label that categorizes the entry to aid in scanning the listings, and the historical data. The topical headings assigned to the entries are based on a standardized list generated for this reference work; they are intended, overall, to reflect both scientific or disciplinary content, and organizational, institutional, and other social aspects that are so important to a full understanding of science in America. Some entries inevitably arc labeled "General or Miscellaneous." (When *two* facets of an historical entry are represented in its label, they are separated by a right-leaning slash [/]. Two such facets can relate as major and subordinate, or as coordinate topics.) Among the types of information are the founding of societies and other institutions and deaths

Preface ix

of some leading scientists (the beginnings of selected careers are in the separate scientist cohort listings), issuance of important or representative publications, and beginnings of some periodicals, as well as scientific activities and discoveries, occasional characterizations of a time period or trend, activities of government and other agencies, occurrences that were "firsts" and thus prelude to similar events, singular happenings that were particular to the times, occasions for scientific recognition (chiefly conferral of the Nobel Prize).

Research Guide

The chronology is followed by an essay and a bibliography on reference works for the history of science in the United States. Readers are directed to the introductory pages of that part of this work for an indication of its content. However, it is important to point out that users will *not* find the secondary historical literature on the subject there. What they will find is guidance to such literature as well as direction to a number of other types of information sources, including a short section on electronic access.

Users

This work is intended for a wide array of users. High school students and adult readers will find the chronology easily accessible for fact-finding, browsing, or reading. Students and scholars in the history of science will find it equally helpful both in regard to individual entries and as a reminder of, or initial orientation to, important features of successive historical eras; it can help to reaffirm or to establish a sense of the temporal relations of historical events. The research guide likewise addresses the concerns of individuals who want to find additional data, who seek means of orientation to the secondary literature, or who contemplate more extensive research on a particular topic.

Acknowledgments

I am grateful to Kennie Lyman, formerly at Garland Publishing, with whose help and advice I was able to conceptualize and undertake this work, one for which I had long sensed a need. Only my wife, Priscilla, knows how many hours of effort separated that original formulation and the project's completion. Her love and her understanding in this and all things are heartily acknowledged.

History of Science in the United States

1

Chronology

ca.1493. GENERAL OR MISCELLANEOUS / NATURAL HISTORY
As early as this date, the exchange of plants and animals took place between Europe and the Americas.

1518. MEDICINE / DISEASE
An epidemic of smallpox among the native population of Hispaniola marked the arrival of the disease in the new world.

1526. NATURAL HISTORY
Gonzalo Ferdandez de Oviedo y Valdes (1478-1557) published his *De la natural hystoria de las India* (Natural History of the West Indies) (Toledo, Spain), of which important portions were available by midcentury to English readers. Oviedo's was the first systematic rendering of new world plants and animals, based on first-hand knowledge during his years as overseer of mines in Hispaniola.

1588. NATURAL HISTORY
Thomas Hariot (1560-1621), who with the painter John White accompanied Sir Walter Raleigh in the settlement of Roanoke, published *A Briefe and True Report of the New Found Land of Virginia...* (London). Richard Hakluyt arranged for an edition with White's paintings to be published in Frankfurt (1590) with a London edition in 1593. Hariot's work carried the earmarks of a vehicle for advancing the interests of the colony, conveying information on the geology and natural history of the region, including mineral resources and medical and other useful products. Hariot's *Report* is significant largely as the first English

publication giving a first-hand account of the objects of natural history and their uses from the New World.

1590. NATURAL HISTORY
Historia Natural y Moral de las Indias by Jose Acosta (1539-1600) was published in four volumes (Seville), with an English translation in 1604, as *The Naturall and Morall History of the East and West Indies...* (London). Acosta, a Jesuit, was for almost twenty years situated in Peru (1569-1588) and in writing on natural history reports his observations, without recourse to special scientific knowledge or system.

1607. GENERAL OR MISCELLANEOUS / TECHNOLOGY AND INVENTION
Settlers at Jamestown, Virginia, produced glass objects, including beads that were traded with the Indians. The following year, the London Company sent glass workers to establish glass furnaces.

1610. BOTANY
Lawrence Bohune (?-1621), one of the first physicians in the English colonies, arrived in Virginia. He established there the first colonial botanical garden and experimented on the medicinal uses of native plants. The garden did not survive its founder.

1610. BOTANY
Samuel de Champlain (1567-1635) established a garden at Montreal, for experiments on European and American botanical specimens. He sent to France seeds for a great many plants, trees, and shrubs from the St. Lawrence Valley and northern New England. By the 1620s, the Paris garden contained substantial numbers of North American specimens.

1612. AGRICULTURE
In Virginia, John Rolfe (1585-1622) succeeded in planting and harvesting a crop of native tobacco. In 1614, he began to ship to England the crop that became the colony's chief export.

1612. GEOGRAPHY AND CARTOGRAPHY
Captain John Smith (1579/80-1631) published *A Map of Virginia with a Description of the Countrey...* (Oxford).

1616. GEOGRAPHY AND CARTOGRAPHY
Captain John Smith (1579/80-1631) published *A Description of New England...* (London), including a map.

1619. TECHNOLOGY AND INVENTION / ORGANIZATIONS—INDUSTRY
The first English colonial ironworks was established at Falling Creek, Virginia.

1626. TECHNOLOGY AND INVENTION / ORGANIZATIONS—INDUSTRY
The first flour mill in the colonies was constructed by the Dutch at New Amsterdam.

1634. GENERAL OR MISCELLANEOUS / NATURAL HISTORY
William Wood, who lived in Massachusetts from 1629-1633 (probably at Saugus [Lynn]), published *New England's Prospects, A True, Lively, and Experimentall Description of That Part of America, Commonly Called New England* (London), with accounts of the geography, flora, and fauna, and of the Indians. Though sound in its observations, it is not the work of a scientist but of a traveler.

1635. BOTANY
The publication of Paris physician Jacques Philippe Cornuti (1606-1651), *Canadensium Plantarum...* (Paris), drawing on the resources of the Paris garden, was the first scientific work on North American plants.

Late 1630s to early 1640s. GENERAL OR MISCELLANEOUS / TECHNOLOGY AND INVENTION
John Winthrop, Jr. (1606-1676) was involved in the promotion of salt and ironworks in the Massachusetts Bay Colony and was involved in the search for minerals of utilitarian value.

1637. GENERAL OR MISCELLANEOUS / NATURAL HISTORY
Thomas Morton (ca.1590-ca.1647) published his *New English Canaan or New Canaan Containing an Abstract of New England...* (Amsterdam). While containing comments on the natural world, it was a means to promote English colonization.

1640s. GENERAL OR MISCELLANEOUS / ZOOLOGY
Giles Firmin lectured on anatomy in Boston and in 1647 conducted an autopsy.

1646. GENERAL OR MISCELLANEOUS / MEDICINE, CHEMISTRY
George Starkey (originally, Stirk) (1628-1665) graduated from Harvard College and for a time before emigrating to England practiced medicine in Boston. In England, he published works on alchemy (some of them posthumously) and apparently also practiced medicine.

1660. ASTRONOMY / INSTRUMENTS AND INSTRUMENTATION
As of this date, John Winthrop, Jr. (1606-1676) was using a ten-foot refractor telescope, most likely the largest in the colonies up to that time.

[1660]. GENERAL OR MISCELLANEOUS
Before 1660, the date of the Stuart Restoration, the English settlers in America showed little interest in natural science as such. What was produced reflected a popular rather than scientific view with overtones of promotion of the colonies rather than of science.

1661 (December 18). ORGANIZATIONS—SOCIETIES AND ASSOCIATIONS
John Winthrop, Jr. (1606-1676), governor of Connecticut, was made a charter member of the Royal Society of London. He was in England at the time, on business relating to the Connecticut charter, and was the first colonial member of the Society. During the early years, Winthrop tended to play the part of chief correspondent for the Society in that area of the world.

1662 (July 16). TECHNOLOGY AND INVENTION
John Winthrop, Jr. (1606-1676), in England at the time, gave his and the first of any colonist's paper to the Royal Society. Titled "A Description of ye Artifice and making of Tarr and Pitch in New England, and ye Materialle of wch it is made," it described the means by which tar was produced from pitch pine.

1663 to 1783. ORGANIZATIONS—SOCIETIES AND ASSOCIATIONS
During this period, fifty-three colonials in British North America were made fellows of the Royal Society, of whom twenty were honored especially because of their positions as governor. Of the non-governors,

the distribution, within the geographical area of the future United States, was ten in New England, three in the Middle colonies, and four in the South. (The remainder were in Canada and the West Indies.)

1669. GENERAL OR MISCELLANEOUS / NATURAL HISTORY
John Winthrop, Jr. (1606-1676), governor of Connecticut, made the first notable donation to the Royal Society of New World plants, shells, a fish, and mineral and ore products.

1672. NATURAL HISTORY / MEDICINE
John Josselyn (ca.1608-1675) published his *New England's Rarities Discovered: In Birds, Beasts, Fishes, Serpents, and Plants of that Country, Together with the Physical and Chyrurgical Remedies wherewith the Natives constantly use to Cure their Distempers, Wounds and Sores...* (London). Two years later, he published *An Account of two Voyages to New England* (London), with a second edition in 1675. Josselyn visited the colonies in 1638-1639 and 1663-1671, and produced the most comprehensive work on New England natural history, and especially the botanical aspects, during the colonial period. Throughout his work, Josselyn, a physician, shows a particular interest in medicinal and other uses of natural specimens and evidence of his own lack of specific training or familiarity with advances in the study of natural history in England at the time.

1676 (April 5). GENERAL OR MISCELLANEOUS
John Winthrop, Jr. (b.1606) died in Boston, Massachusetts. He was the earliest significant student of science in the British North American colonies.

1678. MEDICINE / DISEASE
Thomas Thacher (1620-1678), Boston minister published the broadside *A Brief Rule To Guide the Common-People of New-England How to Order Themselves and Theirs in the Small Pocks, or Measels* (Boston). It was the first medical work published in North America.

1678. NATURAL HISTORY
John Banister (1650-1692) went to Virginia where he undertook scientific studies of the area. Banister had the patronage of various persons in England, most of whom were members of the Royal Society.

He collected specimens and prepared manuscripts and drawings that went to England but published nothing on his own, dying at a young age. Banister was an informed and up-to-date naturalist and the results of his efforts found their way into the works of others (e.g., John Ray's *Historia Plantarum*, 1686-1704, included notices and a catalogue of plants gathered by Banister in Virginia). In consequence, his overall achievement for the natural history of the colonies exceeded that of any of his predecessors or immediate successors.

Late 1600s. GENERAL OR MISCELLANEOUS
During the final decades of the seventeenth century, a limited but lively interest in science grew up in New England, which included an acceptance of the work of Copernicus and Galileo and the methodology of science that incorporated observation and experiment. These developments were incorporated into the colonists' established theological views as a means of learning the ways and the laws of God and God's creation.

1680. ASTRONOMY
With a telescope donated to Harvard College by John Winthrop, Jr. (1606-1676) in the winter of 1671/72, Thomas Brattle (1658-1713) made observations of the comet of this year which later were referenced by Isaac Newton in his *Principia Mathematica*. The observations had been printed in the local almanac for 1681.

1683. ASTRONOMY
Kometographia, or A Discourse Concerning Comets, by Increase Mather (1639-1723), was published (Boston). Drawing on his own observations of Halley's Comet in 1682, and also related literature, Mather recognized comets as natural phenomena but also endorsed the idea that they were associated with calamitous occurrences in human history.

1683. ORGANIZATIONS—SOCIETIES AND ASSOCIATIONS
A Philosophical Society was established at Boston for purposes of philosophical experiment and natural history studies, probably the first colonial scientific society. Among the members was Increase Mather (1639-1723), its apparent head. The group ceased to meet after about 1688.

1688. NATURAL HISTORY
John Clayton (1657-1725), a man with scientific credentials who had resided as minister 1684-1686 in Virginia, began a series of presentations to the Royal Society on what he had seen and learned while in the colony. The communications from Clayton spanned the years 1688-1694 and most were published in the Society's *Philosophical Transactions*. They covered the climate, geography, flora and fauna (especially the birds), and natives of Virginia.

1696. GENERAL OR MISCELLANEOUS
William Byrd II (1674-1744), Virginia's most sustained contributor to the Royal Society during the first half of the eighteenth century, was elected a fellow of the Royal Society. Byrd's interests were eclectic and often of a practical nature but had the effect of helping to promote interest in the study of the new world.

1696. NATURAL HISTORY
Hugh Jones (?-1701/02) was sent to Maryland, as a member of the clergy but also as a naturalist successor to John Banister (1650-1692). Jones was sent with the support and with provision of equipment by participants in the London Temple Coffee House Botany Club. He supplied his sponsors with specimens and descriptive information until his death in 1702, the most significant portions going to James Petiver.

1698. NATURAL HISTORY
William Vernon (?-1706), with the endorsement of the Royal Society and incentive of payment of his expenses plus a fee from the governor of Maryland, went to that colony with the purpose of studying its natural history. He remained there only for a few months, but returned substantial collections. During approximately the same period, physician David Krieg (?-ca.1712) also was in Maryland, with the support of James Petiver, and returned with specimens and illustrations.

[1700]. GENERAL OR MISCELLANEOUS
There is little evidence of interest in science in colonial New York prior to the eighteenth century.

1701. NATURAL HISTORY
Hannah Williams, in South Carolina, sent the first of several shipments of natural history specimens (in this instance, butterflies) to James Petiver in England. After the initial receipt, Petiver sent collecting supplies and instructions.

1708. NATURAL PHILOSOPHY
During a trip to England, James Logan (1674-1751) purchased the first copy of Isaac Newton's *Principia* known to be in the American colonies.

1709. GENERAL OR MISCELLANEOUS / NATURAL HISTORY
John Lawson's *A New Voyage to Carolina* was published, and after his death in 1711 it was republished with the title *The History of Carolina: Containing the Exact Description and Natural History of the Country...* (London, 1714), a work that was strongest in its non-science aspects and characteristic as an account of travels. Lawson is distinguished as the single significant scientist in North Carolina during the colonial period. First arriving in the colony in 1700, in 1709 he was appointed its Surveyor General. He also served as a natural history collector for James Petiver.

1711. GENERAL OR MISCELLANEOUS
Cotton Mather (1662/63-1727/28) began the accumulation of information on American phenomena, with the intention of presenting it to the Royal Society. In November of 1712, he began a series of letters, transmitting information and observations on a wide range of natural, medical, and philosophical matters, for the Society. The series of communications over a dozen years took the title *Curiosa Americana*, and included Mather's own observations as well as reports derived from others.

1711. ORGANIZATIONS—ACADEMIC / NATURAL PHILOSOPHY
College of William and Mary appointed a person named LeFevre as professor of natural philosophy. He was dismissed the following year for alleged moral misconduct. His successor was Hugh Jones (ca.1692-1760).

1712. NATURAL HISTORY
Mark Catesby (1683-1749) first arrived in the British colonies, in Virginia, where he undertook natural history collecting, returning for a time to England ca.1719.

1716 (September 24). BOTANY / GENETICS
In a letter to James Petiver, Cotton Mather (1662/63-1727/28) gave what later [1935] came to be recognized as the first account of hybridization in plants, recounting events in the garden of a friend.

1717. GENERAL OR MISCELLANEOUS / ASTRONOMY
In this year Pennsylvania Governor William Keith (1680-1749) reported to the Royal Society on a solar eclipse at Philadelphia, the first communication from that colony to the Society since sporadic communications in the late seventeenth century.

1718. GENERAL OR MISCELLANEOUS / NATURAL HISTORY
James Petiver, English sponsor of several natural history collectors and observers in the colonies, died in London.

1719. GENERAL OR MISCELLANEOUS
Paul Dudley (1675-1751), Massachusetts jurist, sent his first communication to the Royal Society, on the making and values of maple syrup. Over the years he wrote on a variety of topics, including the height of Niagara Falls, ambergris, botanical items, and others, and when his contributions ended in 1736, they represented the longest series of any colonialist up to that time.

1720. CHEMISTRY
The earliest extant detailed chemical analysis by a resident of the colonies, Thomas Robie's (1688/89-1729) study of a piece of a Cambridge (Massachusetts) oak, was published as an "Account of a large Quantity of Alcalious Salt produced by burning rotten Wood" in the Royal Society's *Philosophical Transactions*, volume 31, no. 366 (September-December 1720): 121-24. Robie was assisted in this study by Samuel Danforth (1696-1777).

1721. GENERAL OR MISCELLANEOUS / NATURAL PHILOSOPHY
Cotton Mather (1662/63-1727/28) published *The Christian Philosopher: A Collection of the Best Discoveries in Nature, with Religious Improvements* (London). It incorporated the first extended exposition by a colonist of Isaac Newton's contributions in *Principia Mathematica*.

1721-1722. MEDICINE / DISEASE
During a smallpox epidemic in Boston, experiments in inoculation were initiated by Cotton Mather (1662/63-1727/28) and carried out by Zabdiel Boylston (1680-1766), to determine the efficacy of the procedure. Boylston was the first colonist to engage in such practice. A great public controversy surrounded the event, but the experiments and their results helped to promote the use of inoculation.

1722-1726. NATURAL HISTORY
Mark Catesby (1683-1749) returned to the British Colonies, sponsored by William Sherard, Sir Hans Sloane, and others, where he explored and accumulated specimens, notes, and illustrations in South Carolina and Florida (Georgia), going in 1725 to the Bahamas.

1723 (March). GENERAL OR MISCELLANEOUS / PERIODICALS AND PUBLISHING
In the issue for February 28-March 7, 1722/23, the *Boston News-Letter* proposed to print items relating to natural occurrences as well as politics and foreign affairs. The paper had, in fact, previously run selected items of agricultural and scientific interest.

1724. BOTANY / GENETICS
Paul Dudley (1675-1751) published a notable work on New England garden vegetables and fruit trees, a very early horticultural account. The paper, "Observations on Some of the Plants of New England, with Remarkable Instances of the Nature and Power of Vegetation," *Philosophical Transactions of Royal Society* 33 (October-December 1724): 194-200, also included comments on cross-fertilization of corn.

1725. PERIODICALS AND PUBLISHING / ASTRONOMY
Boston physician Nathaniel Ames (1708-1764) began publication of *Astronomical Diary and Almanac*; it continued until the period of the Revolution.

1726. MEDICINE / DISEASE
Zabdiel Boylston (1680-1766), who had joined Cotton Mather (1662/63-1727/28) in support of inoculation during the smallpox outbreak in 1721-1722, published *An Historical Account of the Small-Pox Inoculated in New England* (London). The work recounted his experiences and gave clinical and statistical accounts.

1727. GENERAL OR MISCELLANEOUS / GEOLOGY
The earthquake of this date evoked twenty-seven published responses, most of which were sermons.

1727. ORGANIZATIONS—ACADEMIC / MATHEMATICS, NATURAL PHILOSOPHY
London merchant Thomas Hollis's funds and apparatus contributed to Harvard College led to establishment of the Hollis professorship in mathematics and natural philosophy. Isaac Greenwood (1702-1745) was first to hold the professorial appointment, assuming the duties in February 1728. He was removed from the office in 1738 for intemperance. Prior to his appointment, in January 1727 Greenwood began the first course of public lectures on science in New England, having brought demonstration apparatus with him on his recent return from England. Greenwood described his program in *An Experimental Course in Mechanical Philosophy...* (Boston, 1726).

1727. ORGANIZATIONS—SOCIETIES AND ASSOCIATIONS
Benjamin Franklin's (1706-1790) Junto was formed in Philadelphia. A society for self-improvement and inquiry, it was concerned with literary, political, and scientific questions, but contentious debate or the seeking of victory in opinion was prohibited. The group may have lasted for a decade or so, but its history is uncertain.

1728. ORGANIZATIONS—SOCIETIES AND ASSOCIATIONS
In a letter to William Douglass (ca.1691-1752), Cadwallader Colden (1688-1776) suggested formation of a society for the promotion of knowledge. This was apparently the first such suggestion for the creation of an intercolonial scientific society. It was suggested by Colden that the Massachusetts colony assume the leadership for the proposed body.

1728. GENERAL OR MISCELLANEOUS / RELIGION AND THEOLOGY
After Cotton Mather (b.1662/63), who died in this year, the influence of the religious perspective in New England science began to fade considerably and was replaced by the mechanistic world view of the Enlightenment.

1729. MATHEMATICS
The first American publication on mathematics appeared, Isaac Greenwood's (1702-1745) *Arithmetick Vulgar and Decimal: With the Application Thereof to a Variety of Cases in Trade and Commerce* (Boston). It was prepared for use in Harvard College.

1729-1747. NATURAL HISTORY
The two volumes of Mark Catesby's (1683-1749) *The Natural History of Carolina, Florida, and the Bahama Islands* (London) was published in parts (1731-1743, 1729-1747). The completed work contained over 200 colored illustrations by Catesby, who also wrote the text. His work is most notable for its ornithological content.

1730s and 1740s. GENERAL OR MISCELLANEOUS / NATURAL HISTORY
With the aid of persons in England, such as the London Quaker merchant Peter Collinson and Mark Catesby (1683-1749), an interchange among colonial scientists began to develop. From these developments also came an integration of colonialists in the international effort to study natural history.

1730 (October). ASTRONOMY / INSTRUMENTS AND INSTRUMENTATION
Thomas Godfrey (1704-1749), of Philadelphia, produced a double reflecting quadrant, used for measurement of elevation and an aid to determination of latitude in navigation. Coming to be known as Hadley's quadrant, for John Hadley, independent inventor of the instrument in England, it rapidly came into widespread use. James Logan (1674-1751) promoted Godfrey's claims in "An Account of Mr. T. Godfrey's Improvement of Davis's Quadrant," *Philosophical Transactions of Royal Society* 38 (1733-1734): 441-450.

1730. BOTANY
As of this date, John Bartram (1699-1777) had established a modest garden on his farm at Kingsessing, Pennsylvania, where he cultivated

American plants from various locations. Around 1734, Bartram began to provide specimens and seeds to Peter Collinson in London. In turn, Collinson assisted Bartram to make contacts with other potential patrons and purchasers, in the identification of specimens, and by sending Bartram seeds and plants not native to America.

1731. ORGANIZATIONS—LIBRARIES
The Library Company of Philadelphia, founded in this year, soon had the best scientific book collection in the colonies. The group also collected natural history specimens and scientific apparatus.

[1733]. GENERAL OR MISCELLANEOUS / PERIODICALS AND PUBLISHING
After this time, publication of almanacs and newspapers, with a certain level of scientific content, spread beyond their earlier base in New England.

1734. GENERAL OR MISCELLANEOUS / MEDICINE
South Carolinian William Bull (1710-1791) became the first colonial to earn a medical degree in Europe.

1735. TECHNOLOGY AND INVENTION
Roland Houghton, of Boston, patented a theodolite, one of the very few mechanical devices to be patented in the colonial period.

1736. BOTANY
John Bartram (1699-1777) went on his first botanical and collecting trip, to the head of the Schuykill River.

1736. BOTANY / GENETICS
Through the intervention of Peter Collinson, James Logan (1674-1751) published "Some Experiments concerning the Impregnation of the Seeds of Plants," in the Royal Society's *Philosophical Transactions* 39 (January-March 1736): 192-195. The paper had been sent to Collinson as a letter. Johann Friedrich Gronovius assisted in publication of a fuller version, *Experimenta et Meletemata de Plantarum generatione* (Leiden, 1739), and John Fothergill was responsible for an English-language edition of the latter as *Experiments and Considerations on the Generation of Plants* (London, 1747). Logan's work reported on his germination and

hybridization experiments and observations on maize, a first account of the sexual mechanisms of reproduction in that plant.

1736. GENERAL OR MISCELLANEOUS
An act for the encouragement of silk culture was passed in South Carolina.

1736. MEDICINE / DISEASE
William Douglass (ca.1691-1752), European-educated physician, published *The Practical History of a New Epidemical Eruptive Miliary Fever, with an Angina Ulcosculosa, which prevailed in Boston, New England, in the Years 1735 and 1736* (Boston), an accomplished clinical account of scarlet fever.

1736. ORGANIZATIONS—SOCIETIES AND ASSOCIATIONS / MEDICINE
A medical society was formed in Boston, William Douglass (ca.1691-1752) taking a leading role. This apparently was the earliest such medical establishment in the American colonies.

1738. ORGANIZATIONS—ACADEMIC / MATHEMATICS, NATURAL PHILOSOPHY
John Winthrop (1714-1779) became professor of natural philosophy and mathematics in Harvard College, succeeding Isaac Greenwood (1702-1745).

1739. BOTANY
Based upon a manuscript "Catalogue of Plants, Fruits, and Trees Native to Virginia" of John Clayton (1694-1773), *Flora Virginica* was published by Johann Friedrich Gronovius in Leiden; an addition to the *Flora* was published by Gronovius in 1743, drawing on Clayton's contributed materials. A later revision of the work, by Gronovius's son (Leiden, 1762), preempted Clayton's own and superior attempt at an update. As published, the work of Clayton was cast in the Linnaean system of classification.

1739 (April). ASTRONOMY
The first colonial observations of sunspots appeared in Harvard Professor John Winthrop's (1714-1779) manuscript "Notes on Sunspots."

1743. BOTANY
Linnaeus published a description of plants by Cadwallader Colden (1688-1776) as "Plantae Coldenghamiae" in *Acta Societatis regiae scientarium Upsaliensis* 4 (1743): 81-136 and 5 (1744-1750): 47-82.

1743. ORGANIZATIONS—SOCIETIES AND ASSOCIATIONS
Benjamin Franklin (1706-1790) proposed intercolonial cooperation as a means to promote science, recognizing that the colonies had reached a point of advance when such interests could be pursued. The plan appeared in his broadside *Proposal for Promoting Useful Knowledge among the British Plantations in America*; the scheme appears to have been worked out jointly with John Bartram (1699-1777), and the suggested name was The American Philosophical Society. Thomas Bond (1712-1784) also was among the initiators of the Society. At least by April 1744, the Society was formed, in Philadelphia, and the first president was Thomas Hopkinson (1709-1751), while Franklin served as secretary. It was inactive by about 1747 but was revived in the 1760s.

1743-1751. GENERAL OR MISCELLANEOUS / ELECTRICITY AND ELECTRONICS
Archibald Spencer (?-1760), public lecturer on science, was active during this time period. Benjamin Franklin (1706-1790) heard him lecture in Boston in 1743 and was influenced by Spencer in the pursuit of electrical studies. Franklin later assisted and encouraged his friend, Ebenezer Kinnersley (1711-1778), to conduct scientific lectures throughout the colonies, which he did during the years 1749-1753, having available apparatus that drew on Franklin's own experience. Expositions on electricity were the significant part of Kinnersley's repertoire.

1744. ASTRONOMY
Thomas Clap (1703-1767), who became president of Yale College in 1745, published *Conjectures upon the Nature and Motion of Meteors which are above the Atmosphere* (Boston, and republished posthumously, Norwich, Connecticut, 1781).

ca.1744. MEDICINE / DISEASE
John Mitchell (1711-1768) wrote a paper on supposed yellow fever outbreaks in Virginia in the years 1737, 1741, and 1742, drawing upon

his dissections of five casualties of the epidemic. Not published until early in the next century, it circulated in manuscript and has become his best-known work, but later historians have cast doubt on whether it was in fact yellow fever he described.

1745. NATURAL PHILOSOPHY
Cadwallader Colden (1688-1776) published *An Explication of the First Causes of Action in Matter, and, of the Cause of Gravitation* (New York). It was reprinted the following year in London; a German translation was published in Hamburg in 1748 and a French edition appeared in Paris in 1751. In 1751, an enlarged edition as *The Principles of Action in Matter* was published in London in 1751. In spite of this extensive distribution, Colden's rational and nonexperimental efforts to explicate what he considered his discovery of the cause of gravity either was not understood or was heavily criticized by contemporaries. The bold work, in fact, showed the author's lack of knowledge of the best related works at the time, while he pretended to build on the contributions of Isaac Newton.

1746. ORGANIZATIONS—ACADEMIC / PHYSICS
Harvard College Professor John Winthrop (1714-1779) gave the earliest demonstration of electricity and magnetism in what has been called the first experimental physics laboratory in the colonies.

1748. NATURAL HISTORY
Through the efforts of Peter Collinson, two manuscripts by John Mitchell (1711-1768), "Dissertatio brevis de principiis botanicorum et zoologicorum" and "Nova genera plantarem Virginiensum" were published in *Acta Physico-Medica Academiae ... Leopoldina ... Ephemerides* (Nuremberg), appendix to volume 8. Written in 1738, the "Dissertatio" was a significant early work on the principles of taxonomy.

1748 (September). BOTANY
Sponsored by the Royal Swedish Academy of Sciences, Pehr Kalm (1716-1779) arrived in Philadelphia on an expedition to find plants having practical value that could be grown in Scandinavia. Kalm had studied with Linnaeus. He carried out explorations and collecting in Pennsylvania, New Jersey, New York, and Canada, and was befriended by Benjamin Franklin (1706-1790), John Bartram (1699-1777), and

Cadwallader Colden (1688-1776). He remained in North America until 1751.

[1750]. GENERAL OR MISCELLANEOUS
By the middle of the eighteenth century, the center for colonial science was at Philadelphia. By the 1750s, an effective network of communication had developed among colonial scientists and connected with the scientific community across the Atlantic.

1751. GENERAL OR MISCELLANEOUS / NATURAL HISTORY
John Bartram (1699-1777) published *Observations on the Inhabitants, Climate, Soil, Rivers, Productions, Animals and Other Matters ... from Pensilvania to Onondago, Oswego and the Lake Ontario, in Canada* (London).

1751. PHYSICS / ELECTRICITY AND ELECTRONICS
A series of communications to the Royal Society by Benjamin Franklin (1706-1790) (the first in 1749) was published in London, through the efforts of Peter Collinson, as *Experiments and Observations on Electricity, Made at Philadelphia in America*. A French translation was published in 1752. Supplements appeared in 1753 and 1754. The work included Franklin's reports and ideas on pointed conductors, grounding, the Leyden jar, conservation of charge, and his theory of electrical action. He developed the idea of electricity as a single fluid and used the terms positive and negative. In 1752, his suggestion that electricity and lightning were the same was proven in Europe and in June by Franklin himself, with his famous kite experiment. (In the next year, George Wilhelm Richmann was killed while carrying out a similar experiment at St. Petersburg, Russia.)

1752. GENERAL OR MISCELLANEOUS
Along with the mother country, the American colonies began to use the Gregorian calendar. In the adjustment, the 14th of September immediately followed the 3rd.

1752. NATURAL HISTORY
Alexander Garden (1730-1791) arrived in Charleston, South Carolina, where he took up the practice of medicine and remained until his return to Great Britain in 1783. While resident in the colonies, he was active

in promoting the interests of the Royal Society of Arts and in the study of natural history. Most of his work was published by others, including John Ellis, his chief supporter in England.

1752. ORGANIZATIONS—HOSPITALS
The Pennsylvania Hospital in Philadelphia was opened. It received a collection of anatomical drawings and casts and medical books from the English physician John Fothergill.

1753. BOTANY
Linnaeus's *Species Plantarum* (Stockholm), published this year, included some 700 North American species.

1753. GENERAL OR MISCELLANEOUS
Sir Hans Sloane, a central figure in the support and promotion of natural studies of the British North American colonies, died in London.

1753 (November 30). AWARDS AND PRIZES / ELECTRICITY AND ELECTRONICS
Benjamin Franklin (1706-1790) was awarded the Copley Medal, the Royal Society of London's highest honor, for his work on electricity. In 1756, Franklin was elected a member of the Royal Society.

1755. EARTH SCIENCES—GENERAL / SEISMOLOGY
Professor John Winthrop (1714-1779) published *A Lecture on Earthquakes: Read in the Chapel of Harvard College ... November 26, 1755. On Occasion of the Great Earthquake which Shook New-England the Week Before* (Boston) in the wake of the great Lisbon earthquake that was felt in the colonies. Winthrop took a naturalistic approach and his publication was in response to one by the Reverend Thomas Prince, a sermon. Prince sought to keep alive the idea that earthquakes had a divine connection, but he appended ideas on the causation of earthquakes by lightning and the suggestion that lightning rods may have been a contributing factor, a contention that Winthrop refuted. Winthrop's remarks included the important idea that earthquakes were characterized by wave-like motion.

1755. GEOGRAPHY AND CARTOGRAPHY
Lewis Evans (1700-1756) published "A General Map of the Middle British Colonies in America," with an accompanying booklet, *Analysis of a General Map of the Middle British Colonies ... and a Description of the Face of the Country...* (Philadelphia), that gave supplementary description of the geography and geology of the region. His commentary suggested what later became recognized physiographic and geologic regions.

1755. GEOGRAPHY AND CARTOGRAPHY
John Mitchell (1711-1768), then a resident of England, published *Map of the British and French Dominions in North America* (London). Authenticated as a production done at the request of the Board of Trade and Plantations and based on extensive survey and other sources, it became the authority on colonial geography.

1759. ASTRONOMY
John Winthrop (1714-1779) published *Two Lectures on Comets* (Boston).

1759. ORGANIZATIONS—SOCIETIES AND ASSOCIATIONS
William Small (1734-1775), professor of mathematics and philosophy at the College of William and Mary, was instrumental in establishing a Williamsburg society, in imitation of the Royal Society of Arts (London), directed toward scientific experimentation and related activities. In 1764, Small went back to England.

1760. GOVERNMENT—STATE / MEDICINE
The first law in the colonies for the examination and licensing of physicians was enacted in New York.

ca.1760. BOTANY
The system of classification of Linnaeus, based on the sexual characters of plants, was acceptable to many in the American colonies by about this date, though it was not universally acclaimed in Europe.

1761. PHYSICS / ELECTRICITY AND ELECTRONICS
Ebenezer Kinnersley (1711-1778) published the results of his work on the electrical air thermometer, and a theory of electrical repulsion, in

"New Experiments in Electricity," *Philosophical Transactions of Royal Society* 53 (1763): 84-97.

1761 (June 6). ASTRONOMY
The transit of Venus before the sun that occurred on this date was the object of much interest in Europe and the colonies, with the intention to use the observations to determine the solar parallax, and thus to help calculate the distance from earth and from the planets to the sun. Harvard Professor John Winthrop (1714-1779) led an expedition to St. John, Newfoundland, for the observations, the chief contribution from the British colonies. He went with support from the colonial government of Massachusetts, and carried instruments belonging to Harvard College. Winthrop published the result in *Relation of a Voyage from Boston to Newfoundland, for the Observation of the Transit of Venus, June 6, 1761* ... (Boston, 1761).

1761-1764. GENERAL OR MISCELLANEOUS / NATURAL HISTORY
Adam Kuhn (1741-1817), who would become professor of materia medica and botany at the College of Philadelphia (later part of the University of Pennsylvania), studied with Linnaeus at the University of Uppsala, the only colonialist to do so.

1763. GENERAL OR MISCELLANEOUS / EXPLORATION AND SURVEYING
Jeremiah Dixon (1733-1779) and Charles Mason (1728-1786), British astronomers, on the recommendation of astronomer royal Nathaniel Bliss, were hired to survey the boundary between Pennsylvania and Maryland, long an uncertain location. The work took almost five years but resulted in the termination of the disagreement about the boundary, and the establishment of what came to be called the "Mason and Dixon Line."

1764 (January 24). ORGANIZATIONS—ACADEMIC / INSTRUMENTS AND INSTRUMENTATION
Harvard College's Harvard Hall burned and with it the institution's collection of scientific instruments. Professor John Winthrop (1714-1779) undertook to replenish the loss and among those who most assisted was Benjamin Franklin (1706-1790), then in England.

1765. BOTANY
John Bartram (1699-1777) was named Royal Botanist, carrying a compensation of fifty pounds a year. (The previous year, William Young had been given the title of Botanist to the King and Queen at a yearly salary of three hundred pounds.) In consequence of his appointment, Bartram undertook an exploratory trip to Florida, accompanied by his son William. This was John's final excursion.

1765. EARTH SCIENCES—GENERAL / SEISMOLOGY
Before this date, the *Philosophical Transactions of the Royal Society* published nearly a dozen accounts of earthquakes by Americans.

1766. BOTANY
Jane Colden (1724-1766), the first recognized woman botanist in America, died. She was the daughter of Cadwallader Colden (1688-1776), whose botanical interests she assumed. As early as 1755, she had begun her studies and prepared a "Flora of New York," with illustrations; the manuscript is in the British Museum. The novelty of a woman working at natural history generated interest and comment, Peter Collinson remarking that he knew of no other who was competent in the system of Linnaeus.

1766. GENERAL OR MISCELLANEOUS
John Winthrop (1714-1779) was elected to membership in the Royal Society of London.

1766. ORGANIZATIONS—ACADEMIC / MEDICINE
The first medical school in the colonies was begun in Philadelphia, largely through the efforts of John Morgan (1735-1789) and in association with the College of Philadelphia (University of Pennsylvania). The early faculty included Morgan (theory and practice of medicine), William Shippen, Jr. (1736-1808) (anatomy), Adam Kuhn (1741-1817) (botany and materia medica), and Benjamin Rush (1746-1813) (chemistry). The first graduates completed their studies in 1768.

1766. PALEONTOLOGY / ZOOLOGY
What eventually were designated as mastodon bones from Big Bone Lick, on the Ohio River, were produced by Indian agent George

Croghan, and they came into the hands of English scientists. Although bones had been found in various places earlier, this event helped to generate wider interest.

1766 (December). ORGANIZATIONS—SOCIETIES AND ASSOCIATIONS
An informal group that had been meeting in Philadelphia for several months took the name American Society for Promoting and Propagating Useful Knowledge, Held in Philadelphia.

1767. ASTRONOMY
John Winthrop (1714-1779) published "Cogitata de Cometis," *Philosophical Transactions of Royal Society* 57:132-154. Submitted at the time of his election to the Society, the work included Winthrop's calculations of the mass and density of comets.

1768. GENERAL OR MISCELLANEOUS / NATURAL HISTORY
Peter Collinson (b.1693), English patron and promoter of natural history studies in the colonies, died in Mill Hill, Middlesex, England. Especially interested in botany, he was responsible for the introduction to England of some fifty American plants. Collinson's active involvement with the colonies began about 1730 and he came to be the preeminent English promoter of science in America. Among his special beneficiaries was John Bartram (1699-1779), but there were many others. He was London agent for the Library Company of Philadelphia.

1768 (December 20). ORGANIZATIONS—SOCIETIES AND ASSOCIATIONS
Union was effected between the American Society Held at Philadelphia for Promoting Useful Knowledge (a subsequent variation on the name taken in December 1766), and the American Philosophical Society, the latter a recent resurrection of the society of that name established in 1743. The new group was called the American Philosophical Society, held at Philadelphia, for Promoting Useful Knowledge, and held its first meeting on 2 January 1769. Benjamin Franklin (1706-1790) was elected president even though he was not resident in the colonies at the time; Franklin held the office until his death.

1769 (June 3). ASTRONOMY
At least twenty-two American observations or accounts of the transit of Venus of this date were eventually published. They appeared in

newspapers and in scholarly journals. Half of these were done with the sponsorship of the American Philosophical Society and occurred in the Philadelphia region; for this event, the legislature of Pennsylvania granted the Society nearly 450 pounds, including funds for an observatory at Philadelphia. They were published in the Society's *Transactions* (1771). Despite efforts by John Winthrop (1714-1779) and others, the political events in Massachusetts that had suspended the legislature for its opposition to the Townshend Acts made it impossible to appropriate funds to support an expedition to Lake Superior to observe the transit of Venus. John Winthrop did publish *Two Lectures on the Parallax and Distance of the Sun, as Deducible from the Transit of Venus, Read in Holden-Chapel at Harvard College in Cambridge, New England, in March 1769* (Boston). This work was based on lectures to prepare his students for the transit. Winthrop's own observations of the event were published in the *Philosophical Transactions of Royal Society*.

1769 (August 1). ORGANIZATIONS—ACADEMIC / CHEMISTRY
Benjamin Rush (1746-1813) was appointed professor of chemistry in the College of Philadelphia (later the University of Pennsylvania Medical College). This appointment was a milestone in the teaching of chemistry in the United States. The following year he published *Syllabus of a Course of Lectures on Chemistry* (Philadelphia).

1770. INSTRUMENTS AND INSTRUMENTATION
As of this time, American crafters were able to provide most instruments for pedagogical purposes, but the colonies still depended on Europe for more precise and sophisticated scientific apparatus. This state of affairs continued for some years after independence.

1771. INSTRUMENTS AND INSTRUMENTATION / ASTRONOMY
David Rittenhouse (1732-1796) completed construction of an orrery, begun in 1767, the best such piece of apparatus produced in the colonies.

1771 (February). PERIODICALS AND PUBLISHING
The American Philosophical Society published the first volume of its *Transactions*, which thereafter tended to replace the Royal Society's *Philosophical Transactions* as outlet for American scientific productions. A primary content in the first volume was reports on observations and calculations for the 1769 transit of Venus.

1772. ZOOLOGY—HUMAN / PHYSIOLOGY
Virginian James McClurg (1746?-1823) published his studies of bile and liver in *Experiments upon the Human Bile: And Reflections on the Biliary Secretions* (London).

1773 (May 13). ORGANIZATIONS—SOCIETIES AND ASSOCIATIONS
The Virginia Society for Promoting Useful Knowledge was established at Williamsburg. Botanist John Clayton (1694-1773) served as first president. There is no record of a meeting of the society after 15 June 1774.

1773-1778. GENERAL OR MISCELLANEOUS / NATURAL HISTORY
Naturalist William Bartram (1739-1823) traveled in the southeastern United States, under the patronage of London doctor John Fothergill.

ca.1774. AWARDS AND PRIZES / TECHNOLOGY AND INVENTION
John Hobday received a gold medal from the Virginia Society for Promoting Useful Knowledge for his invention of a threshing machine, the first award by an American society for a useful invention.

ca.1774. ORGANIZATIONS—MUSEUMS / ANATOMY
Dr. Abraham Chovet (1704-1790) established a wax anatomical museum in Philadelphia.

1775. GENERAL OR MISCELLANEOUS / NATURAL HISTORY
Bernard Romans (ca.1720-ca.1784) published *A Concise Natural History of East and West Florida* (New York), which included an appendix on sailing directions.

1775. TECHNOLOGY AND INVENTION / MILITARY AND WAR
David Bushnell (ca.1742-1824) invented a submarine, called the American Turtle, which was propelled by foot and hand power and intended for use against the British. This was the most significant device to be produced during the war, even though it did not succeed in its military mission to attach a bomb to the hulls of ships.

1775 (July). MILITARY AND WAR / MEDICINE
Congress appointed Dr. Benjamin Church (1734-1778?) as director general for the military medical service. He was succeeded shortly thereafter by Dr. John Morgan (1735-1789).

Revolutionary War Period. GENERAL OR MISCELLANEOUS
Overall, the effects of the war on science in America were more negative than positive. During the war period, Benjamin Franklin's (1706-1790) diplomatic position in Paris made it possible for him to serve as the best bridge between American scientists and Great Britain.

1776. GENERAL OR MISCELLANEOUS / METEOROLOGY AND CLIMATOLOGY, MEDICINE
Lionel Chalmers (1715-1777) published *An Account of the Weather and Diseases of South-Carolina*, in two volumes (London). Chalmers had been a medical associate of John Lining (1708-1760), who earlier had done similar work (including detailed measurements of his own bodily functions) that was published in the Royal Society's *Philosophical Transactions* in 1743 and 1745.

ca.1776. GENERAL OR MISCELLANEOUS / MEDICINE
Of approximately 3500 physicians in the colonies in the period of the Revolution, no more than 5% had medical degrees.

1777. MILITARY AND WAR / ENGINEERING AND APPLIED SCIENCE
Congress commissioned several French officers as engineers in the American army. One of these men, Louis Lebeque de Presle Duportail, was made commander of the Corps of Engineers and Companies of Sappers and Miners. In 1782, of the fourteen engineering officers, only one was an American.

1778. PHARMACOLOGY AND PHARMACY
Dr. William Brown (1752-1792) published *Pharmacopoeia Simpliciorum et Efficaciorum* (Philadelphia), based on an Edinburgh work, but the first pharmacopoeia published in America. Brown at the time was physician general of the army's Middle Department.

1779 (May 3). GENERAL OR MISCELLANEOUS / ASTRONOMY, NATURAL PHILOSOPHY
Astronomer and natural philosopher John Winthrop (b.1714) died at Cambridge, Massachusetts.

1780. ORGANIZATIONS—SOCIETIES AND ASSOCIATIONS
The Philadelphia Humane Society was formed to deal with respiratory concerns. This was the first such organization to be established in the country.

1780 (May 4). ORGANIZATIONS—SOCIETIES AND ASSOCIATIONS
The Massachusetts legislature chartered the American Academy of Arts and Sciences. The first meeting of the Academy was on 30 May 1780 in the Philosophy Chamber of Harvard College. John Adams, who had been most active in conceiving the Academy, served as president from 1791 to 1814, following the term of first president James Bowdoin (1726-1790).

1780 (October 27). ASTRONOMY
An expedition went out to observe the solar eclipse, of this date, at Penobscot Bay (in the state of Maine), headed by Harvard professor Samuel Williams (1743-1817). Carried out with the support of Harvard University, the American Academy of Arts and Sciences, and the Massachusetts legislature, the British forces in control of the observation area cooperated to make the enterprise feasible. The results were published in 1785, in the first volume of the *Memoirs* of the American Academy. It is probable that Williams' report included the first notice of what later were named Baily's beads (light phenomenon that appear along the edge of the moon during the eclipse).

1780-1830. GENERAL OR MISCELLANEOUS
In this period, the United States evolved out of its earlier role of provider of scientific specimens and data for Europe and became a functioning if lesser collaborator.

1781. GENERAL OR MISCELLANEOUS / PERIODICALS AND PUBLISHING
Joseph Willard (1738-1804) published an account of the longitude of Cambridge, Massachusetts, in the *Philosophical Transactions of Royal Society* (1781); another contribution by an American did not appear in

the *Transactions* until Thomas Say (1787-1834) published a letter in 1819. After the War, other changes in publishing also occurred so that, thereafter, scientific writings seldom appeared in newspapers but were printed in magazines instead.

1784. AGRICULTURE
John Beale Bordley (1727-1804) drew upon his own agricultural experiments and familiarity with up-to-date knowledge from England in publishing *A Summary View of the Courses of Crops, in the Husbandry of England and Maryland* (Philadelphia).

1784. GEOGRAPHY AND CARTOGRAPHY
Jedidiah Morse (1761-1826) published the earliest geography in the United States, a work for use in the schools entitled *Geography Made Easy* (New Haven); by 1819, it had gone through twenty editions.

1784. METEOROLOGY AND CLIMATOLOGY
Harvard College accepted the offer of a thermometer, barometer, hygrometer, and variation needle from the Meteorological Society of Mannheim. The gift helped to promote interest in weather observation on the part of Professor Samuel Williams (1743-1817), who collected data at the request of the Society. It also led to an increase in the weather records submitted to the American Academy of Arts and Sciences.

1784. TECHNOLOGY AND INVENTION
Bifocal eyeglasses were devised by Benjamin Franklin (1706-1790).

1784 (June 24). GENERAL OR MISCELLANEOUS / AERONAUTICS
What apparently was the first balloon ascent in America, with a human passenger, took place in Baltimore. It was carried out by Peter Carnes (1749-1794). (This was less than a year after the Montgolfier brothers made their first public demonstration in France.)

1785. BOTANY
The first systematic work on trees and shrubs to be produced in the United States, Humphry Marshall's (1722-1801) *Arbustrum [i.e., Arbustum] Americanum: The American Grove; or, An Alphabetical*

Catalogue of Forest Trees and Shrubs, Natives of the American United States, was published in Philadelphia.

1785. BOTANY
Andre Michaux (1746-1802) was sent to America by the French government to gather botanical items (and particularly trees) of possible value to the French. To effect his mission, he traveled through the country and established two botanical gardens (in New Jersey and in Charleston).

1785. BOTANY
Manasseh Cutler (1742-1823) published "An Account of Some of the Vegetable Productions, Naturally Growing in This Part of America, Botanically Arranged," *Memoirs of American Academy of Arts and Sciences* 1. It was a significant attempt to study systematically the botanical life of New England.

1785. GENERAL OR MISCELLANEOUS / AERONAUTICS
American loyalist Dr. John Jeffries (1744/45-1819) accompanied Jean Pierre Blanchard on one of the first balloon flights across the English Channel. He procured upper air samples which were analyzed by Henry Cavendish. Jeffries later returned to the United States.

1785. GENERAL OR MISCELLANEOUS / NATURAL HISTORY
Thomas Jefferson (1743-1826) published *Notes on the State of Virginia, written in the Year 1781, Somewhat Corrected and Enlarged in the Winter of 1782, for the Use of a Foreigner of Distinction, in answer to Certain Queries proposed by Him Respecting ... 1. Its Boundaries; 2. Rivers; 3. Seaports; 4. Mountains* ... (Paris, 1784-1785). The content included material on geography, flora and fauna, topography, ethnology, and economic and other matters. Among the things that Jefferson's work did was to answer the assertion of Comte de Buffon that life forms had degenerated in the American environment. The work appeared in other editions in England, France, Germany, and the United States; the latter was published in Philadelphia in 1788.

1785. PERIODICALS AND PUBLISHING
The first volume of the *Memoirs* of the American Academy of Arts and Sciences was published with the contents in three sections for astronomy,

natural philosophy and natural history, and medicine. The influence of Harvard College was strongly evident in the volume.

1785. TECHNOLOGY AND INVENTION
Oliver Evans (1755-1819) developed a mechanized flour mill, possibly the first such automated manufacturing process.

1785 (February 11). ORGANIZATIONS—SOCIETIES AND ASSOCIATIONS / AGRICULTURE
The Philadelphia Society for Promoting Agriculture held its initial meeting, the earliest such American society devoted to agriculture. John Beale Bordley (1727-1804), author of a book on agricultural practices in England and Maryland, published the previous year, played a significant role in forming the Society. Established with the intention to promote agricultural production, a central goal of the Society became the testing and disseminating of agricultural knowledge developed in England.

ca.1785. ORGANIZATIONS—SOCIETIES AND ASSOCIATIONS
The Humane Society of Massachusetts was formed. Dr. Henry Moyes (1750-1807), a blind lecturer on scientific topics from Great Britain, and Benjamin Waterhouse (1754-1846) organized the society. As with similar groups in Europe, the Massachusetts organization was especially concerned with methods to deal with drowning victims and other persons whose breathing was stopped.

1786. OCEANOGRAPHY
Benjamin Franklin's (1706-1790) study on the Gulf Stream, and related items, were published in the American Philosophical Society's *Transactions*, volume 2.

1786. ORGANIZATIONS—MUSEUMS / NATURAL HISTORY
Charles Willson Peale (1741-1827) opened the first natural history museum in the United States, in his home in Philadelphia. In 1794, it was moved to the headquarters of the American Philosophical Society, and in 1802 to the Pennsylvania State House (i.e., Independence Hall). The museum came to include, among other accessions, specimens collected by the Lewis and Clark expedition that were entrusted to the museum by President Thomas Jefferson. Peale gave up management of the museum to his son Rubens in 1809.

1786. PHYSICS
David Rittenhouse (1732-1796) published "An Account of Some Experiments on Magnetism," *Transactions of American Philosophical Society* 2:178-181. In the paper, he gave an early and clear statement of the molecular theory of magnetism.

1787. GENERAL OR MISCELLANEOUS / CHEMISTRY, NATURAL PHILOSOPHY
Benjamin Rush (1746-1813) published *Syllabus of Lectures, Containing the Application of the Principles of Natural Philosophy and Chemistry, to Domestic and Culinary Purposes* (Philadelphia). This publication grew out of Rush's lecture series to the Young Ladies Academy.

1787. GENERAL OR MISCELLANEOUS / NATURAL HISTORY, EARTH SCIENCES—GENERAL
During the years 1783-1784, Johann David Schopf (1752-1800), a German physician, traveled in the United States, from New York to Charleston and to St. Augustine and on to Pittsburgh, and collected natural history information. In 1787, he published two related works, on American materia medica, and on the geology of eastern North America: *Materia Medica Americana Potissimum Regni Vegetabilis* (Erlangen), and *Beyträge zur Mineralogischen Kenntnis des Ostlichen Theils von Nordamerica und seiner Geburge* (Erlangen). The latter, the earliest general work on American geology, apparently did not circulate to any degree in the United States and had little influence on the course of American geological studies. (Schopf is better known for his travel account published in German in 1785.)

1787. ZOOLOGY—HUMAN
Samuel S. Smith (1750-1819) published *Essay on the Causes of the Variety of Complexion and Figure in the Human Species* (Philadelphia), the first such work in the new American nation. He argued for a single creation of humans (in keeping with the Biblical account) and racial differences in terms of natural causes, especially climate and social factors. A second and enlarged edition appeared in 1810.

1787 (August). TECHNOLOGY AND INVENTION
John Fitch (1743-1798) had a steamboat in operation on the Delaware River during this time, the first successful one. It did not prove to be commercially feasible.

1788. ASTRONOMY / INSTRUMENTS AND INSTRUMENTATION
Joseph Pope (1750-1826), in Boston, completed construction of an orrery. Harvard College acquired the instrument by way of a lottery conducted to raise the 450 pounds which it cost.

1788. BOTANY
Thomas Walter's (ca.1740-1789) *Flora Caroliniana* (London) was published. This was the earliest work on the plants of an area of the country to be done by an American.

1788. CHEMISTRY
A traveler from France observed in this year that Aaron Dexter (1750-1829) was using a textbook by Antoine F. de Fourcroy to teach the new French chemistry to students in Harvard College. (The theory of oxygen was the basis of the new ideas.) When Joseph Priestley (1733-1804) published his *Considerations on the Doctrine of Phlogiston, and Decomposition of Water* (Philadelphia) in 1796, he admitted that the system of Lavoisier was by then widely known and universally taught in the schools.

1788. MATHEMATICS
Nicholas Pike (1743-1819) published *A New and Complete System of Arithmetic, Composed for the Use of the Citizens of the United States* (Newbury-Port).

1788. MEDICINE
The earliest volume of medical papers published in the country was produced by the Medical Society of New Haven County as *Cases and Observations* (New Haven).

1789. GEOGRAPHY AND CARTOGRAPHY
Jedidiah Morse (1761-1826) published *The American Geography; or A View of the Present Situation of the United States of America* (Elizabethtown, N.J.). It was reprinted in Great Britain and translated

into German and Dutch. Subsequent and expanded editions appeared. The work drew extensively on published sources and on information provided to Morse, including responses to questionnaires; the maps were prepared by others. It included historical and political as well as geological and topographic aspects and material on climate and on plants and animals, but Morse was not himself a field worker. A new, two-volume edition appeared in 1793 as *The American Universal Geography: Or a View of the Present State of All the Empires, Kingdoms, States and Republics in the Known World, and of the United States of America in Particular* (Boston).

1789 (November). GENERAL OR MISCELLANEOUS
The American Philosophical Society held its first meeting in its new Philosophical Hall. In 1794, Charles Willson Peale (1741-1827) leased an unutilized part of the building which he used for his museum and for his family. In 1802 the museum moved to the State House where it had larger quarters.

1789 (November). GENERAL OR MISCELLANEOUS / TECHNOLOGY AND INVENTION
Samuel Slater (1768-1835) arrived in the United States from England, bringing his knowledge of textile machinery that he later applied in the mills of Almy and Brown in Pawtucket, Rhode Island.

1790. AGRICULTURE
Reverend Samuel Deane (1733-1814) published *The New England Farmer; or Georgical Dictionary* (Worcester), with the encouragement of the American Academy of Arts and Sciences. Deane had carried out agricultural experiments and also drew upon American and European background knowledge.

1790. GOVERNMENT—FEDERAL / TECHNOLOGY AND INVENTION
The first patent act was passed by Congress, acting under provisions of Article I, Section 8 of the U.S. Constitution. The secretary of state, secretary of war, and the attorney general were made a board to decide on patents. Administration was placed in the state department and its secretary, Thomas Jefferson. A revision of the law in 1793 put the secretary of state in charge but it was essentially a registering and fee

collecting function and did not involve decision-making regarding the acceptability of an invention.

1790 (April 17). GENERAL OR MISCELLANEOUS / PHYSICS
Benjamin Franklin (b.1706) died at Philadelphia.

1791. BOTANY
William Bartram (1739-1823) published *Travels Through North & South Carolina, Georgia, East and West Florida* ... (Philadelphia). The work was an account of his botanical explorations but written in a luxurious style that became a source for European Romantic authors such as Chateaubriand, Wordsworth, and Coleridge.

1792. GOVERNMENT—FEDERAL
Astronomer and natural philosopher David Rittenhouse (1732-1796) was appointed as the first director of the U.S. Mint.

1792. ORGANIZATIONS—SOCIETIES AND ASSOCIATIONS / AGRICULTURE
The Massachusetts Society for Promoting Agriculture was incorporated. It had its beginnings in 1785 in a special committee appointed by the American Academy of Arts and Sciences, and its early members generally were not directly engaged in farming.

1792. ORGANIZATIONS—SOCIETIES AND ASSOCIATIONS / CHEMISTRY
The Chemical Society of Philadelphia was established under the leadership of James Woodhouse (1770-1809). It was likely the first organized chemical society in the world, although John Penington, a student of Benjamin Rush (1746-1813), is said to have established a chemical society in 1789. Among the Philadelphia society's accomplishments was the promotion of the new chemistry of Lavoisier. In 1811, it was succeeded by the Columbian Chemical Society.

1793. GENERAL OR MISCELLANEOUS / CHEMISTRY
Chemist Thomas Cooper (1759-1839) came to the United States from England, accompanied by two sons of Joseph Priestley (1733-1804), who came the next year.

1793. TECHNOLOGY AND INVENTION
Eli Whitney (1765-1825) invented the cotton gin.

1794. CHEMISTRY
Samuel L. Mitchill (1764-1831) published *Nomenclature of the New Chemistry* (New York). Mitchill explained to his readers that the new ideas were prevalent to the degree that any chemically knowledgeable person should know them and promoted them as a new truth in chemistry. (In 1801, Mitchill also published *Explanation of the Synopsis of Chemical Nomenclature and Arrangement* [New York].)

1794. GOVERNMENT—FEDERAL / MILITARY AND WAR
The Springfield Armory was made a national facility by Congress.

1794 (June 4). GENERAL OR MISCELLANEOUS / CHEMISTRY
Chemist and natural philosopher Joseph Priestley (1733-1804) moved from England to the United States, arriving in New York on this date. Soon thereafter, he settled at Northumberland, Pennsylvania.

1795. GENERAL OR MISCELLANEOUS
David Rittenhouse (1732-1796) was elected a member of the Royal Society of London.

1795. ORGANIZATIONS—ACADEMIC / CHEMISTRY
John Maclean (1771-1814) was appointed professor of chemistry and natural history at Princeton University (then College of New Jersey). This was the first such American professorship in chemistry not attached to a medical school.

1795. TECHNOLOGY AND INVENTION
Oliver Evans (1755-1819) published *The Young Mill-Wright & Miller's Guide* (Philadelphia). Through its many editions, it became a favorite source for young mechanics.

1796. FUNDS AND FUNDING
Benjamin Thompson (Count Rumford) (1753-1814), gave $5,000 in stock to the American Academy of Arts and Sciences to establish a biennial medal and premium for American work in heat and light. (At the same time, he also established a Rumford medal with the Royal Society of London.) Probably the earliest research endowment, in 1832 the Academy received legal permission to make use of the fund more

flexible in its support of research. The first prize was not given until 1839.

1797. CHEMISTRY
John Maclean (1771-1814) published *Two Lectures on Combustion* (Philadelphia). It helped to promote the antiphlogiston views that Maclean had adopted before coming to the United States in 1795, and directly challenged Joseph Priestley's (1733-1804) refutation of the new theory of oxygen. A discussion of the phlogiston question followed its publication, among Maclean, Joseph Priestley, James Woodhouse (1770-1809), and Samuel L. Mitchill (1764-1831), in the *Medical Repository*.

1797. PERIODICALS AND PUBLISHING / MEDICINE
The *Medical Repository* began publication in New York as the first such independent or commercial publication in the United States to be devoted to science (as well as medicine). The journal included developments in Europe as well as in the United States. Samuel Latham Mitchill (1764-1831) was a founder and served as editor until 1813. Issuance continued until 1824.

1797. ZOOLOGY / ENTOMOLOGY
The earliest book on American insects was published. This was *The Natural History of the Rarer Lepidopterous Insects of Georgia* (London), prepared by British naturalist James Edward Smith. It included the illustrations and notes of John Abbot (1751-1840/41), with descriptions of new species by Smith based on Abbot's drawings.

1797-1800. GEOLOGY / MINERALOGY AND CRYSTALLOGRAPHY
Samuel L. Mitchill (1764-1831) published "A Sketch of the Mineralogical and Geological History of the State of New York ...," *Medical Repository* 1 (1797-1798): 293-314, 445-452 and 3 (1799-1800): 325-335. This work was based on a survey that Mitchill undertook on behalf of the Society for the Promotion of Agriculture, Arts, and Manufactures.

1797-1807. BOTANY
The largest collection of American plants of the time was accumulated by the Philadelphia botanist Benjamin Smith Barton (1766-1815).

1798. ANTHROPOLOGY AND ETHNOLOGY
Benjamin Smith Barton (1766-1815) published *New Views of the Origin of the Tribes and Nations in America* (Philadelphia). Though a work of only limited success, it reflected the point of view that such origins were to be found in linguistic studies and included tables of comparative words from various new and old world sources. This view of linguistic studies was shared by Thomas Jefferson (1743-1826), who made the point in his *Notes on the State of Virginia,* written in 1782.

1798. ORGANIZATIONS—SOCIETIES AND ASSOCIATIONS / GEOLOGY
A brief-tenured American Mineralogical Society was founded in New York, the first American geological society. Samuel L. Mitchill (1764-1831) was a leading member of the Society.

1798. PHYSICS
Benjamin Thompson (Count Rumford) (1753-1814) published *Enquiry Concerning the Source of Heat Which Is Excited by Friction.* Reporting on his experiments with boring cannons, he argued against the caloric theory of heat in favor of one based on motion.

1798 (July 16). GOVERNMENT—FEDERAL / PUBLIC HEALTH
A law that provided for the deduction of a fee from the wages of merchant seaman for the support of their medical care was enacted by Congress. The U.S. Public Health Service found its origins in the act through the hospitals that were established.

1799. ORGANIZATIONS—SOCIETIES AND ASSOCIATIONS
The Connecticut Academy of Arts and Sciences was established as the first state academy. Its *Memoirs* began publication in 1810 and included papers on science.

1799. PALEONTOLOGY / ZOOLOGY
On 10 March 1797, Thomas Jefferson (1743-1826) presented a paper on the megalonyx to the American Philosophical Society. It was published as "A Memoir on the Discovery of Certain Bones of a Quadruped of the Clawed Kind in the Western Parts of Virginia," *Transactions of American Philosophical Society* 4:255-256. Caspar Wistar (1761-1818) also published an account of the megalonyx in the same volume. In 1822, it was named Megalonyx jeffersoni by a French naturalist.

Chronology

1800. MEDICINE / DISEASE
Benjamin Waterhouse (1754-1846) vaccinated his son for smallpox, the first American physician to use the procedure.

ca.1800. GENERAL OR MISCELLANEOUS / CHEMISTRY
John Griscom (1774-1852) began public subscription lectures in New York on science (especially chemistry). This was the start of Griscom's lecturing that continued for a number of years.

1801. ENGINEERING AND APPLIED SCIENCE
The first modern suspension bridge was built by Pennsylvanian James Finley.

1801. MATHEMATICS
Jared Mansfield (1759-1830) published *Essays, Mathematical and Physical* (New Haven), which has been referred to as the first work of original research in mathematics by an American.

1801. PALEONTOLOGY / ZOOLOGY
Charles Willson Peale (1741-1827) and his son Rembrandt Peale (1778-1860) excavated and assembled the almost complete skeletal remains of two mastodons found in Orange County, New York. One was added to the father's Philadelphia natural history museum, while the son took the other on tour in 1802 to New York and to London. In 1814, Rembrandt Peale established a museum in Baltimore and the second mastodon was displayed there.

1801 (September 1). ORGANIZATIONS—BOTANICAL GARDENS
David Hosack (1769-1835) purchased twenty acres on Manhattan on which he subsequently developed his Elgin Botanic Garden. From 1809 to 1811, Frederick Pursh (1774-1820) served as gardener. In 1811, the Garden became the property of the Regents of the State of New York; in 1814, it was attached to Columbia University but was not maintained.

1801 (December 10). CHEMISTRY
Robert Hare (1781-1858) presented to the Chemical Society of Philadelphia his paper, "Memoir of the Supply and Application of the Blow-Pipe." It presented the twenty-year-old scientist's discovery relating to the intense production of heat with his oxyhydrogen

blow-pipe, progenitor of the welding torch. The Society published the paper the following year, and accounts of the invention also appeared in Tilloch's *Philosophical Magazine* and in the *Annales de Chemie*. In 1839 the American Academy of Arts and Sciences awarded him the first of its Rumford Medals.

1801-1809. GENERAL OR MISCELLANEOUS
With Thomas Jefferson (1743-1826) in office during these years, the country was headed by the president who did most to promote the interests of science.

1802. GENERAL OR MISCELLANEOUS / ASTRONOMY
The third edition of Nathaniel Bowditch's (1773-1838) *New American Practical Navigator* (Newburyport, Mass.) was published. Originally a corrected and expanded version of a British work by John Hamilton Moore, from this time until the twentieth century it carried Bowditch's name. It became a standard guide to navigation, the most popular ever done.

1802. ORGANIZATIONS—ACADEMIC / CHEMISTRY, NATURAL HISTORY
Benjamin Silliman (1779-1864) was given the new chair of chemistry and natural history in Yale College. In preparation for his duties, he went for a time to study in Philadelphia and in 1805 went to Great Britain for further study and purchase of books and apparatus. In the year of Silliman's appointment, there were about 21 academic scientists in the United States, and this group has been said to constitute all of the full-time positions in American science. However, the years thereafter saw a dramatic increase in the number of academic positions.

1802. ORGANIZATIONS—ACADEMIC / ENGINEERING AND APPLIED SCIENCE
The United States Military Academy (West Point) was established on March 16 by Congress and officially opened on July 4. Its programs emphasized education of officers for engineering and related activities.

1802. ORGANIZATIONS—ACADEMIC / MATHEMATICS
It was not until this year that Harvard College required knowledge of arithmetic for admission.

1802. ORGANIZATIONS—INDUSTRY / CHEMISTRY
DuPont chemical company was founded in Delaware for the manufacture of gunpowder. Until after World War I, 75 percent of the country's explosives were made by this company.

1802. TECHNOLOGY AND INVENTION
A propeller-driven steamboat was constructed by John Stevens (1749-1838). In 1804, it became operational on the Hudson River.

1803. BOTANY
Benjamin Smith Barton (1766-1815) published *Elements of Botany: Or Outlines of the Natural History of Vegetables* (Philadelphia), the first botanical textbook produced in the United States. The work included illustrations based on drawings by William Bartram (1739-1823). Three editions appeared during Barton's lifetime and a sixth was published in 1836.

1803. BOTANY
The *North American Flora* (*Flora Boreali-Americana, sisten caracteres plantarum quas in America Septentrionali collegit et detexit*, Paris, 2 volumes) of Andre Michaux (1746-1802), the first overall portrayal of American botany, was published through the efforts of his son Francois A. Michaux (1770-1855).

1803. GENERAL OR MISCELLANEOUS
Samuel Miller's (1769-1850) two-volume *A Brief Retrospect of the Eighteenth Century* (New York) concluded that contributions to literature and science in America could not be favorably compared to those of Europe. He explained the deficit in terms of lack of the necessary cultural institutions and leisure for learning, historic dependence on Great Britain, and similar factors.

1803. GENERAL OR MISCELLANEOUS / NATURAL HISTORY
Constantin-F. Chasseboeuf, Count de Volney (1757-1820) published a work on America in French in this year that was translated by Charles Brockden Brown and published as *View of the Soil and Climate of the United States of America* (Philadelphia, 1804). Approximately one-third of the work was on geography and geology and made important contributions to knowledge of the region of the Mississippi valley.

1803. ZOOLOGY / PHYSIOLOGY
John Richardson Young (1782-1804) prepared a medical dissertation, *An Experimental Inquiry, Into the Principles of Nutrition, and the Digestive Process* (Philadelphia) for his degree at the University of Pennsylvania. There, Young demonstrated that gastric juice is a component of normal gastric secretion, that it is simultaneous with the secretion of saliva, and is acidic. His dissertation was evidence of experimental skill and insight and has been noted as a significant American work in the physiology of digestion, but more recent evaluation comparing it to knowledge of the time finds it less remarkable. Young died in his twenty-second year.

1804. GENERAL OR MISCELLANEOUS / PERIODICALS AND PUBLISHING
In a listing of 1,338 titles in *The Catalogue of All the Books Printed in the United States*, which was issued by booksellers of Boston, no more than 20 could be classified as science (not counting medicine). From this and other contemporary sources, chemistry appeared as the most popular of the sciences. (These assessments are taken from John Greene, *American Science in the Age of Jefferson* [1984].)

1804. PERIODICALS AND PUBLISHING / MATHEMATICS
The *Mathematical Correspondent*, the first such journal in the United States, was begun. It was edited in New York by George Baron and later by Robert Adrain (1775-1843).

1804 (February 6). GENERAL OR MISCELLANEOUS / CHEMISTRY
Chemist and theologian Joseph Priestley (b.1733) died at Northumberland, Pennsylvania.

1804 (October). EXPLORATION AND SURVEYING / EARTH SCIENCES—GENERAL
A scientific expedition to explore the lower Red River and Ouachita River began. The government-sponsored venture was headed by William Dunbar (1749-1810). It returned successfully to Natchez, Mississippi in January 1805, among its accomplishments the first report of the mineral wells at Hot Springs, Arkansas.

1804-1806. GOVERNMENT—FEDERAL / EXPLORATION AND SURVEYING
Consequent to the Louisiana Purchase in 1803, President Thomas Jefferson (1743-1826) organized the government-sponsored Lewis and

Clark expedition to explore the western country to the northwest Pacific. Meriwether Lewis (1774-1809) and William Clark (1770-1838) departed along the Missouri River on 21 May 1804; by 7 November 1805, they reached the Pacific Ocean near the mouth of the Columbia River in the Oregon area. They collected information on the geography, native Americans, natural history, and specimens of minerals, plants, and animals, and the expedition returned to St. Louis in September 1806. In Washington, the expedition was largely coordinated by Jefferson, who had contributed detailed instructions on the selection and use of scientific equipment. Lewis, who was to prepare the report, died in 1809 and it was finally published in 1814 by Nicholas Biddle and Paul Allen as *History of the Expedition under the Command of Captains Lewis and Clark, to the Sources of the Missouri, Thence across the Rocky Mountains and down the River Columbia to the Pacific Ocean, Performed during the Years 1804-5-6* (Philadelphia). The botanical specimens were, for the most part, included in Frederick Pursh's (1774-1820) *Flora Americae Septentrionalis* (London, 1814). The zoological items were deposited in the museum of Charles Willson Peale (1741-1827) and were published by several naturalists, including Alexander Wilson (1766-1813), George Ord (1781-1866), Constantine S. Rafinesque (1783-1840), and Thomas Say (1787-1834). With Jefferson's urging and intervention, the journals and natural history, ethnographic, and other data were deposited with the American Philosophical Society.

1804-1809. PERIODICALS AND PUBLISHING
The *Philadelphia Medical and Physical Journal* was published, edited by Benjamin S. Barton (1766-1815).

1805. ORGANIZATIONS—SOCIETIES AND ASSOCIATIONS / BOTANY
The Charleston Botanic Garden and Society was established in South Carolina as the first botanical society in the country. It had as its primary objective the study of native plants.

1806. BOTANY / AGRICULTURE
Bernard M'Mahon [McMahon] (1775-1816) published the first American gardening book, *The American Gardener's Calendar: Adapted to the Climate and Seasons of the United States* (Philadelphia). The work achieved eleven editions.

1806. ORGANIZATIONS—SOCIETIES AND ASSOCIATIONS / NATURAL HISTORY
The Linnaean Society of Philadelphia was formed by Benjamin Smith Barton (1766-1815). Initially called the Philadelphia Botanical Society, it took the name Linnaean in 1807. It endured only for a few years.

1806. ZOOLOGY / ENTOMOLOGY
Frederick V. Melsheimer (1749-1814) published *Catalogue of Insects of Pennsylvania* (Hanover, Penn.). It was the earliest systematic work on the entomology of a region of the country.

1807. GOVERNMENT—FEDERAL / GEOPHYSICS AND GEODESY
With the initiative of President Thomas Jefferson (1743-1826), Congress authorized funds for a coast survey and the plan of Ferdinand R. Hassler (1770-1843) was chosen. No work took place immediately, but in 1811 Hassler was sent to England to buy instruments; he did not return until 1815. Fieldwork began in 1816 but in 1818 the work was limited to army or naval personnel; Hassler's involvement for the time ended and the work of the Survey essentially ceased until the early 1830s.

1807 (August 17-21). TECHNOLOGY AND INVENTION
A steamboat, Robert Fulton's (1765-1815) *Clermont*, made a roundtrip on the Hudson River, from New York City to Albany, at the rate of about five miles per hour. This successful outing demonstrated the commercial viability of steam navigation. The vessel used a paddlewheel powered by a Watt engine; financial support came from Robert R. Livingston.

1807 (December 14). ASTRONOMY / CHEMISTRY
A meteor exploded over Weston, Connecticut. Benjamin Silliman (1779-1864) and James L. Kingsley (1778-1852) gathered fragments, one of which weighed six pounds. Silliman conducted chemical analyses and he and Kingsley published the results of their studies in the *Transactions of American Philosophical Society*, vol. 6 (1809), which was republished in the *Memoirs of Connecticut Academy of Arts and Sciences*, 1810. In 1815, Nathaniel Bowditch (1773-1838) published "An Estimate of the Height, Direction, Velocity and Magnitude of the Meteor, That Exploded over Weston in Connecticut, December 14, 1807 ...," *Memoirs of American Academy of Arts and Sciences* 3, part 1.

Bowditch's studies indicated that the original meteor had weighed some six million tons.

1808. MATHEMATICS
Robert Adrain (1775-1843) published "Research Concerning the Probabilities of the Errors Which Appear in Making Observations," *Analyst* 1:93-109. This was the earliest exposition of the exponential law of error; Carl F. Gauss produced a similar result the next year and it bears his name.

1808. PERIODICALS AND PUBLISHING / MATHEMATICS
Robert Adrain (1775-1843) published *The Analyst, or Mathematical Companion*, a periodical produced in Philadelphia. Adrain was an important contributor; others who published there included Nathaniel Bowditch (1773-1838) and Robert Patterson (1743-1824). The publication did not endure beyond the first volume.

1808-1814. ZOOLOGY / ORNITHOLOGY
Alexander Wilson (1766-1813) published his illustrated *American Ornithology; or The Natural History of the Birds of the United States* (Philadelphia). Seven volumes appeared before his death; volumes 8 and 9 were completed by his friend George Ord (1781-1866). In the work, Wilson included illustrations and descriptions for 264 species, of which 48 were new to the American scene. Among the resources that he had available, in addition to his own collections, were the specimens gathered by the Lewis and Clark expedition.

1809. GEOLOGY
William Maclure (1763-1840) published "Observations on the Geology of the United States, Explanatory of a Geological Map," *Transactions of American Philosophical Society* 6:411-428. It included a color map. An expanded version appeared in 1817 as *Observations on the Geology of the United States* (Philadelphia), to accompany a revised geological map. A summary of the geology of the area east of the Mississippi River, Maclure's endeavor was the first geological survey of the region.

1809. MEDICINE / SURGERY
Ephraim McDowell (1771-1830) carried out the first ovariotomy. The operation was done in Kentucky, although McDowell had had medical training in Edinburgh.

1810. CHEMISTRY
Thomas Cooper (1759-1839) prepared potassium for the first time in the United States. He used a procedure devised in 1808 by Guy-Lussac and Thenard, involving the heating of potash with iron in a gun barrel.

1810. ORGANIZATIONS—ACADEMIC / MINERALOGY AND CRYSTALLOGRAPHY
George Gibbs (1776-1833), who had accumulated a substantial mineralogical collection by purchases in Europe, the most extensive such collection in the country at the time, agreed to loan some parts of it to Yale College if they would provide for its display. This arrangement with Gibbs was made by Benjamin Silliman (1779-1864). As deposited at Yale in 1811, it included 12,000 specimens.

1810 (October 1). ORGANIZATIONS—FAIRS AND EXPOSITIONS / AGRICULTURE
The Berkshire Cattle Show opened in Pittsfield, Massachusetts, where Elkanah Watson (1758-1842) had put two Merino sheep on exhibit in 1807. Organized by Watson, the cattle show soon led to establishment of the Berkshire Agricultural Society, the first permanent agricultural association in the country. These efforts laid the foundation for agricultural fairs in the United States.

1810-1814. PERIODICALS AND PUBLISHING
The *American Medical and Philosophical Register* was published, having been founded by David Hosack (1769-1835) and John W. Francis (1789-1861).

1810-1814. PERIODICALS AND PUBLISHING / MINERALOGY AND CRYSTALLOGRAPHY
Archibald Bruce (1777-1818) founded and edited the *American Mineralogical Journal* in New York City. This was the earliest American journal devoted to a particular field of science (other than mathematics or medicine). Only one volume appeared.

1811. BOTANY
Benjamin Waterhouse (1754-1846) published *The Botanist* (Boston), a work that had appeared serially in the *Monthly Anthology and Boston Review*.

1811. MEDICINE / ANATOMY
Caspar Wistar (1761-1818) published *A System of Anatomy for the Use of Students of Medicine* (Philadelphia), in two volumes. It eventually had nine editions.

1812. MEDICINE / PSYCHIATRY
Benjamin Rush (1746-1813) published *Medical Inquiries and Observations Upon the Diseases of the Mind* (Philadelphia). It was the earliest work on psychiatry by a native-born American.

1812. ORGANIZATIONS—SOCIETIES AND ASSOCIATIONS
The Academy of Natural Sciences of Philadelphia was formed. Gerard Troost (1776-1850) was the first president. In 1817, William Maclure (1763-1840) became president and was reelected every year until his death in 1840. In addition to his leadership, Maclure also supported the Academy financially.

1812. PERIODICALS AND PUBLISHING / MEDICINE
The *New England Journal of Medicine and Surgery* began publication. It was founded by John Gorham (1783-1829), James Jackson (1777-1867), and John C. Warren (1778-1856), the earliest medical journal in the Boston area.

1812-1814. PERIODICALS AND PUBLISHING
The *Emporium of Arts and Sciences*, a Philadelphia-based serial first directed by John R. Coxe (1773-1864) and in 1813 by Thomas Cooper (1759-1839), appeared during this period. It was especially directed to the application of scientific knowledge to technological interests, under Cooper stressing applied chemistry.

1813. BOTANY
Gotthilf Henry Ernest Muhlenberg (1753-1815) published *Catalogus Plantarum Americae Septentrionalis, Huc Usque Cognitarum Indigenarum et Circum; or, A Catalogue of the Hitherto Known Native*

and *Naturalized Plants of North America Arranged According to the Sexual System of Linnaeus* (Lancaster, Penn.).

1813 (April 19). GENERAL OR MISCELLANEOUS / CHEMISTRY, MEDICINE
Physician and chemist Benjamin Rush (b.1746) died at Philadelphia, Pennsylvania.

1814. BOTANY
German botanist Frederick Pursh (1774-1820) published *Flora Americae Septentrionalis: Or, A Systematic Arrangement and Description of the Plants of North America* (London). Written while in England after a period of work and collecting in the United States, he drew upon a number of collections and included the chief account of the specimens gathered by the Lewis and Clark expedition, a fact that irritated patriotic American botanists. The work signaled a new era in the study of American botany.

1814. ORGANIZATIONS—SOCIETIES AND ASSOCIATIONS
The Literary and Philosophical Society of Charleston, South Carolina was established. Among its primary interests was natural history and the sciences generally. Stephen Elliott (1771-1830) became first president.

1814. ORGANIZATIONS—SOCIETIES AND ASSOCIATIONS / NATURAL HISTORY
The New England Society for the Promotion of Natural History was established in Boston, at a December meeting at the home of Jacob Bigelow (1786-1879). (The first president was John Davis, 1761-1847.) In the next year, it became the Linnaean Society of New England and was incorporated in 1820. It went out of existence about 1822, and in 1823 Harvard University took charge of the Society's museum collections, which fell to neglect. With the formation in 1830 of the Boston Society of Natural History, the earlier Society's collections went there.

1814. TECHNOLOGY AND INVENTION / ORGANIZATIONS—INDUSTRY
The first factory building incorporating both power cotton spinning and weaving machinery was opened by Francis Cabot Lowell (1775-1817) in Waltham, Massachusetts.

1814 (May 4). GENERAL OR MISCELLANEOUS
DeWitt Clinton (1769-1828) delivered an address on the progress of science in America. It was published the next year in the *Transactions* of the Literary and Philosophical Society of New York, which was founded in 1814.

1814 (August 21). GENERAL OR MISCELLANEOUS / PHYSICS
Physicist Benjamin Thompson (Count Rumford) (b.1753) died at Auteuil (near Paris), France.

1815. BOTANY
When Gotthilf Henry Ernest Muhlenberg (b.1753) died in this year, his herbarium in excess of 5000 specimens was the largest in the country. It subsequently was procured by the American Philosophical Society.

1815. TECHNOLOGY AND INVENTION
The first steam-powered warship was constructed in the United States. It was conceived by Robert Fulton and called *Demologos*, or *Fulton the First*. Fulton died 24 February 1815 and the War of 1812 was over before it was completed.

1815. TECHNOLOGY AND INVENTION / CHEMISTRY
Thomas Cooper (1759-1839) published *A Practical Treatise on Dyeing and Calicoe Printing* (Philadelphia).

1816. GEOLOGY / MINERALOGY AND CRYSTALLOGRAPHY
Parker Cleaveland (1780-1858) published *An Elementary Treatise on Mineralogy and Geology* (Boston), the first significant textbook on mineralogy in the country. There was a second edition, in two volumes, in 1822. The work incorporated data widely solicited from American correspondents and had a theoretical grounding in European ideas.

1816. ORGANIZATIONS—SOCIETIES AND ASSOCIATIONS
The Columbian Institute for the Promotion of Arts and Sciences was established in Washington, D.C. It was an organization devoted to all areas of knowledge, including ambitious plans for science and related subjects. In 1818, it received a federal charter and land for a botanical garden. By the late 1820s, the Institute was no longer active and the

botanical garden seems never to have been actualized. The charter expired in 1838.

1816. ZOOLOGY / CONCHOLOGY AND MALACOLOGY
Thomas Say (1787-1834) published "Conchology" in the British Nicholson's *Encyclopedia of Arts and Sciences* (American edition, Philadelphia, 1816-1817). It was the first paper on native shells by an American author.

1816 (June 11). TECHNOLOGY AND INVENTION / ORGANIZATIONS—INDUSTRY
The Gas Light Company of Baltimore, using coal gas, was established, making the city the first to have such a facility for street illumination.

1816-1824. BOTANY
Stephen Elliott (1771-1830) issued *Sketch of the Botany of South Carolina and Georgia* (Charleston), in two volumes.

1817. BOTANY
Amos Eaton (1776-1842) published *Manual of Botany for the Northern States* (Albany), a field guide. In 1840, there was an eighth edition.

1817. ORGANIZATIONS—ACADEMIC / ENGINEERING AND APPLIED SCIENCE
The United States Military Academy at West Point underwent a reorganization that followed the example of the Ecole Polytechnique with greater stress on engineering and cognate sciences.

1817. ORGANIZATIONS—SOCIETIES AND ASSOCIATIONS / NATURAL HISTORY
The Lyceum of Natural History of New York was founded. Samuel L. Mitchill (1764-1831) served as the first president and in 1823 the Lyceum's published *Annals* began. In January 1876, it became the New York Academy of Sciences.

1817. PERIODICALS AND PUBLISHING / NATURAL HISTORY
The Academy of Natural Sciences of Philadelphia began publication of its *Journal*.

1817-1825. GENERAL OR MISCELLANEOUS / ENGINEERING AND APPLIED SCIENCE
The Erie Canal was constructed with state funding. Measuring 363 miles and connecting New York City with the Great Lakes, it was completed on October 26, 1825. The Canal extended from Albany to Buffalo.

1817-1828. ZOOLOGY / ENTOMOLOGY
Thomas Say (1787-1834) published *American Entomology; or Descriptions of the Insects of North America* (Philadelphia). The first part was issued in 1817, and volumes 1-2 were published in 1824 and volume 3 in 1828. The work included illustrations by Titian R. Peale (1799-1885) as well as by Charles Lesueur (1778-1846). The lasting value of Say's entomological work is indicated in John L. LeConte's (1825-1883) two-volume *Complete Works of Thomas Say on the Entomology of North America* (New York, 1859).

1818. BOTANY
Thomas Nuttall (1786-1859) published *The Genera of North American Plants, and Catalogue of the Species to the Year 1817* (Philadelphia), in two volumes. In this work, the author drew upon his prior explorations in the West and produced an early overall view of American botany.

1818. GEOLOGY
Amos Eaton (1776-1842) published *Index to the Geology of the Northern States, with a Transverse Section from Catskill Mountain to the Atlantic* (Leicester, Mass.). A second edition appeared in 1820, published at Troy, N.Y. In this work, Eaton addressed questions of stratigraphy and nomenclature for the geology of America.

1818. GEOLOGY / MINERALOGY AND CRYSTALLOGRAPHY
James Freeman Dana (1793-1827) and Samuel L. Dana (1795-1868) published "Outlines of the Mineralogy and Geology of Boston and Its Vicinity, with a Geological Map," *Memoirs of American Academy of Arts and Sciences* 4:129-224, the first such account of that part of the country.

1818. ZOOLOGY—HUMAN
William C. Wells (1757-1817), American born but long resident in England, published "An Account of a Female of the White Race of

Mankind, Part of Whose Skin Resembles That of a Negro; With Some Observations on the Causes of the Differences in Colour and Form between the White and Negro Races of Men" as an appendix to his *Two Essays: One upon Single Vision with Two Eyes; the Other on Dew...* (London). In the "Account," Wells anticipated later ideas of natural selection.

1818 (July). PERIODICALS AND PUBLISHING
The American Journal of Science and Arts was founded by Benjamin Silliman (1779-1864). During its early years, mineralogy and geology were about one-third to one-half of the articles. The journal's circulation was about 500 to 1000 subscribers, and in 1826 Silliman had to take over full financial responsibility for the journal.

1818-1829. ORGANIZATIONS—ACADEMIC / MATHEMATICS, PHYSICAL SCIENCES—GENERAL
Harvard University professor John Farrar (1779-1853) translated a series of French mathematical, physical, and astronomical texts. In the process, he significantly changed the direction of the science curriculum in Harvard College and elsewhere through the introduction of French mathematics. The series came to be known as the Cambridge Mathematics and Cambridge Natural Philosophy.

1819. CHEMISTRY / INSTRUMENTS AND INSTRUMENTATION
Robert Hare (1781-1858) invented a galvanic device, the calorimotor.

1819. ORGANIZATIONS—MUSEUMS
The Western Museum in Cincinnati was apparently operational by the end of this year. Among those who took part in its establishment was Daniel Drake (1785-1852), while John James Audubon (1785-1851) was recruited in the preparation of the displays. The museum was based on subscribed funds as well as contributions of natural and historical specimens of various kinds. It flourished as an institution for science only for a few years before it passed to commercial interests.

1819. ORGANIZATIONS—SOCIETIES AND ASSOCIATIONS / GEOLOGY
The American Geological Society was sanctioned by the Connecticut state legislature and was situated in New Haven. George Gibbs (1776-1833) and Benjamin Silliman (1779-1864) had promoted the idea

for the Society. At its first meeting, at Yale University on September 7, 1819, William Maclure (1763-1840) was named president, while the vice presidents were Gibbs, Silliman, Parker Cleaveland (1780-1858), Stephen Elliott (1771-1830), Robert Gilmor, Jr. (1774-1848) (Baltimore), Samuel Brown (1769-1830) (Kentucky), and Robert Hare (1781-1858). The group never flourished and its last apparent meeting was in 1828.

1819. PERIODICALS AND PUBLISHING / AGRICULTURE
The *American Farmer* began publication. It was edited by John Stuart Skinner (1788-1851) in Baltimore.

1819. TECHNOLOGY AND INVENTION
The American *Savannah* made the first crossing of the Atlantic Ocean by a steamboat. During 87 percent of the voyage, however, the ship moved under sail.

1819 (April 15). GENERAL OR MISCELLANEOUS / TECHNOLOGY AND INVENTION
Inventor Oliver Evans (b.1755) died in New York City.

1819-1820. CHEMISTRY
John Gorham's (1783-1829) *The Elements of Chemical Science* (Boston), in two volumes, appeared as the earliest original textbook on the subject to be authored by an American.

1819-1820. EXPLORATION AND SURVEYING / NATURAL HISTORY
Major Stephen H. Long (1784-1864) led an expedition to the Rocky Mountains. In addition to military personnel who carried out topographical and astronomical work, the expedition included naturalists Edwin James (1797-1861) (botany), Thomas Say (1787-1834) (zoology), and Titian R. Peale (1799-1885). James prepared the account of the expedition, which was published in 1823. The botanical results of the expedition were published by John Torrey (1796-1873), who received specimens from James.

1820. ORGANIZATIONS—SOCIETIES AND ASSOCIATIONS / ARCHAEOLOGY
The founding of the American Antiquarian Society in this year had as a primary interest the study of early humans and their archaeological remains in North America.

1820. PERIODICALS AND PUBLISHING / MEDICINE
In answer to a challenge from Great Britain (asserted by Sidney Smith in the *Edinburgh Review*) that denigrated the contributions made by Americans to various fields, the *Philadelphia Journal of the Medical and Physical Sciences* was established by Nathaniel Chapman (1780-1853). Chapman was especially concerned with the indictment of American medicine and his journal was directed at that field.

1820. PHARMACOLOGY AND PHARMACY
The first meeting on American pharmacy, the United States Pharmacopoeial Convention, was held in the capitol in Washington, D.C.

1820. ZOOLOGY / ICHTHYOLOGY AND PISCICULTURE
The descriptive study of fishes in the United States began, in a sense, through Constantine Rafinesque's (1783-1840) work on fish of the Ohio River region. See his *Ichthyologia Ohiensis* (Lexington, Ky., 1820).

1820s. GENERAL OR MISCELLANEOUS / BOTANY
During this time, botany became, and, for much of the remainder of the century, was the most popular science for recreational and general educational purposes.

1821. CHEMISTRY / INSTRUMENTS AND INSTRUMENTATION
Robert Hare (1781-1858) invented the deflagrator for the production of high levels of electric current.

1822. ZOOLOGY / ANATOMY
John C. Warren (1778-1856) published *Comparative View of the Sensorial and Nervous Systems in Men and Animals* (Boston). It was based on extensive examinations of animal and human brains and was the earliest American work on comparative anatomy. Appended was an illustrated "Account of the Crania of Some of the Aborigines of the United States."

1822-1823. PERIODICALS AND PUBLISHING
The *Western Quarterly Reporter of Medical, Surgical and Natural Science* was published at Cincinnati, the first journal devoted to science west of the Alleghenies.

1822-1824. GEOLOGY
Amos Eaton (1776-1842) engaged in a survey of the Erie Canal route, which was financed by Stephen Van Rensselaer. In 1824, Eaton published the outcome in *A Geological and Agricultural Survey of the District Adjoining the Erie Canal* (Albany).

1823. GOVERNMENT—STATE / GEOLOGY
The first state-sponsored geological survey was carried out by Denison Olmsted (1791-1859) in North Carolina.

1824. BOTANY
John Torrey (1796-1873) published *Flora of the Northern and Middle Sections of the United States Or A Systematic Arrangement and Description of All the Plants Hitherto Discovered in the United States North of Virginia* (New York), an attempt to encompass the entire range of botany of North America in one work. An intended second volume never appeared.

1824. ORGANIZATIONS—ACADEMIC
The Rensselaer School, the first such institution for study of science and engineering in the United States, was founded at Troy, N.Y. In 1851, it took the name Rensselaer Polytechnic Institute. The school was founded by Stephen Van Rensselaer, at the suggestion of Amos Eaton (1776-1842) who directed it as senior professor. A one-year course of study was offered.

1824. ORGANIZATIONS—SOCIETIES AND ASSOCIATIONS / TECHNOLOGY AND INVENTION
The Franklin Institute of the State of Pennsylvania for the Promotion of the Mechanic Arts was founded in Philadelphia. The Institute's *Journal* began publication in 1826.

1825. GENERAL OR MISCELLANEOUS
The so-called "boatload of knowledge," Charles Lesueur (1778-1846), William Maclure (1763-1840), Thomas Say (1787-1834), and Gerard Troost (1776-1850), with Robert Dale Owen (1801-1877), went to New Harmony, Indiana, with plans to establish a model community.

1825. METEOROLOGY AND CLIMATOLOGY
A state weather service was established in New York. (Pennsylvania established such a service in 1837.)

1825. ORGANIZATIONS—ACADEMIC / MEDICINE
Robley Dunglison (1798-1869) was appointed to teach medicine at the University of Virginia as the first full-time professor of medicine in the country.

1825. PERIODICALS AND PUBLISHING / MATHEMATICS
The *Mathematical Diary* was begun by Robert Adrain (1775-1843). Publication continued until 1832.

1825. ZOOLOGY—HUMAN / PHYSIOLOGY
William Beaumont (1785-1853) began his experimental studies of human digestion, using as his subject a French Canadian trapper, Alexis St. Martin, whom Beaumont had begun treating in 1822 when he suffered an abdominal gunshot wound. The injury did not close (i.e., it resulted in formation of a fistula). Continuing his studies until 1833, Beaumont sent specimens of St. Martin's gastric juice to several American chemists. The result of his studies was published in *Experiments and Observations on the Gastric Juice and the Physiology of Digestion* (Plattsburgh, N.Y., 1833). The work was widely noted, a German translation appeared in 1834, and helped to establish the chemical character of digestive processes.

1825 (January 8). GENERAL OR MISCELLANEOUS / TECHNOLOGY AND INVENTION
Inventor Eli Whitney (b.1765) died in New Haven, Connecticut.

1825 (December 6). GOVERNMENT—FEDERAL / ASTRONOMY
President John Quincy Adams (1767-1848) prepared the first annual presidential message to Congress. Among his requests were for a

national university and the construction of a national astronomical observatory at Washington.

1825-1826. ZOOLOGY / MAMMALOGY
Richard Harlan (1796-1843) published *Fauna Americana: Being a Description of the Mammiferous Animals Inhabiting North America* (Philadelphia, 1825), a work that never came to completion. A pioneering systematic effort in American zoology, but based substantively on the works of others, it was criticized by his contemporaries. An important feature of the work was its inclusion of fossils. The next year, a similar but less derivative work by John D. Godman (1794-1830), entitled *American Natural History* (Philadelphia, 1826-1828), began publication. A bitter dispute erupted between the two authors.

1825-1833. ZOOLOGY / ORNITHOLOGY
Charles Lucien Bonaparte (1803-1857) published *American Ornithology or the Natural History of Birds Inhabiting the United States, Not Given by Wilson* (Philadelphia), in four volumes. For volume one, Titian Ramsay Peale (1799-1885) prepared all but one of the plates and also collected many of the specimens. In 1826, Bonaparte published *Observations on the Nomenclature of Wilson's "Ornithology"* (Philadelphia).

1826. BOTANY
William Darlington (1782-1863) published *Florula Cestrica* (West Chester, Penn.). An expanded work, *Flora Cestrica* (West Chester, Penn.), appeared in 1837 with a second edition in 1853. Both works dealt with plants in the vicinity of West Chester, Pennsylvania.

1826. GENERAL OR MISCELLANEOUS / EDUCATION IN SCIENCE
Before this date, science in the schools of New York state included only natural philosophy, astronomy, and chemistry. By 1830, some schools had begun to introduce geology, mineralogy, mechanics, natural history, and botany.

1826. GEOLOGY
The Philadelphia Society for Promoting Agriculture published a geological map of the area by Gerard Troost (1776-1850).

1826. GEOLOGY
Ebenezer Emmons (1799-1863) published *Manual of Mineralogy and Geology* (Albany).

1826 (July 4). GENERAL OR MISCELLANEOUS
Thomas Jefferson (b.1743) died at Monticello.

1826 (October 7). TECHNOLOGY AND INVENTION
The first railroad in the country was completed at Quincy, Massachusetts. The three-mile long metal rail, serving the granite quarry, was designed for horse-drawn vehicles.

1827. ORGANIZATIONS—SOCIETIES AND ASSOCIATIONS / BOTANY
The Pennsylvania Horticultural Society was established as the first of its kind in the country.

1827. PHYSICS / ELECTRICITY AND ELECTRONICS
Joseph Henry (1797-1878) began his studies of electricity and magnetism.

1827-1838. ZOOLOGY / ORNITHOLOGY
John James Audubon's (1785-1851) *The Birds of America*, a four-volume work consisting of 435 aquatint copper engravings, depicting 1065 birds, was issued (Edinburgh and London). The text relating to the illustrations was published in five volumes as *Ornithological Biography* (Edinburgh, 1831-1839). An American edition of *The Birds of America*, produced at New York 1840-1844, included the text but the illustrations were not up to the standard (or size) of the British originals.

1828. CHEMISTRY
Robert Hare (1781-1858) published *Compendium of the Course of Chemical Instruction* ... (Philadelphia), a textbook with more than 200 copper plate engravings of apparatus.

1828. GEOLOGY / PALEONTOLOGY
Lardner Vanuxem (1792-1848), in his work on the Atlantic coastal plain, was the first to correlate America strata with the Cretaceous in Europe. He used fossil evidence in reaching his conclusions.

1828. ORGANIZATIONS—ACADEMIC
The influential *Yale [University] Report of 1828*, among other aspects, promoted the expansion of natural science teaching in the college curriculum.

1828-1830. BOTANY / MEDICINE
Constantine Samuel Rafinesque (1783-1840) published *Medical Flora, or Manual of the Medical Botany of the United States of North America* (Philadelphia), in two volumes. It included 100 woodcut plates by the author.

1829. ASTRONOMY
The first volume of Nathaniel Bowditch's (1773-1838) translation and commentary on Pierre-Simon Laplace's *Mecanique Celeste* appeared, though his work on the project had been completed in 1818. Subsequent volumes appeared 1832, 1834, and 1839, the last after his death in 1838. (Because of its timing, Bowditch was unable to translate Laplace's fifth volume.)

1829. BOTANY
Almira Hart Lincoln Phelps (1793-1884) published the most popular botanical work in the nineteenth century, *Familiar Lectures on Botany* (Hartford, Conn.). It had 29 editions and sold some 375,000 copies.

1829. GENERAL OR MISCELLANEOUS
When John Quincy Adams (1767-1848) left the presidency, he was the last chief executive to have a broad and working knowledge of science.

1829. MEDICINE / PATHOLOGY
William E. Horner (1793-1853) published *A Treatise on Pathological Anatomy* (Philadelphia), the first such work produced in the United States.

1829. TECHNOLOGY AND INVENTION
Jacob Bigelow (1786-1879) published *Elements of Technology* (Boston), based on his lectures as Rumford Professor on the Application of Science to the Useful Arts in Harvard University. This publication introduced the word "technology" in its modern usage.

1829 (June 27). ORGANIZATIONS—RESEARCH INSTITUTIONS
James Louis Macie Smithson (b.1765) died in Genoa, Italy. The Englishman's first beneficiary died in 1835 and the residue of Smithsonian's estate, by terms of his will, went to the United States to establish a Smithsonian Institution "for the increase & diffusion of knowledge among men." The proceeds were in excess of $500,000.

1830. GOVERNMENT—FEDERAL / ASTRONOMY
A Depot of Charts and Instruments was formed in the U.S. Navy, with Lieutenant L.M. Goldsborough in charge, and with the intent to supply navigational materials to the naval fleet. Goldsborough was succeeded in 1833 by Lieutenant Charles Wilkes (1798-1877) who in turn was succeeded (in 1836) by James M. Gilliss (1811-1865), and in carrying out its mission the Depot established an observatory. In an act of August 31, 1842, money was appropriated for a building that established the Naval Observatory in all but name. When Matthew Fontaine Maury (1806-1873) took over the Depot in 1842, he diverted the work away from astronomy and toward oceanography and meteorology; Maury remained in charge until 1861.

1830 (February 9). ORGANIZATIONS—SOCIETIES AND ASSOCIATIONS / NATURAL HISTORY
The Boston Society of Natural History was established; it was incorporated in 1831.

1830-1831. CHEMISTRY
Benjamin Silliman (1779-1864) published *Elements of Chemistry* (New Haven, Conn.), in two volumes.

1830-1834. ZOOLOGY / CONCHOLOGY AND MALACOLOGY
Thomas Say (1787-1834) published *American Conchology* (New Harmony, Ind.) in six volumes. A seventh volume was published posthumously in 1838 by Timothy A. Conrad (1803-1877). In 1858, William G. Binney (1833-1909) edited *Complete Writings of Thomas Say on the Conchology of the United States* (New York).

1830-1870. GENERAL OR MISCELLANEOUS / EDUCATION IN SCIENCE
During this period, the most widespread teaching of science at the secondary level was in chemistry, natural philosophy, and astronomy. Of the natural history subjects, botany was most common.

1830-1880. GENERAL OR MISCELLANEOUS / GEOLOGY
George P. Merrill, in his *The First Hundred Years of American Geology* called these years in geology the "Era of State Surveys."

1831. CHEMISTRY
Chloroform was discovered by Samuel Guthrie (1782-1848).

1831. METEOROLOGY AND CLIMATOLOGY
William C. Redfield (1789-1857) published "Remarks on the Prevailing Storms of the Atlantic Coast of the North American States," *American Journal of Science* 20:17-51. The paper included his important theory on storms, that the winds blow counterclockwise about a center moving in the direction of prevailing winds. Redfield had begun to formulate his idea a decade earlier, with the observation of tree damage in Massachusetts and Connecticut from a hurricane on 3 September 1821.

1831. PHYSICS / ELECTRICITY AND ELECTRONICS
Joseph Henry (1797-1878) discovered that changes in magnetic fields induce electricity. Henry made this discovery of electromagnetic induction prior to and independently of Michael Faraday. Faraday, however, won credit because he was the first to publish.

1831 (July). PERIODICALS AND PUBLISHING / GEOLOGY
George William Featherstonhaugh (1780-1866) began issuance of *Monthly American Journal of Geology and Natural Science*. Only one volume was published.

1831 (July). TECHNOLOGY AND INVENTION
The earliest version of Cyrus McCormick's (1809-1884) grain reaper was shown. McCormick took a patent on the horse-drawn reaper in 1834.

1831-1866. GOVERNMENT—FEDERAL / EXPLORATION AND SURVEYING
During this period, the Topographical Engineers existed as a separate unit of the U.S. Army. (In 1838, the Corps of Engineers was assigned defense fortification, while the Topographical Bureau took on responsibility for civil projects.) The termination of separate status for the Topographical Engineers in 1866 is an indication of the waning role of the Army in surveying and exploration.

1832. GOVERNMENT—FEDERAL / GEOPHYSICS AND GEODESY
The U.S. Coast Survey, inactive since 1818, was reauthorized by Congress, under the terms of the law of 1807. Ferdinand R. Hassler (1770-1843) again took charge of the Survey. Aimed at the unpopular and failed attempts of former President John Quincy Adams, the legislation prohibited the establishment of a permanent astronomical observatory.

1832. PALEONTOLOGY
Jacob Green (1790-1841) published *Monograph of the Trilobites of North America* (Philadelphia).

1832. ZOOLOGY—HUMAN / PHYSIOLOGY
Robley Dunglison (1798-1869) published the two-volume work, *Human Physiology* (Philadelphia), which had an eighth edition in 1856.

1832 (July 13). EXPLORATION AND SURVEYING
The source of the Mississippi River at Lake Itasca, Minnesota, was found by an exploring party led by Henry R. Schoolcraft (1793-1864).

1832 and 1834. ZOOLOGY / ORNITHOLOGY
Thomas Nuttall (1786-1859) published *A Manual of the Ornithology of the United States and Canada: The Land Birds* (Cambridge, Mass., 1832) and *A Manual of the Ornithology of the United States and Canada: The Water Birds* (Boston, Mass., 1834).

1832-1835. MINERALOGY AND CRYSTALLOGRAPHY
Charles U. Shepard (1804-1886) published *A Treatise on Mineralogy* (New Haven, Conn., 1832, 1835).

1832-1861. PALEONTOLOGY
Timothy Abbott Conrad (1803-1877) published three significant paleontological works during this period: *Fossil Shells of the Tertiary Formations of North America* (Philadelphia, 1832-1835); *Eocene Fossils of Claiborne, With Observations on This Formation in the United States, and Geological Map of Alabama* (Philadelphia, 1835); and *Fossils of the Medial Tertiary of the United States* (Philadelphia, 1838-1861). Conrad made an early American case for the use of organic remains in dating geological strata, views which he presented in "Observations on the Tertiary Strata of the Atlantic Coast," *American Journal of Science*, vol. 28 (1835):104-111, 280-282.

1833. BOTANY
Almira Hart Lincoln Phelps (1793-1884) published a popular *Botany for Beginners* (Hartford, Conn.). It had numerous editions and by 1867 had sold some 270,000 copies.

1833. GEOLOGY
As state geologist since 1830, Edward Hitchcock (1793-1864) published the final report, *Geology of Massachusetts* (Amherst).

1834. BOTANY / EXPLORATION AND SURVEYING
Thomas Nuttall (1786-1859), as a member of Nathaniel Jarvis Wyeth's expedition to Oregon, was the earliest accomplished and informed botanist to carry out collecting and study across the continent.

1834. GOVERNMENT—FEDERAL / GEOLOGY
George William Featherstonhaugh (1780-1866) became the first geologist in the employ of the United States government; he was engaged to examine the mineralogy and geology of the Ozark Mountains and later (1835) the Green Bay, Wisconsin area.

1834. ORGANIZATIONS—INDUSTRY / CHEMISTRY
Samuel Luther Dana (1795-1868) began work for the Merrimac Print Works to improve their bleaching and dyeing operations. He remained in their employ until 1868.

1834. PALEONTOLOGY
Samuel George Morton (1799-1851) published on the fossils collected by the Lewis and Clark expedition in *Synopsis of the Organic Remains of the Cretaceous Group of the United States* (Philadelphia).

1835. ORGANIZATIONS—ACADEMIC / ENGINEERING AND APPLIED SCIENCE
The first degrees in civil engineering in the country were awarded by Rensselaer Institute.

1835. ORGANIZATIONS—SOCIETIES AND ASSOCIATIONS
The Western Academy of Natural Sciences was established at Cincinnati.

mid-1830s to mid-1880s. GENERAL OR MISCELLANEOUS / BOTANY
These years were the highpoint for amateur botany in the United States.

1836. BOTANY
Asa Gray (1810-1888) published *Elements of Botany* (New York). The work was a significant departure from earlier textbooks; it was up-to-date on morphological and physiological aspects, and it promoted natural (as opposed to the Linnaean) classification. In 1842, Gray produced *The Botanical Text-Book for Colleges, Schools and Private Students* (New York) as a successor to the *Elements*.

1836. GOVERNMENT—FEDERAL / PHYSICS
Alexander Dallas Bache's (1806-1867) report on studies of steam boiler explosions, conducted in association with the Franklin Institute and published in its *Journal* (new series volume 17), represented an early example of government interest in the practical issues of scientific investigation. The research was carried out at the request of the U.S. Treasury Department by a committee of the Franklin Institute, chaired by Bache.

1836. GOVERNMENT—FEDERAL / TECHNOLOGY AND INVENTION
The Commissioner of Patents was established as a permanent office with staff. The legislation also required that inventions be tested for novelty and usefulness. Henry L. Ellsworth (1791-1858) was the first commissioner.

1836. GOVERNMENT—STATE / GEOLOGY, NATURAL HISTORY
The Natural History Survey of New York was established by the legislature and continued until 1843. Among the principals involved were Timothy A. Conrad (1803-1877), Ebenezer Emmons (1799-1863), James Hall, Jr. (1811-1898), William Williams Mather (1804-1859), and Lardner Vanuxem (1792-1848). The reports were issued between 1842 and 1894 though most appeared before about 1855. Those delayed were especially James Hall's reports on paleontology. One important change introduced by the Survey was to redirect American geologists away from a concern with correlations to English strata and toward the development of an American geological column.

1836. TECHNOLOGY AND INVENTION
Samuel Colt (1814-1862) patented the revolver named for him.

1836. ZOOLOGY / CONCHOLOGY AND MALACOLOGY
Isaac Lea (1792-1886) published *Synopsis of the Family of Naiades* (Philadelphia and London), which helped to bring order to the mollusks Unionidae. A fourth edition appeared in 1870.

1836. GENERAL OR MISCELLANEOUS / EDUCATION IN SCIENCE, CHEMISTRY
Charles Thomas Jackson (1805-1880) gave up the practice of medicine in 1836. He set up in Boston a laboratory for chemical analysis and it became one of the first instructional facilities of its kind in the country. About the same time, James Curtis Booth (1810-1888), returning from scientific study in Germany, established a student chemistry laboratory in Philadelphia.

1837. GOVERNMENT—STATE / GEOLOGY
The Indiana Geological Survey was established and David Dale Owen (1807-1860) was appointed as state geologist.

1837. GOVERNMENT—STATE / GEOLOGY, NATURAL HISTORY
The Massachusetts legislature authorized a geological and natural history survey of the state.

1837. MINERALOGY AND CRYSTALLOGRAPHY
James Dwight Dana (1813-1895) published *A System of Mineralogy* (New Haven), which became a leading textbook. In his edition of 1844, Dana added a classification by chemical features, and in the 3rd edition of 1850 he went fully with the chemical arrangement rather than the original scheme modeled on natural history that drew upon physical characteristics.

1837. ORGANIZATIONS—OBSERVATORIES
Sears C. Walker (1805-1853) established an observatory associated with Central High School, Philadelphia.

1837. TECHNOLOGY AND INVENTION / ELECTRICITY AND ELECTRONICS
Samuel F.B. Morse (1791-1872) patented his ideas for a telegraph. Joseph Henry (1797-1878), who advised Morse, had worked out the basic concepts for a telegraph in 1835.

1837-1840. GOVERNMENT—STATE / GEOLOGY
The earliest geological survey of Ohio was conducted under the direction of William Williams Mather (1804-1859).

1838. ORGANIZATIONS—ACADEMIC / ASTRONOMY
The first permanent astronomical observatory in the country was established by Professor Albert Hopkins (1807-1872) at Williams College.

1838. PERIODICALS AND PUBLISHING
The American Philosophical Society initiated publication of its *Proceedings*.

1838 (August)-1842 (June). GOVERNMENT—FEDERAL / EXPLORATION AND SURVEYING
The United States Exploring Expedition, with six ships, was carried out under the command of Lieutenant Charles Wilkes (1798-1877). Authorization for the Expedition was signed into law by President Andrew Jackson on May 14, 1836 and the overall cost of the enterprise reached more than $900,000. The Expedition covered some 85,000 miles, carrying out scientific study and collecting in Latin America, Antarctica, the islands of the central Pacific, and the northwestern coast

of North America. A portion of the Antarctic continent came to be called Wilkes Land. Wilkes published a report on the Expedition in 1844 and other reports were done by the scientific staff and by a number of scientists over a period of some thirty years. Some planned volumes never were realized.

1838-1843. BOTANY
Asa Gray (1810-1888) and John Torrey (1796-1873) published the uncompleted *Flora of North America* (New York), in two volumes. The work helped replace the Linnaean with a natural classification for American botany and established the use of type specimens in the taxonomy of American plants.

1839. AWARDS AND PRIZES / CHEMISTRY
Robert Hare (1781-1858) was awarded the first Rumford Medal by the American Academy of Arts and Sciences for his 1801 invention of the oxyhydrogen blowpipe. The prize had been established in 1796 by Count Rumford.

1839. GOVERNMENT—FEDERAL / AGRICULTURE
The Patent Office was given $1,000 by Congress for the purposes of collecting agricultural statistics. This was the first time the federal government gave funds earmarked for agricultural purposes.

1839. ORGANIZATIONS / EDUCATION IN SCIENCE
The Lowell Institute lectures in Boston were initiated with a series of twelve lectures on geology by Benjamin Silliman, Sr. (1779-1864).

1839. ORGANIZATIONS—OBSERVATORIES
Harvard College Observatory was founded with William Cranch Bond (1789-1859) at its head.

1839. ORGANIZATIONS—SOCIETIES AND ASSOCIATIONS / MATHEMATICS
The American Statistical Association was founded with an orientation to applied mathematical studies.

1839. TECHNOLOGY AND INVENTION / CHEMISTRY
Vulcanization of rubber (making the state of the substance constant through changes in temperature) was discovered by Charles Goodyear (1800-1860).

1839. ZOOLOGY—HUMAN / ANATOMY
Samuel George Morton (1799-1851) published *Crania Americana; or, a Comparative View of the Skulls of Various Aboriginal Nations of North and South America* (Philadelphia and London). In this work, and in others, Morton tended to the conclusion that the different races were not related.

1839 (May 11). GENERAL OR MISCELLANEOUS / CHEMISTRY
Thomas Cooper (born 1759) died in Columbia, South Carolina. Arriving in the United States in 1793, he played an important role in the dissemination of chemical knowledge, including the preparation of annotated American editions of English textbooks.

1839-1840. TECHNOLOGY AND INVENTION / ASTRONOMY, PHOTOGRAPHY
On October 7, 1839 New York mechanic Alexander Simon Wolcott (?-1844) took the first photographic portrait. By December, John William Draper (1811-1882) also produced a daguerreotype portrait and he published an account in *Philosophical Magazine* (1840), the first received in Europe. During the winter, Draper produced the first photograph of the moon, which was divulged to the New York Lyceum of Natural History on 23 March 1840.

1840. GENERAL OR MISCELLANEOUS / BOTANY
George Engelmann (1809-1884) met Asa Gray (1810-1888). Thereafter, Engelmann became an important agent in St. Louis for the transmittal eastward of botanical knowledge of the west.

1840. ZOOLOGY / CONCHOLOGY AND MALACOLOGY
Augustus Addison Gould (1805-1866) published "Results of an Examination of the Species of Shells of Massachusetts and Their Geographical Distribution," *Boston Journal of Natural History* 3:483-494. This was a pioneering work in the United States on geographical distribution.

1840 (March 23). GENERAL OR MISCELLANEOUS / GEOLOGY
Geologist William Maclure (b.1763) died in St. Angelo, Mexico.

1840 (April 2). ORGANIZATIONS—SOCIETIES AND ASSOCIATIONS / GEOLOGY, NATURAL HISTORY
The initial meeting of the Association of American Geologists took place in Philadelphia (in 1842, the phrase "and Naturalists" was added to the name of the organization). Plans for the body began as early as 1838, at a meeting of the New York Board of Geologists at the Albany home of Ebenezer Emmons (1799-1863). Edward Hitchcock (1793-1864) was presiding officer at the first meeting and Benjamin Silliman (1779-1864) was president in 1841 and 1842. It was the predecessor of the American Association for the Advancement of Science.

1840 (May 15). ORGANIZATIONS—SOCIETIES AND ASSOCIATIONS
The National Institution for the Promotion of Science was founded in Washington, as a successor to the Columbian Institute, with Secretary of War Joel R. Poinsett taking the lead. In 1841, it was given responsibility by the U.S. Navy for the collections of the Wilkes Expedition, but by about 1843 this was taken away. In 1842, the organization was incorporated by Congress as the National Institute. The Institute, with its political and amateur base, clashed with the emerging professional leadership in American science and with the Association of American Geologists and Naturalists. It never was strong and within a few years became inactive, although it was revived in 1855 as essentially a local organization. When its charter expired in 1862, its library and museum were transferred to the Smithsonian Institution.

[1840]. ORGANIZATIONS—ACADEMIC / ENGINEERING AND APPLIED SCIENCE
After the decade of the 1830s, West Point began to recede from its previous emphasis on engineering and by 1860 it was more emphatically devoted to military concerns.

1841. CHEMISTRY
John White Webster (1793-1850) produced two American editions of Justus von Liebig's *Organic Chemistry in Its Application to Agriculture and Physiology* (Cambridge, Mass.).

1841. CHEMISTRY
John William Draper (1811-1882) developed the proposition that only rays that are absorbed can produce chemical change. It came to be known as the Grotthuss-Draper law (his name teamed with a prior but apparently unknown promulgator of the same idea in 1817).

1841. EXPLORATION AND SURVEYING / ARCHAEOLOGY
John Lloyd Stephens (1805-1852) published *Incidents of Travel in Central America, Chiapas, and Yucatan* (New York), illustrated by Frederick Catherwood. The Stephens and Catherwood expedition began in 1839 to find the ruins of Copan. Their work resulted in the revelation of Copan as the remains of the Maya civilization, then not commonly known in Europe.

1841. GEOLOGY
Ebenezer Emmons' (1799-1863) assertion that geological strata found in the Taconic Mountains constituted a separate geological period set off a disagreement, the Taconic Controversy, that lasted for decades thereafter. Emmons' first formal presentation of the hypothesis apparently appeared in an 1841 paper given to the Association of American Geologists and Naturalists. In 1842, the first full account of the Taconic system was given in his report on the *Geology ... of the Second Geological District of New York* (Albany).

1841. METEOROLOGY AND CLIMATOLOGY
James Pollard Espy (1785-1860) published *The Philosophy of Storms* (Boston).

1841. ZOOLOGY / CONCHOLOGY AND MALACOLOGY
Augustus Addison Gould (1805-1866) published *Report on the Invertebrates of Massachusetts* (Cambridge, Mass.) as part of the work of the state geological and natural history survey. It was the first book in the country to attempt a study of all of the mollusks in a geographical region.

1841-1842. ORGANIZATIONS—ACADEMIC / MATHEMATICS
James Joseph Sylvester (1814-1897) served briefly as professor of mathematics at the University of Virginia, leaving the post in part because of anti-Semitism.

1841-1850. ZOOLOGY
Nicholas M. Hentz (1797-1856) presented a series of papers on spiders, "Descriptions and Figures of the Araneides of the United States," to the Boston Society of Natural History, which subsequently were published in the Society's *Journal*.

1842. GEOLOGY
Henry Darwin Rogers (1808-1866) and William Barton Rogers (1804-1882) presented the first significant American work on geological theory, published as "On the Physical Structure of the Appalachian Chain, as Exemplifying the Laws Which Have Regulated the Elevation of Great Mountain Chains Generally," *Reports of the Meetings of Association of American Geologists and Naturalists* (1843): 474-531.

1842. GOVERNMENT—FEDERAL / ASTRONOMY
Congress appropriated funds for what became a permanent Navy Observatory, in association with the Depot of Charts and Instruments. In this same year, Matthew Fontaine Maury (1806-1873) was put in charge of the Depot and he became head of the new observatory when the building was completed in 1844.

1842. GOVERNMENT—FEDERAL / TECHNOLOGY AND INVENTION
The U.S. Congress appropriated $30,000 to test Samuel F.B. Morse's (1791-1872) telegraph by constructing a line between Baltimore and Washington.

1842. ORGANIZATIONS—SOCIETIES AND ASSOCIATIONS / ANTHROPOLOGY AND ETHNOLOGY
The American Ethnological Society was established.

1842. ORGANIZATIONS—SOCIETIES AND ASSOCIATIONS / ASTRONOMY
The Cincinnati Astronomical Society was established. In November 1843, John Quincy Adams (1767-1848) participated in the laying of the cornerstone for the Society's refracting telescope, the largest in the Americas. It resulted from the widespread solicitation and promotion by Ormsby MacKnight Mitchel (1809-1862).

1842. PERIODICALS AND PUBLISHING
Benjamin Peirce (1809-1880) established *Cambridge Miscellany of Mathematics, Physics and Astronomy*. It ceased publication the next year.

1842. ZOOLOGY
John Edwards Holbrook (1794-1871) published the second edition of *North American Herpetology; or, a Description of the Reptiles Inhabiting the United States* (Philadelphia), in five volumes. The first edition, published 1836-1840, was replaced by the new one.

1842 (March 30). MEDICINE / SURGERY
Crawford Williamson Long (1815-1878) made the first surgical use of ether, but he did not reveal the fact in publication until 1849.

1842 (May). GOVERNMENT—FEDERAL / EXPLORATION AND SURVEYING
Colonel John Fremont (1813-1890) began an expedition to explore the Rocky Mountains in southern Wyoming. Subsequent explorations followed and in 1845 he published *The Report of the Exploring Expedition to the Rocky Mountains in the Year 1842 and to Oregon and North California in the Years 1843-44* (Washington). The report had the effect of contributing to awareness and interest in the West.

1842-1849. BOTANY
Thomas Nuttall (1786-1859) published *The North American Sylva* (Philadelphia), in three volumes. It was initiated as an appendix to Francois Andre Michaux's (1770-1855) *North America Sylva* and also published separately.

1843. BOTANY
John Torrey (1796-1873) published *Flora of the State of New York* (Albany) in two volumes, part of the survey of the state.

1843. GOVERNMENT—FEDERAL / GEOPHYSICS AND GEODESY
Alexander Dallas Bache (1806-1867) became superintendent of the U.S. Coast Survey. Bache remained head of the Survey until his death. During his tenure, more physical scientists found employment with the Survey than in any other place.

1843. MEDICINE
Oliver Wendell Holmes (1809-1894) addressed the problem of puerperal (childbed) fever by advising a routine of personal cleanliness among physicians. His views were presented in his "The Contagiousness of Puerperal Fever," *New England Quarterly Journal of Medicine and Surgery* 1:503-530.

1843. METEOROLOGY AND CLIMATOLOGY
Elias Loomis (1811-1889) presented to the American Philosophical Society a paper that introduced to meteorology the idea of graphic illustration, that showed deviation from the normal barometer. The paper was published as "On Two Storms Which Were Experienced Throughout the United States in the Month of February, 1842," *Transactions of American Philosophical Society* 9 (1846): 161-184.

1843-1857. GOVERNMENT—FEDERAL / METEOROLOGY AND CLIMATOLOGY
James Pollard Espy (1785-1860) published reports on meteorology in the capacity of a paid meteorologist of the federal government (1843, 1850, 1851, 1857). The first report was to the army surgeon general, the next two to the secretary of the navy, and the last directly to Congress.

1844. CHEMISTRY
Eben Norton Horsford (1818-1893) began two years of study with the German chemist, Justus von Liebig. As the second American to study with Liebig, he played an important role in the transfer of German chemistry to the United States, beginning with letters published in the Albany *Cultivator* that brought Liebig's work to the attention of farmers. During a period of study from 1840-1844, John Lawrence Smith (1818-1883) had spent some time with Liebig, the first American to do so.

1844. MEDICINE
Connecticut dentist Horace Wells (1815-1848) undertook to use nitrous oxide as an anaesthetic in the extraction of teeth. A subsequent demonstration at the Harvard Medical School failed, and in the wake of the controversy over priority in the discovery of anaesthesia, Wells committed suicide.

1844. PHILOSOPHY
Samuel Tyler (1809-1877), a lawyer, published *Discourse on the Baconian Philosophy* (Baltimore). In this work, the widely professed philosophy of American science in the early nineteenth century was discussed and defended by its chief American promoter. As late as 1877, Tyler undertook a new edition at the request of Joseph Henry (1797-1878). In the work, Tyler equated scientific methodology with classification, a view widely shared.

1844. PHYSICS / INSTRUMENTS AND INSTRUMENTATION
John William Draper (1811-1882) produced what apparently was the first photograph (daguerreotype) of a diffraction spectrum. The diffraction grating had been engraved by Joseph Saxton (1799-1873).

1844. ZOOLOGY—HUMAN
Josiah Nott (1804-1873) of Alabama most likely was the first American to argue for the separate creations of the races of humans.

1844 (April). GENERAL OR MISCELLANEOUS
The National Institute for the Promotion of Science, directed by Joel Poinsett, Francis Markoe, Jr., and others, called the first national scientific conference in the country. Several hundred persons with scientific interest attended. One effect of the congress, however, was to alert the rising professional scientific community to the need to differentiate its interests from the larger supportive population.

1844 (May 24). TECHNOLOGY AND INVENTION / ELECTRICITY AND ELECTRONICS
The well-known message, "What hath God wrought?" was telegraphed by Samuel F.B. Morse (1791-1872) from Washington to Baltimore.

1845. BOTANY
Alphonso Wood (1810-1881) published *A Class-Book of Botany* (Boston, Mass.) in response to the *Botanical Textbook* of Asa Gray (1810-1888), which had failed to provide users with a field guide. Wood's work relied on a natural classification rather than the older Linnaean scheme.

mid-1840s. GENERAL OR MISCELLANEOUS
By this period, Boston was beginning to replace Philadelphia as the leading center for American scientists.

1845. ORGANIZATIONS—ACADEMIC
The Naval School was established. In 1850-51, it was reorganized as the U.S. Naval Academy.

1845. ORGANIZATIONS—SOCIETIES AND ASSOCIATIONS / TECHNOLOGY AND INVENTION
The National Association of Inventors was established but endured only a couple of years.

1845. PERIODICALS AND PUBLISHING
Scientific American, a technology publication of popular interest, was founded in New York by Rufus Porter (1792-1884). In 1846, it was taken over by Orson Munn (1824-1907) and Alfred Beach (1826-1896).

1846. CHEMISTRY
Charles Thomas Jackson (1805-1880) claimed priority over C.F. Schonbein for the discovery of guncotton.

1846. ORGANIZATIONS—RESEARCH INSTITUTIONS
The Smithsonian Institution was founded at Washington, D.C., by a Congressional act, on 10 August. Joseph Henry (1797-1878) was chosen as the first secretary (director) on 3 December.

1846. TECHNOLOGY AND INVENTION
The lock-stitch sewing machine was patented by Elias Howe (1819-1867).

1846. ZOOLOGY / MICROBIOLOGY AND MICROSCOPY
Joseph Leidy (1823-1891) identified parasitic *Trichina spiralis* in hogs.

1846 (June). ZOOLOGY / ORNITHOLOGY
The Academy of Natural Sciences of Philadelphia received a bird collection of 10,000 European specimens from Thomas B. Wilson (1807-1865), leading to the claim that it was the best ornithological collection in the world.

1846 (July). PERIODICALS AND PUBLISHING / ASTRONOMY
Ormsby M. Mitchel (1809-1862) began publication of the monthly *Sidereal Messenger*. It was the first serial in the United States devoted specifically to astronomy. It ceased publication in October 1848.

1846 (August 19). ORGANIZATIONS—ACADEMIC
The Corporation of Yale University enacted a plan to establish professorships in agricultural and applied chemistry. The so-called Yale School of Applied Chemistry opened on 1 November 1847 and in 1861 became the Sheffield Scientific School. The promoters of the scheme were Benjamin Silliman, Sr. (1779-1864), Benjamin Silliman, Jr. (1816-1885), and John Pitkin Norton (1822-1852). When established, it was the first scientific school affiliated with a college or university.

1846 (October 3). GENERAL OR MISCELLANEOUS / NATURAL HISTORY
Louis Agassiz (1807-1873) came to the United States to lecture at the Lowell Institute, Boston. In 1847, he became professor in the Lawrence Scientific School at Harvard University and continued in that position until his death.

1846 (October 16). MEDICINE
William Thomas Green Morton (1819-1868) at the Massachusetts General Hospital demonstrated the general anaesthetic use of sulphuric ether; John Collins Warren (1778-1856) performed the surgical operation. Morton's former teacher, Charles Thomas Jackson (1805-1880), subsequently claimed priority for the discovery. (Morton was granted a patent, on which Jackson's name also was included, on November 12, 1846.)

1847. ASTRONOMY
Benjamin Peirce (1809-1880) published a list of all the known orbits of comets in *American Almanac and Repository of Useful Knowledge*.

1847. CHEMISTRY
Benjamin Silliman, Jr. (1816-1885) published *First Principles of Chemistry* (Philadelphia / Boston). T. Sterry Hunt (1826-1892) did the section on organic chemistry.

1847. GOVERNMENT—FEDERAL / OCEANOGRAPHY
Matthew Fontaine Maury (1806-1873), at the U.S. Navy Depot of Charts and Instruments, began publication of *Wind and Current Charts*, with the issuance of those for the North Atlantic. The intent of the project was to promote maritime commerce and Maury distributed the charts free in exchange for ships' logs of winds and currents.

1847. ORGANIZATIONS—ACADEMIC
The Lawrence Scientific School was established at Harvard University.

1847. ORGANIZATIONS—OBSERVATORIES / INSTRUMENTS AND INSTRUMENTATION
The 15-inch refractor installed at the Harvard College Observatory was then (with one other) the largest in the world.

1847. ORGANIZATIONS—SOCIETIES AND ASSOCIATIONS
The Association of American Geologists and Naturalists, meeting in Boston, determined to transform itself into an American Association for the Advancement of Science, the first general scientific society to be truly national in scope. The first meeting of the new society was held in Philadelphia in September 1848, where a constitution was adopted and William C. Redfield (1789-1857) was elected as president. There were 461 members during the formative year. Two divisions were established: (1) general physics, mathematics, chemistry, civil engineering, applied sciences; and (2) natural history, geology, physiology, medicine. The Association began publication of *Proceedings* from the outset.

1847. ORGANIZATIONS—SOCIETIES AND ASSOCIATIONS / MEDICINE
The American Medical Association was established.

1847. PALEONTOLOGY / ZOOLOGY
Joseph Leidy (1823-1891) described the fossil of the horse in North America.

1847. ZOOLOGY
Jeffries Wyman (1814-1874) and Thomas S. Savage (1804-1880) published "Notice of the External Characters and Habits of Troglodytes gorilla, a New Species of Ourang From the Gaboon River. Osteology of

the Same," *Boston Journal of Natural History* 5:417-442. This was the first detailed description of the gorilla.

1847 (October 1). ASTRONOMY
Maria Mitchell (1818-1889) discovered a new telescopic comet. For this achievement, she was given a gold medal by the king of Denmark.

1847-1851. ASTRONOMY / PHOTOGRAPHY
George Phillips Bond (1825-1865) carried out pioneering daguerreotype work with stars, at the Harvard College Observatory. Working with his father, William Cranch Bond (1789-1859), a result was the earliest stellar photograph (1850).

1847-1894. PALEONTOLOGY
James Hall, Jr. (1811-1898) published *New York State Natural History Survey: Paleontology* (Albany), 8 volumes in 13.

1848. ASTRONOMY
Hyperion, the eighth moon of Saturn, was discovered by George Phillips Bond (1825-1865).

1848. BOTANY
Asa Gray (1810-1888) published his *Manual of the Botany of the Northern United States* (Boston). It included all known plants of the region, William S. Sullivant (1803-1873) contributing the sections on mosses and liverworts. There were five editions of the work during Gray's lifetime.

1848. GENERAL OR MISCELLANEOUS / ASTRONOMY
Astronomer Benjamin Apthorp Gould (1824-1896) received a doctorate at the University of Gottingen, having studied under C.F. Gauss.

1848. GENERAL OR MISCELLANEOUS / CHEMISTRY
Frederick Augustus Genth (1820-1893) came to the United States from Germany, having earned a doctorate at the University of Marburg in 1845. He was at the time unusually well prepared among analytical chemists in the United States.

1848. GEOPHYSICS AND GEODESY / INSTRUMENTS AND INSTRUMENTATION
John Locke (1792-1856) invented the electro-chronograph or magnetic clock. In use by the U.S. Coast Survey, it contributed to the accuracy of longitude determinations.

1848. MICROBIOLOGY AND MICROSCOPY / ZOOLOGY
Joseph Leidy (1823-1891) began studies on the micro-forms of life in animals. In 1853, he published "A Flora and Fauna Within Living Animals," *Smithsonian Contributions to Knowledge*, 5.

1848. ORGANIZATIONS—SOCIETIES AND ASSOCIATIONS
The first woman, Maria Mitchell (1818-1889), was elected to the American Academy of Arts and Sciences.

1848. ORGANIZATIONS—SOCIETIES AND ASSOCIATIONS / ENGINEERING AND APPLIED SCIENCE
The Boston Society of Civil Engineers was established, the earliest such organization in the country.

1848. PERIODICALS AND PUBLISHING / ARCHAEOLOGY
Ephraim G. Squier (1821-1888) and Edwin H. Davis (1811-1888) published *Ancient Monuments of the Mississippi Valley*. It was the first of the Smithsonian Contributions to Knowledge and a start for the program of the Institution's Secretary Joseph Henry (1797-1878) to promote research. The Contributions series continued until 1910.

1848. ZOOLOGY
Louis Agassiz (1807-1873) and Augustus Addison Gould (1805-1866) published *Principles of Zoology* (Boston). Most of the illustrations were done by Louis Francois de Pourtales (1823/24-1880). A German edition appeared in 1851.

1848 and 1854. ORGANIZATIONS—MUSEUMS / NATURAL HISTORY
The collections of the natural history museum established by Charles Willson Peale (1741-1827) in Philadelphia in 1786 were sold.

1849. ASTRONOMY
A paper by Daniel Kirkwood (1814-1895) in the *Proceedings* of the American Association for the Advancement of Science generated interest in the topic of planetary rotation.

1849. GEOGRAPHY AND CARTOGRAPHY
Arnold Henri Guyot (1807-1884) published *Earth and Man, or Lectures on Comparative Physical Geography in Its Relation to the History of Mankind* (Boston), based on his Lowell Institute lectures.

1849. GOVERNMENT—FEDERAL
The U.S. Department of the Interior was established by Congress. The function of the department was originally a general oversight of various governmental activities and facilities (e.g., federal lands, Indian affairs, patents, public buildings, pensions, the census). With the development of the governmental structure, Interior came to concentrate on federal natural resources.

1849. GOVERNMENT—FEDERAL / ASTRONOMY
The *Nautical Almanac* was established by the U.S. Navy; it was nominally connected with the Naval Observatory but situated in Cambridge, Massachusetts. Naval officer Charles Henry Davis (1807-1877) was made the first head of the Almanac office. The first volume of *The American Ephemeris and Nautical Almanac* was issued in 1852.

1849. GOVERNMENT—FEDERAL / TECHNOLOGY AND INVENTION
The U.S. Congress granted to Charles G. Page (1812-1868) $20,000 for development of a wet-cell-battery-powered locomotive, although the action did not set a precedent. Page eventually devised a prototype but it was very expensive and not dependable.

1849. INFORMATION ACCESS
An international exchange of scientific periodicals was organized by the Smithsonian Institution.

1849. METEOROLOGY AND CLIMATOLOGY
The Smithsonian Institution, under Joseph Henry (1797-1878), began the establishment of a national organization for daily collection of

meteorological data. The program included a system of telegraph stations.

1849. PERIODICALS AND PUBLISHING / ASTRONOMY
The *Astronomical Journal* was established by Benjamin A. Gould (1824-1896). It was intended to serve the needs and interests of astronomers and to further research rather than educational interests. Publication was suspended from 1861 to 1886.

1849-1852. ASTRONOMY
An astronomical observatory was established and operated in Santiago, Chile, by Navy Lieutenant James M. Gilliss (1811-1865). The motivating purpose for the observatory was to compile simultaneous observational data on Venus and Mars in the northern and southern hemispheres, the northern aspect to be carried out at the Naval Observatory. The end purpose was to determine the distance of the earth from the sun. Unfortunately, Matthew Fontaine Maury (1806-1873), head of the Naval Observatory, failed to make the corresponding observations. When the government of Chile took over the southern observatory from Gilliss, it became the earliest such facility in that continent.

1850. ASTRONOMY
William Cranch Bond (1789-1859) and George Phillips Bond (1825-1865) discovered the Crepe (dusky) or C ring of Saturn.

1850. ASTRONOMY / PHOTOGRAPHY
The earliest clear daguerreotype of the moon was produced by William Cranch Bond (1789-1859), ten years after John William Draper (1811-1882) achieved the first but less definitive image.

1850. CHEMISTRY / AGRICULTURE
John Pitkin Norton (1822-1852) published the textbook, *Elements of Scientific Agriculture* (Albany, N.Y.).

1850. GENERAL OR MISCELLANEOUS / CHEMISTRY
Harvard University chemistry professor John White Webster (b.1793) was executed for murder.

1850 and 1854. MEDICINE / DISEASE
Daniel Drake (1785-1852) published a two-volume work, *A Systematic Treatise, Historical, Etiological and Practical, on the Principal Diseases of the Interior Valley of North America, as They Appear in the Caucasian, African, Indian, and Esquimaux Varieties of Its Population* (Cincinnati).

1850. ORGANIZATIONS—INDUSTRY / CHEMISTRY
William P. Blake (1825-1910) became what was probably the first full-time American college-educated chemist to work for an industrial concern (a Baltimore chemical manufacturer).

1850. PERIODICALS AND PUBLISHING
David A. Wells (1828-1898) began publication of the *Annual of Scientific Discovery*, which served the need for summaries and listing of American and European developments and publications. It continued until 1871.

mid-19th century. GENERAL OR MISCELLANEOUS
During the middle decades of the nineteenth century, an informal leadership group, known as the Lazzaroni, undertook to direct American science, especially in regard to its organization and standards of work and conduct. The core members of the group were Alexander Dallas Bache (1806-1867), Joseph Henry (1797-1878), Benjamin Peirce (1809-1880), and Louis Agassiz (1807-1873). Benjamin A. Gould (1824-1896) and Oliver Wolcott Gibbs (1822-1908) were less central but also within the circle; others also were in varying ways associated with the group. The name itself was current as of the mid-1850s. By the early 1860s, the group, in any self-conscious sense, was no longer active.

ca.1850. ZOOLOGY / PHYSIOLOGY
John Call Dalton (1825-1889), having graduated from Harvard Medical School (1847) and visited Paris (1850) where he attended lectures of Claude Bernard, gave up the practice of medicine and became the first American to devote his career to teaching and research in physiology.

1850-1856. MICROBIOLOGY AND MICROSCOPY / INSTRUMENTS AND INSTRUMENTATION
John William Draper (1811-1882) engaged in what probably was the first microphotographic work, producing photographs of slides through a microscope. These were published in his 1856 work, *Human Physiology, Statical and Dynamical; or, The Conditions and Course of the Life of Man* (New York).

1851. CHEMISTRY
Frederick Augustus Genth (1820-1893) published an account of the earliest definite knowledge of crystallized salts of the cobalt ammines. This appeared in *Keller-Tiedemann's Nordamerikan Monatsbericht* 2:8-12.

1851. EXPLORATION AND SURVEYING
Lieutenant William Lewis Herndon (1813-1857) undertook an expedition to explore the Amazon River.

1851. INSTRUMENTS AND INSTRUMENTATION
Joseph Saxton (1799-1873) won a gold medal at the Great Exhibition in London for his precision balances.

1851. ORGANIZATIONS—ACADEMIC
The Lawrence Scientific School at Harvard University conferred its first degrees. Of the three recipients, one was geologist and naturalist Joseph LeConte (1823-1901), later a professor at the University of California.

1851. ORGANIZATIONS—SOCIETIES AND ASSOCIATIONS / GEOGRAPHY AND CARTOGRAPHY
The American Geographical and Statistical Society was established in New York. In 1852, the Society began publishing a *Bulletin*. In 1871, the organization became the American Geographical Society of New York.

1851. RELIGION AND THEOLOGY / GEOLOGY
Edward Hitchcock (1793-1864) published *Religion of Geology and Its Connected Sciences* (Boston).

1851 (January 27). GENERAL OR MISCELLANEOUS / NATURAL HISTORY
Naturalist and artist John James Audubon (b.1785) died in New York City.

1851-1857. ANTHROPOLOGY AND ETHNOLOGY
Henry Rowe Schoolcraft (1793-1864) published *Historical and Statistical Information Respecting the History, Condition and Prospects of the Indian Tribes of the United States*, in six volumes (1-5, Washington, D.C.; 6, Philadelphia).

1852. GENERAL OR MISCELLANEOUS / GEOPHYSICS AND GEODESY
Benjamin Peirce (1809-1880) took charge of longitude observations for the U.S. Coast Survey. He continued that work until he became superintendent of the Survey in 1867. During these connections, he continued to hold his position as professor at Harvard University.

1852. GEOLOGY
David Dale Owen (1807-1860) published the results of his geological studies in Minnesota, Iowa, and into the South Dakota Badlands in the country's most lavish geological report up to that time. The report included maps and plates prepared by Owen and his brother Richard Owen (1810-1890). The publication was *Report of a Geological Survey of Wisconsin, Iowa and Minnesota* (Philadelphia), in two volumes, prepared for the U.S. Treasury Department.

1852. ORGANIZATIONS—ACADEMIC
The Chandler School of Science and Arts began at Dartmouth College, based on a gift from Abiel Chandler (1777-1851).

1852. ORGANIZATIONS—SOCIETIES AND ASSOCIATIONS / ENGINEERING AND APPLIED SCIENCE
The American Society of Engineers and Architects was established as the first organization of engineers on the national level; it later became the American Society of Civil Engineers. Formed by engineers in the vicinity of New York City, the Society went into an extended period of inactivity about 1855.

1852. ORGANIZATIONS—SOCIETIES AND ASSOCIATIONS /
PHARMACOLOGY AND PHARMACY
The American Pharmaceutical Association was established.

1852. ASTRONOMY / INSTRUMENTS AND INSTRUMENTATION
The Alvan Clark and Sons company was formed in Cambridge, Massachusetts, and became one of the leading telescope-making firms in the world. Of American observatories constructed during the second half of the century, nearly all had an equatorial refracting telescope and other apparatus made by Clark and Sons.

ca.1852. TECHNOLOGY AND INVENTION
A steelmaking process was devised by William Kelly (1811-1888) (in Eddyville, Kentucky) that was similar to the Bessemer process developed in England in 1856.

1852-1859. GENERAL OR MISCELLANEOUS / ASTRONOMY
During the 1850s, a public controversy revolved around the Dudley Observatory in Albany, New York, which was established in 1852. A complex affair, it involved the emerging professional scientific community in the persons of Benjamin Apthorp Gould (1824-1896), Alexander D. Bache (1806-1867), Benjamin Peirce (1809-1880), and Joseph Henry (1797-1878) (the group also constituting the so-called Lazzaroni), who were made a Scientific Council in 1855, and the Observatory's trustees. Gould served as head of the Observatory and his personality was a major factor in the contentious situation, but the issues also were ones of relative authority between professional and lay persons in the affairs of a scientific institution. On 3 January 1859, Gould was forcibly removed from his home and directorship.

1853. GEOLOGY
Jules Marcou (1824-1898) published *A Geological Map of the United States and the British Provinces of North America* (Boston). Subsequent editions appeared in 1855 and 1858 (published in Europe) and all were controversial with American geologists.

1853. METEOROLOGY AND CLIMATOLOGY
Matthew Fontaine Maury (1806-1873) organized the first international meteorological meeting, at Brussels. It dealt with meteorology at sea

only, in spite of Maury's hopes that it also would involve land weather as well.

1853. METEOROLOGY AND CLIMATOLOGY
James H. Coffin (1806-1873) delineated three wind zones in the northern hemisphere. His conclusions were given in his "Winds of the Northern Hemisphere," *Smithsonian Contributions to Knowledge*, 6.

1853. ORGANIZATIONS—SOCIETIES AND ASSOCIATIONS
The New Orleans Academy of Sciences was established.

1853. ORGANIZATIONS—SOCIETIES AND ASSOCIATIONS
The organization that came to be known in 1868 as the California Academy of Sciences was founded at San Francisco as the California Academy of Natural Sciences.

1853. ORGANIZATIONS—SOCIETIES AND ASSOCIATIONS / NATURAL HISTORY
The Elliott Society of Natural History was established in Charleston, South Carolina. It was named for Stephen Elliott (1771-1830).

1853 (May). EXPLORATION AND SURVEYING
Elisha Kent Kane (1820-1857), with support from New York merchant Henry Grinnell (1799-1874) and the U.S. Navy, undertook an expedition to the Arctic. Before their ship became ice bound and was abandoned in summer 1855, the party reached further north than any previous outing. It was not possible to bring back the natural history materials collected but other scientific data did get into print. (Grinnell, with the Navy, had supported a similar expedition in 1850 but scientific work was not a conspicuous part of it.)

1853 (Summer)-1855 (October). GOVERNMENT—FEDERAL / EXPLORATION AND SURVEYING
The U.S. Navy's North Pacific Exploring and Surveying Expedition was carried out under the command first of Cadwalader Ringgold (1802-1867) and later John Rodgers (1812-1882) (the Ringgold-Rodgers expedition). It went to South Africa, Australia, the Coral Sea, Hong Kong, Japan, and the Aleutian Islands. Though economic purposes were primary, scientific study also was a significant part of the intent,

including oceanography, astronomy, and natural history. Civilian scientists were part of the corps.

1854. CHEMISTRY
Josiah Parsons Cooke, Jr. (1827-1894) published his widely noted first paper, "The Numerical Relation Between the Atomic Weights and Some Thoughts on the Classification of the Chemical Elements," *Memoirs of American Academy of Arts and Sciences*, new series 5 (1855): 235-257, 412 and *American Journal of Science*, 2nd series 17 (1854): 387-407. He proposed an arrangement of the elements in six series.

1854. GEOLOGY / MINING
Josiah Dwight Whitney (1819-1896) published *The Metallic Wealth of the United States* (Philadelphia).

1854. GOVERNMENT—FEDERAL / ENTOMOLOGY
Townend Glover (1813-1883) was the first federal entomologist, holding a position in the Bureau of Agriculture from this year until 1878.

1854. GOVERNMENT—STATE / ENTOMOLOGY
Asa Fitch (1809-1879) became the first person to receive appointment as full-time state entomologist, holding that position in New York state until 1870. Thaddeus W. Harris (1795-1856) had been the first to be paid by a state for entomological work when he was compensated at $175 for his 1841 report on the injurious insects of Massachusetts. During the years 1855-1872, Fitch published a series of fourteen "Reports on the Noxious, Beneficial and Other Insects of the State of New York" in *Transactions of New York State Agricultural Society*.

1854. GOVERNMENT—STATE / GEOLOGY
David Dale Owen (1807-1860) was appointed Kentucky state geologist.

1854. PALEONTOLOGY / BOTANY
Leo Lesquereux (1806-1889), who was the earliest student of fossil plants in the United States, published his first work on the subject. It was on Carboniferous fossils of Pennsylvania.

1854. PHYSICS
Based on his experiments, the spectral identification of elements was proposed by David Alter (1807-1881).

1855. CHEMISTRY / GEOLOGY
Benjamin Silliman, Jr. (1816-1885) prepared *Report on the Rock Oil, or Petroleum, From Venango County, Pennsylvania* (New Haven) as a consultant to individuals interested in drilling in that location. In his chemical analysis, Silliman used fractional distillation methods to derive the component parts and made recommendations for their use. The report played an important part in the promotion of the petroleum industry.

1855. OCEANOGRAPHY
Matthew Fontaine Maury (1806-1873) published *Physical Geography of the Sea* (New York), the earliest oceanography textbook. Although the work was reprinted repeatedly and translated into several languages, it was widely criticized by the scientific community on account of its lack of organization, errors, and unfounded generalizations.

1855. ORGANIZATIONS—INDUSTRY / CHEMISTRY
The first chemist regularly employed by an American tanner began work.

1855. PHYSICS / MATHEMATICS
Benjamin Peirce (1809-1880) published *A System of Analytic Mechanics* (Boston), approaching the subject from the perspective of mathematical theory.

1856. CHEMISTRY
Frederick Augustus Genth (1820-1893) and Oliver Wolcott Gibbs (1822-1908) published a significant chemical work, "Researches on the Ammonia-Cobalt Bases," *American Journal of Science*, 2nd series 23:234, 319, and 24:86.

1856. GEOLOGY
J. Peter Lesley (1819-1903) published *A Manual of Coal and Its Topography* (Philadelphia), based on his studies of Pennsylvania geology.

1856. ORGANIZATIONS—SOCIETIES AND ASSOCIATIONS
The Academy of Science of St. Louis was established.

1856. ORGANIZATIONS—SOCIETIES AND ASSOCIATIONS
The Chicago Academy of Natural Sciences was established. In 1859, it was incorporated as the Chicago Academy of Sciences.

1857. GENERAL OR MISCELLANEOUS / NATURAL HISTORY
Henry Darwin Rogers (1808-1866) became what was probably the first native-born American to assume a professorship in Europe when he was appointed Regius professor of natural history in the University of Glasgow.

1857. NATURAL HISTORY / ZOOLOGY
Louis Agassiz (1807-1873) published the first volume of his *Contributions to the Natural History of the United States* (Boston), which included the "Essay on Classification." When plans were announced in 1855, a ten-volume comprehensive work was contemplated, but only four volumes ever appeared.

1857. PALEONTOLOGY
The existence of Permian fossils in North America was first noted by Fielding Bradford Meek (1817-1876).

1857. ZOOLOGY / PHYSIOLOGY
The pioneering work in America on the kidney, by Charles Edward Isaacs (1811-1860), was published in the *Transactions of New York Academy of Medicine* as "Researches Into the Structure and Physiology of the Kidney," vol. 1:377-435, and "On the Function of the Malpighian Bodies of the Kidney," vol. 1:437-457.

1857 (September). GENERAL OR MISCELLANEOUS / EVOLUTION
Charles Darwin wrote a letter to Asa Gray (1810-1888), relating to Darwin's ideas on evolution by natural selection. In 1858, it was presented to the Linnaean Society, along with other documents, in support of Darwin's priority over Alfred Russel Wallace.

1858. GEOLOGY
Henry Darwin Rogers (1808-1866) published *The Geology of Pennsylvania* (Edinburgh and Philadelphia), in two volumes. He had served as Pennsylvania state geologist in the 1830s.

1858. GEOLOGY
Jules Marcou (1824-1898) published *The Geology of North America* (Zurich). In the work, he was critical of geologists native to the United States.

1858. GOVERNMENT—FEDERAL / NATURAL HISTORY
The natural history collections held by the U.S. Patent Office were transferred to the Smithsonian Institution, with the aid of a Congressional appropriation and annual support thereafter. Prior to this, the support of the Institution (including its museum collections accumulated by Spencer Baird [1823-1887]) had come entirely from the Smithsonian bequest but the federal government support signalled the establishment of a National Museum.

1858. PHYSICS / ACOUSTICS
John LeConte (1818-1891) published "On the Influence of Musical Sounds on the Flame of a Jet of Coal Gas," *American Journal of Science*, 2nd series 23 (1858): 62-67. This work helped in the development of means to visualize sound wave effects.

1858. TECHNOLOGY AND INVENTION / ELECTRICITY AND ELECTRONICS
The first Atlantic telegraph was laid but its functionality did not endure. In 1866, Cyrus W. Field (1819-1892), also involved in the earlier venture, successfully laid a telegraph cable across the Atlantic Ocean.

1858 (May 15). GENERAL OR MISCELLANEOUS / CHEMISTRY
Chemist Robert Hare (b.1781) died in Philadelphia.

1859. BOTANY
As part of the reports of the U.S. and Mexican Boundary Survey, George Engelmann (1809-1884) published *Cactaceae of the Boundary* (Washington, D.C.), with illustrations by Paul Roetter. The work was a significant contribution toward the classification of cactus.

1859. ORGANIZATIONS—BOTANICAL GARDENS
The Missouri Botanical Garden was established in St. Louis by Henry Shaw (1800-1889). George Engelmann (1809-1884) was influential in persuading Shaw to do so.

1859. ORGANIZATIONS—MUSEUMS / ZOOLOGY
The Museum of Comparative Zoology at Harvard opened under the direction of its founder, Louis Agassiz (1807-1873). Of the endowment for the project, a bequest of $50,000 for research came from Francis Calley Gray, $100,000 from the Commonwealth of Massachusetts, and substantial additional funds from private donors.

1859. ORGANIZATIONS—SOCIETIES AND ASSOCIATIONS / ENTOMOLOGY
The Entomological Society of Philadelphia was established. In 1867, it was renamed the American Entomological Society, the earliest national organization in the biological sciences.

1859. ORGANIZATIONS—SOCIETIES AND ASSOCIATIONS / MEDICINE
The American Dental Association was established.

1859. PHYSICS
Benjamin Silliman, Jr. (1816-1885) published *First Principles of Physics, or Natural Philosophy* (Philadelphia).

1859. TECHNOLOGY AND INVENTION / GEOLOGY
The first productive oil well in the world was drilled at Titusville, Pennsylvania by Edwin Drake (1819-1880).

1859. ZOOLOGY—HUMAN / PHYSIOLOGY
John Call Dalton (1825-1889) published *A Treatise on Human Physiology* (Philadelphia), a text for students and medical practitioners. It reached a seventh edition in 1882. The work drew upon Dalton's personal experience in repeating many experimental procedures, such as in digestion and nutrition.

1859 (January 29). GENERAL OR MISCELLANEOUS / ASTRONOMY
Astronomer William Cranch Bond (b.1789) died in Cambridge, Massachusetts.

1859-1860. METEOROLOGY AND CLIMATOLOGY
William Ferrel (1817-1891) published "The Motions of Fluids and Solids Relative to the Earth's Surface," *Mathematical Monthly* 1 and 2 (January 1859-August 1860). This work presented his significant quantitative theory of motion relative to the earth's surface and its currents.

1860. ASTRONOMY
Elias Loomis (1811-1889) originated the first map of the frequency distribution of auroras. This was published in the *American Journal of Science*.

1860. BOTANY
Alvan Wentworth Chapman (1809-1899) published *Flora of the Southern United States* (New York). The volume complemented the work of John Torrey (1796-1873) and Asa Gray (1810-1888) to give a more general geographical coverage for American botany. A third and more complete edition of the *Flora* appeared in 1897.

1860. GOVERNMENT—STATE / GEOLOGY
As of this time, 29 of 33 states had at one time conducted a state geological survey.

1860 (February-April). EVOLUTION
A debate over Charles Darwin's *Origin of Species* took place at four meetings of the Boston Society of Natural History. The principal participants were Louis Agassiz (1807-1873), who argued against Darwin's propositions, and William Barton Rogers (1804-1882). The event was one of the primary early debates on the topic anywhere, and it was generally conceded that Rogers had the upper hand. During the same year, the American Association for the Advancement of Science was criticized in the press for not officially addressing Darwin's ideas at its annual meeting.

1860 (March). EVOLUTION
Asa Gray (1810-1888) published a review of Darwin's *Origin of Species* in the *American Journal of Science*. Darwin thought it was the best commentary to appear up to that time. Later in the year, Gray published a three-part review of the work in the *Atlantic Monthly* which was reprinted by Darwin and circulated in his country.

1861. ENGINEERING AND APPLIED SCIENCE / HYDROLOGY
Andrew A. Humphreys (1810-1883), assisted by Henry L. Abbot (1831-1927), both of the U.S. Army Corps of Topographical Engineers, published *Report upon the Physics and Hydraulics of the Mississippi River*. It required innovative ideas to deal with the data and helped to promote subsequent studies in hydraulics.

1861. GEOLOGY
John Strong Newberry (1822-1892) described the Grand Canyon as the result of erosion in his geological report, in Joseph C. Ives (1828-1868), *Report Upon the Colorado River of the West, Explored in 1857 and 1858* (U.S. Congress, Senate, Executive Document), pp.25, 32, 41-48, 103.

1861. ORGANIZATIONS—ACADEMIC
The Massachusetts Institute of Technology was chartered by the state; instruction began in 1865. In 1862, William Barton Rogers (1804-1882) became the first president and served until 1870 and again from 1878 to 1881.

1861. ORGANIZATIONS—ACADEMIC
The scientific school at Yale University was named the Sheffield Scientific School following a $100,000 gift from Joseph Sheffield.

1861-1875. GEOGRAPHY AND CARTOGRAPHY
Arnold Henri Guyot (1807-1884) prepared a series of textbooks and wall maps that helped to develop geographic education.

1862. ASTRONOMY
George Phillips Bond (1825-1865) published his work on Donati's Comet of 1858 in *Annals of Harvard College Observatory*. In 1865, he became the first American recipient of the gold medal from the Royal Astronomical Society for this work.

1862. ASTRONOMY
Benjamin Apthorp Gould (1824-1896) published the earliest effort to collect in one catalog star positions from different observatories, as *Standard Mean Right Ascensions of Circumpolar and Time Stars Prepared for the Use of U.S. Coast Survey* (Washington, D.C.).

1862. GEOLOGY
James Dwight Dana (1813-1895) published his *Manual of Geology* (Philadelphia and London). Through this and subsequent editions, it became for some years the standard work on American geology and a milestone in the maturation of that science in the United States.

1862. GOVERNMENT—FEDERAL / AGRICULTURE
An independent department of agriculture was established in the federal government, with a noncabinet commissioner at its head. It was an outgrowth of the old agricultural section of the U.S. Patent Office. The new department had among its charges to procure and dispense information of a scientific character and to hire scientists and conduct research. The first commissioner was Isaac Newton (1800-1867) and the first chemist was Charles M. Wetherill (1825-1871) (who stayed only a short time). The Department achieved cabinet status in 1889 along with an expansion of its responsibilities.

1862. PERIODICALS AND PUBLISHING
The Smithsonian Institution began publication of its Smithsonian Miscellaneous Collections series. It came to include pieces such as bibliographies, natural history descriptions, and physical tables.

1862 (July 2). GOVERNMENT—FEDERAL / EDUCATION IN SCIENCE
The Morrill Act (i.e., the federal land-grant college bill), for support of education in agriculture and mechanic arts, was signed by President Abraham Lincoln. Funding was based on the granting of 30,000 acres to each state for each of its senators and representatives. Similar legislation had been passed by Congress in 1859, but Southern influence resulted in its veto by the President.

1863. ASTRONOMY / INSTRUMENTS AND INSTRUMENTATION
Astrophysicist Lewis Morris Rutherfurd (1816-1892) began his work on diffraction gratings. In 1867, he constructed a ruling engine that was operated by a screw in place of levers and was powered by a turbine. The resulting gratings were better than any others then available, and by 1877 he could produce more than 17,296 lines per inch. His product was widely used and was the best available until the time of Henry A. Rowland (1848-1901).

1863. GOVERNMENT—FEDERAL / EXPLORATION AND SURVEYING
The Army's Topographical Engineers was terminated.

1863. ORGANIZATIONS—ACADEMIC
Josiah Willard Gibbs (1839-1903) received a doctorate in engineering from Yale University, one of the earliest Ph.D. degrees in the country. His thesis was titled "On the Form of the Teeth of Wheels in Spur Gearing."

1863. PERIODICALS AND PUBLISHING / ZOOLOGY
The *Bulletin* of the Museum of Comparative Zoology at Harvard began publication. This was the first serial devoted wholly to zoology in the country.

1863 (February 11). GOVERNMENT—FEDERAL / MILITARY AND WAR
The Secretary of the Navy established a Permanent Commission to which questions relating to science were referred. The Commission members were Charles Henry Davis (1807-1877), Joseph Henry (1797-1878), and Alexander Dallas Bache (1806-1867). During the War, the Commission's chief function was to review inventions and other ideas submitted to the Navy. It terminated at the end of the War and nothing of importance emerged from its work.

1863 (March 3). ORGANIZATIONS—SOCIETIES AND ASSOCIATIONS
The bill establishing the National Academy of Sciences was enacted. The details had been worked out at a dinner meeting by Alexander D. Bache (1806-1867), Charles H. Davis (1807-1877), Louis Agassiz (1807-1873), Benjamin Peirce (1809-1880), Benjamin A. Gould (1824-1896), and Massachusetts Senator Henry Wilson on 19 February. Joseph Henry (1797-1878) was not a part of this planning group. The fifty-member Academy had all the original and life members named in the bill of enactment. The first meeting of the Academy took place in New York on April 22, 1863 and a structure of two classes, (1) mathematics and physics, and (2) natural history, was agreed upon. The Academy made Bache the first president.

1863-1864. ZOOLOGY
James Dwight Dana (1813-1895) published a series of articles in the *American Journal of Science* presenting his ideas on "cephalization,"

which related to his theory of the progress of species through the increasing importance of the brain, a manifestation of Dana's interest in finding design in nature. His views also appeared in "On Cephalization," *New Englander* 22 (1863): 495-506. His ideas on the importance of the head region as a gauge of an animal's relative development grew out of work for his 1852 volume on Crustacea, one of the Wilkes Expedition reports.

1864. ASTRONOMY
Hubert Anson Newton (1830-1896) published his result of an examination of all records of the meteor shower of November 1861 and concluded that the periodic event had occurred thirteen times since 932 in a 33.25-year cycle. His results were in the *American Journal of Science* 37:377-389 and 38:53-61.

1864. ASTRONOMY / INSTRUMENTS AND INSTRUMENTATION
Henry Draper (1837-1882) published "On the Construction of a Silvered Glass Telescope 15-1/2 Inches in Aperture, and Its Use in Celestial Photography," *Smithsonian Contributions to Knowledge*, 14, part 2. It became a standard work on telescope construction.

1864. ENVIRONMENT AND CONSERVATION
George Perkins Marsh (1801-1882) published *Man and Nature; or, Physical Geography as Modified by Human Action* (New York and London), perhaps the earliest such work on the threat of humans to the natural world. In 1874, he issued a revised edition as *Earth as Modified by Human Action*.

1864. GENERAL OR MISCELLANEOUS
A falling-out occurred between Louis Agassiz (1807-1873) and his students at Harvard University. Alpheus Hyatt (1838-1902), Edward S. Morse (1838-1925), Alpheus Spring Packard, Jr. (1839-1905), Frederic W. Putnam (1839-1915), Samuel H. Scudder (1837-1911), and Addison E. Verrill (1839-1926) departed from Cambridge.

1864. GENERAL OR MISCELLANEOUS / ASTRONOMY
Cleveland Abbe (1838-1916) was accepted for study at the Pulkovo Observatory, the only American for whom there is evidence of study of science in Russia during the nineteenth century.

1864. GENERAL OR MISCELLANEOUS / ORGANIZATIONS—INDUSTRY
Bessemer steel was manufactured in the United States for the first time, in Michigan.

1864. RELIGION AND THEOLOGY / CHEMISTRY
Josiah Parsons Cooke, Jr.'s (1827-1894) 1860 lectures at the Brooklyn Institute of Arts and Sciences were published as *Religion and Chemistry or Proof of God's Plan in the Atmosphere and the Elements* (New York), with a second edition in 1880.

1864 (November 15). ORGANIZATIONS—ACADEMIC / MINING
The Columbia University School of Mines began operations. The first Engineer of Mines degree was awarded in 1867 and for several decades the School graduated nearly half of the mining engineers in the country. In the 1890s, the School became a more general engineering institution.

1864 (November 24). GENERAL OR MISCELLANEOUS / CHEMISTRY, GEOLOGY
Chemist and geologist Benjamin Silliman (b.1779) died at New Haven, Connecticut.

1865. EXPLORATION AND SURVEYING / GEOLOGY, ZOOLOGY
Geologist and zoologist Louis Agassiz (1807-1873) made an expedition to Brazil.

1865. ORGANIZATIONS—SOCIETIES AND ASSOCIATIONS / SOCIAL SCIENCES—GENERAL
The American Social Science Association was founded in Boston with the purpose of societal improvement (e.g., civil service, prisons, public health, education). It terminated in 1912, replaced by other, more specialized and professionalized societies.

1866. ASTRONOMY
Daniel Kirkwood (1814-1895) published on his discovery of gaps in the distribution of mean distances of asteroids from the sun; the disturbances in the distribution of the orbits were attributed to Jupiter's gravity. The work appeared as "On the Theory of Meteors," *Proceedings of American Association for the Advancement of Science for 1866* (1867): 8-14.

Kirkwood had made the initial discovery as early as 1857, and the phenomenon came to be known as Kirkwood gaps.

1866. ASTRONOMY
Chester Smith Lyman (1814-1890), in this year and again in 1874, made studies of Venus that resulted in the earliest reliable evidence of an atmosphere on that planet. The report of this work appeared in the *American Journal of Science*, 2nd series 43 (1867): 129-130 and 3rd series 9 (1875): 47-48.

1866. BIOLOGY—GENERAL / EVOLUTION
Edward D. Cope (1840-1897) and Alpheus Hyatt (1838-1902) each, and independently of Ernst Haeckel, proposed the idea of biological recapitulation in relation to evolutionary progress.

1866. GOVERNMENT—FEDERAL / ASTRONOMY
The Hydrographic Office was separated from the U.S. Naval Observatory. This allowed the Observatory to direct its efforts more to pure research in astronomy.

1866. ORGANIZATIONS—ACADEMIC / PALEONTOLOGY
Othniel Charles Marsh (1831-1899) was appointed professor of vertebrate paleontology at Yale University, the first such appointment in the United States.

1866. TECHNOLOGY AND INVENTION / ELECTRICITY AND ELECTRONICS
Mahlon Loomis (1826-1886) telegraphed messages between two mountains in West Virginia by way of radio waves. Kites were used to hold the aerials in the air.

1866. ZOOLOGY / MICROBIOLOGY AND MICROSCOPY
Joseph Leidy (1823-1891) identified parasitic hookworm in cats.

ca.1866. GEOLOGY
Benjamin Smith Lyman (1835-1920) developed a means of using structural contour lines to show subsurface geological structure.

Chronology

1867. EXPLORATION AND SURVEYING
James Orton (1830-1877) went on the first of three expeditions to the Andes. This initial excursion was sponsored by Williams College with the support of the Smithsonian Institution, which lent instruments. A second expedition took place in 1873 and a third in 1876. On the latter outing, Orton died while crossing Lake Titicaca in Bolivia. He published an account of his first expedition in *Andes and the Amazon; or, Across the Continent of South America* (New York, 1870); a new edition in 1876 included an account of his 1873 expedition.

1867. GOVERNMENT—FEDERAL / GEOLOGY
Ferdinand V. Hayden (1829-1887) began his Geological and Geographical Survey of the Territories, under the auspices of the U.S. Department of Interior. The survey took the name given above in 1873 and was active until 1879 especially in Nebraska and also in Wyoming, Idaho, and Colorado.

1867. GOVERNMENT—FEDERAL / GEOLOGY
The U.S. Army's Geological Exploration of the Fortieth Parallel began. The territory to be studied was a swath of about 100 miles in width running from California to Colorado and following what amounted to the route of the Central Pacific railroad, with special attention to mining interests. Clarence R. King (1842-1901), a civilian who had promoted the idea of the Survey, was put in charge. The Survey's field work continued until 1872 and publication of its reports in seven volumes was completed in 1880.

1867. GOVERNMENT—FEDERAL / GEOPHYSICS AND GEODESY
Benjamin Peirce (1809-1880) became superintendent of the U.S. Coast Survey. Among his new directions for the agency was to begin a transcontinental triangulation along the 39th parallel to connect the surveys on the Atlantic and Pacific coasts.

1867. ORGANIZATIONS—SOCIETIES AND ASSOCIATIONS
The Peabody Academy of Science was founded at Salem, Massachusetts. It later became the Peabody Museum.

1867. ORGANIZATIONS—SOCIETIES AND ASSOCIATIONS / BOTANY
A group of New York botanists, who had met informally for some years, began a more formal organization. In 1870, it took the name of the Torrey Botanical Club and began to publish what became the *Bulletin of Torrey Botanical Club*; the *Bulletin* achieved a reputation as the leading botanical taxonomic journal in the country.

1867. ORGANIZATIONS—SOCIETIES AND ASSOCIATIONS / ENGINEERING AND APPLIED SCIENCE
The American Society of Civil Engineers, inactive since 1855, was revived. The Society began publication of its *Transactions* in 1873 and was formally chartered in 1877.

1867. PERIODICALS AND PUBLISHING / BIOLOGY--GENERAL
The *American Naturalist* was founded by Alpheus Hyatt (1838-1902), Edward Sylvester Morse (1838-1925), Alpheus Spring Packard, Jr. (1839-1905), and Frederic Ward Putnam (1839-1915), former students of Louis Agassiz (1807-1873). It began as a popular journal but became increasingly technical in content.

1867-1872. ORGANIZATIONS—ACADEMIC / CHEMISTRY
New York University granted five Ph.D.s to students in chemistry. This early attempt to establish the doctor of philosophy degree in the United States was influenced by John William Draper (1811-1882).

1868. CHEMISTRY
Josiah Parsons Cooke, Jr. (1827-1894) published *Principles of Chemical Philosophy* (Cambridge, Mass.). A second revised and enlarged edition appeared in 1887 (Boston). It was an influential textbook.

1868. MATHEMATICS / INSTRUMENTS AND INSTRUMENTATION
Hubert Anson Newton (1830-1896) published *The Metric System of Weights and Measures* (Washington, D.C.).

1868. ORGANIZATIONS—ACADEMIC
Massachusetts Institute of Technology graduated its first students.

1868 and 1871. EVOLUTION
Edward Drinker Cope (1840-1897), a leading figure in the neo-Lamarckian branch of evolutionary studies, published his most significant contributions: "On the Origins of Genera," *Proceedings of Academy of Natural Sciences of Philadelphia* 20 (1868); 242-300; and "The Method of Creation of Organic Forms," *Proceedings of American Philosophical Society* 12 (1871): 229-263.

1869. ASTRONOMY
Jonathan Homer Lane (1819-1880) presented a paper, "On the Theoretical Temperature of the Sun," to the National Academy of Sciences. It was subsequently published in the *American Journal of Science*, 2nd series 50 (1870): 57-74. It was notable for its calculations of the relations of heat and mass in the sun.

1869. CHEMISTRY / TECHNOLOGY AND INVENTION
Celluloid, the first artificial plastic, was developed by John Wesley Hyatt (1837-1920). The discovery was independent of Englishman Alexander Parkes, who demonstrated celluloid as early as 1862. Hyatt patented the process in 1870, which involved combining camphor and collodion with heat and pressure.

1869. EXPLORATION AND SURVEYING / GEOLOGY
John Wesley Powell (1834-1902) carried out an exploration of the canyons of the Green and the Colorado rivers by boat, covering some nine hundred miles and including the Grand Canyon. This was the last such exploration of unknown territory in continental United States.

1869. GOVERNMENT—FEDERAL / EXPLORATION AND SURVEYING, GEOLOGY
Lieutenant George M. Wheeler (1842-1905) began a U.S. Army survey in the West that came to be devoted to the region west of the one hundredth meridian. Originally a topographical survey, in 1871 geological concerns were incorporated.

1869. GOVERNMENT—STATE / PUBLIC HEALTH
The first state board of health was established by Massachusetts.

1869. ORGANIZATIONS—ACADEMIC
Iowa State University at Ames opened as the first land-grant college in the country.

1869. ORGANIZATIONS—MUSEUMS / NATURAL HISTORY
The American Museum of Natural History (New York) was founded.

1869. ORGANIZATIONS—SOCIETIES AND ASSOCIATIONS
The American Union Academy of Literature, Science and Art was established. The organization was a response to creation of the National Academy of Sciences and the omission of John William Draper (1811-1882) from its membership. Draper was first president of the American Union Academy.

1869. PALEONTOLOGY
Joseph Leidy (1823-1891) published "The Extinct Mammalian Fauna of Dakota and Nebraska...," *Journal of Academy of Natural Sciences of Philadelphia*, 2nd series, vol. 7.

1869. TECHNOLOGY AND INVENTION
George Westinghouse (1846-1914) patented the railway air brake.

1869 (May 10). GENERAL OR MISCELLANEOUS
Completion of the first transcontinental railroad was signified by the meeting of two trains of the Union Pacific and the Central Pacific railways in Utah. Construction of the project was authorized by Congress by the Pacific Railway Act of July 1, 1862.

1869 (September 1). METEOROLOGY AND CLIMATOLOGY
Cleveland Abbe (1838-1916) at the Cincinnati Observatory commenced the issuance of weather forecasting bulletins.

1870. ASTRONOMY / PHYSICS
In viewing a solar eclipse at Jerez, Spain, on December 22, 1870, Charles Augustus Young (1834-1908) observed that the sun's dark spectral lines briefly reverse at totality. In this he is credited with discovery of the "reversing layer."

1870. GENERAL OR MISCELLANEOUS / GEOGRAPHY AND CARTOGRAPHY
William Healey Dall (1845-1927) published *Alaska and Its Resources* (Boston).

1870. GENERAL OR MISCELLANEOUS / ORGANIZATIONS—INDUSTRY
Standard Oil Company was established by John D. Rockefeller (1839-1937). In 1972, Standard Oil (New Jersey) took the name Exxon Corporation.

1870. GOVERNMENT—FEDERAL / GEOLOGY
John Wesley Powell (1834-1902) was put in charge of a Congressionally enacted survey that in 1876 became known as the Geographical and Geological Survey of the Rocky Mountain Region. Powell continued at the head until it was ended in 1879. For a time associated with the Smithsonian Institution, in 1874 the Survey was transferred to the Department of Interior.

1870. GOVERNMENT—FEDERAL / METEOROLOGY AND CLIMATOLOGY
Congress effectively assigned to the Army's Signal Service the task of providing meteorological information and weather forecasting, the beginnings of a national weather service, with part of its mission being aid to agriculture and business. At the time, the Signal Service was under the charge of Colonel Albert J. Myer (1829-1880). In 1871, Cleveland Abbe (1838-1916) became associated with the enterprise as a civilian meteorologist.

1870. MATHEMATICS
Benjamin Peirce (1809-1880) published his *Linear Associative Algebra* as a memoir of the National Academy of Sciences. It was issued in lithographed form in a limited edition and has been considered the earliest mathematical work of first importance to appear in the United States. After Peirce's death, his son, Charles Sanders Peirce (1839-1914), published the work in the *American Journal of Mathematics* 4, no. 2 (1881): 97-229; it was published separately in 1882 (New York).

1870. OCEANOGRAPHY
Louis Francois de Pourtales (1823/24-1880) prepared a chart that showed the distribution of ocean-floor sediments from Cape Cod to

Florida. Pourtales published results of his work in "Der Boden des Golfstroms und der Atlantischen Kuste Nord Amerika's," *A. Petermanns Mittheilungen aus J. Perthes geographischer Anstalt* 16.

1870. ORGANIZATIONS / ASTRONOMY
Simon Newcomb (1835-1909) recommended the establishment of a body to plan for observations of the transits of Venus of 1874 and 1882. The result was the Transit of Venus Commission, of which Newcomb was secretary.

1870. ORGANIZATIONS—SOCIETIES AND ASSOCIATIONS
By act of Congress, the limitation on the number of members in the National Academy of Sciences was removed.

1870s. ORGANIZATIONS—ACADEMIC / INSTRUMENTS AND INSTRUMENTATION
Before this date, there were few laboratory facilities available for college students. Microscopes were not very often available in college laboratories for students before 1870.

1870s. ORGANIZATIONS—SOCIETIES AND ASSOCIATIONS / PHILOSOPHY
William James (1842-1910) and Charles Sanders Peirce (1839-1914), in the context of the Metaphysical Club which they founded in Cambridge, Massachusetts, gave early formulation to pragmatism.

1870-1885. ASTRONOMY
With the assistance of the government, Benjamin Apthorp Gould (1824-1896) founded the Argentine National Observatory, at Cordoba. During his years in Argentina Gould carried out mapping of the stars of the southern skies and also assisted in a range of projects that helped to establish science in Argentina.

1870s and 1880s. EVOLUTION
During this period, the neo-Lamarckian approach to evolution, as promoted by Alpheus Hyatt (1838-1902) and Edward Drinker Cope (1840-1897), characterized the American biological community.

1871. ANTHROPOLOGY AND ETHNOLOGY / AGRICULTURE
Edward Palmer (1831-1911) published "Food Products of the North American Indians," *Report of the Commissioner of Agriculture.*

1871. FUNDS AND FUNDING / ORGANIZATIONS—SOCIETIES AND ASSOCIATIONS
The National Academy of Sciences received the residue of the estate of Alexander Dallas Bache, who had died in 1867. The funds were to be used for the support of scientific research.

1871. GOVERNMENT—FEDERAL / ICHTHYOLOGY AND PISCICULTURE
The U.S. Fish Commission was established. Spencer Fullerton Baird (1823-1887) was made head of the Commission, serving without salary. Baird established a research program and later developed a station at Woods Hole, Massachusetts, which was supported by university and private funds. The Commission was significantly involved in economic interests with its hatcheries and work in relation to fish culture. In 1903 the Commission was made a Bureau of Fisheries in the Department of Commerce and Labor, and in 1940 was incorporated into the newly founded Fish and Wildlife Service.

1871. OCEANOGRAPHY / ZOOLOGY
Louis Francois de Pourtales (1823/24-1880) published *Deep-Sea Corals* (Cambridge).

1871. ORGANIZATIONS—ACADEMIC / ENGINEERING AND APPLIED SCIENCE
Stevens Institute of Technology (Hoboken, New Jersey) was established. It became a center of training and research in mechanical engineering.

1871. ORGANIZATIONS—MEDICINE / PHYSIOLOGY
Henry Pickering Bowditch (1840-1911) was appointed an assistant professor in the Harvard Medical School and about this time established the country's first physiological teaching laboratory. In doing so, Bowditch drew upon his own experience as a student in Europe, especially at Leipzig.

1871. ORGANIZATIONS—SOCIETIES AND ASSOCIATIONS /
ENGINEERING AND APPLIED SCIENCE
The American Institute of Mining and Metallurgical Engineers was founded at Wilkes-Barre, Pennsylvania. In 1957, it was reorganized into three constituent societies: Society of Mining Engineers, The Metallurgical Society, and the Society of Petroleum Engineers.

1871. ZOOLOGY
Joel Asaph Allen (1838-1921) published "On the Mammals and Winter Birds of East Florida, with an Examination of Certain Assumed Specific Characters in Birds, and a Sketch of the Bird Faunae of Eastern North America," *Bulletin of Museum of Comparative Zoology* 2:161-450. The work was significant for its contributions to thinking on biogeography.

1871-1872. OCEANOGRAPHY / ZOOLOGY
Louis Agassiz (1807-1873) led a U.S. Coast Survey dredging expedition, aboard the *Hassler*, along the coast of South America. Louis Francois de Pourtales (1823/24-1880) was in charge of the dredging operations but equipment failure limited the success of the venture.

1872. ASTRONOMY / PHOTOGRAPHY
The first photographic spectrum of a star, Vega, was produced by Henry Draper (1837-1882).

1872. GOVERNMENT—FEDERAL / ENVIRONMENT AND CONSERVATION
The Yellowstone National Park in Wyoming was established by Congress as a national preserve.

1872. MEDICINE / NEUROBIOLOGY
Silas Weir Mitchell (1829-1914) published *Injuries of Nerves and Their Consequences* (Philadelphia). This was an expansion of a work he wrote in 1864 with William W. Keen, Jr. (1837-1932) and George R. Morehouse (1829-1905) entitled *Gunshot Wounds and Other Injuries of Nerves* (Philadelphia).

1872. ORGANIZATIONS—SOCIETIES AND ASSOCIATIONS / PUBLIC HEALTH
The American Public Health Association was founded.

1872. PERIODICALS AND PUBLISHING
Popular Science Monthly was established by Edward Livingston Youmans (1821-1887).

1872. ZOOLOGY / ORNITHOLOGY
Elliott Coues (1842-1899) published *Key to North American Birds* (Salem, Mass., and New York), which introduced to zoological interests the artificial key. The work also included a significant revision in taxonomy. The *Key* eventually reached a fifth edition, the last appearing posthumously in 1903.

1872. ZOOLOGY—HUMAN / PHYSIOLOGY
Henry Pickering Bowditch (1840-1911) undertook a study of the growth rates of Boston children, and showed that environmental and nutritional factors were of more likely importance than race in determining size. Results of his study appeared in the eighth *Annual Report of the Massachusetts State Board of Health* (Boston, 1877), pp.275-325, and in *Transactions of American Medical Association* 32 (1881): 371-377.

1872 (October). GENERAL OR MISCELLANEOUS / FUNDS AND FUNDING
English physicist John Tyndall (1820-1893) arrived in the United States to undertake a lecture tour, having been invited by a group of twenty-five American scientists, to promote the values of scientific research. He arrived in New York and remained until the early part of 1873. The profits of the lectures were given to endow the Tyndall Fund that supported study of science by Americans in Europe.

1872-1874. ZOOLOGY
Zoologist and oceanographer Alexander Agassiz (1835-1910) published "Revision of the Echini," *Illustrated Catalogue of the Museum of Comparative Zoology at Harvard College*, 7.

1873. FUNDS AND FUNDING / ORGANIZATIONS—SOCIETIES AND ASSOCIATIONS
The American Association for the Advancement of Science received its first endowment for the support of research. It was in the amount of $1000, donated by New York philanthropist Elizabeth Thompson (1821-1899).

1873. GENERAL OR MISCELLANEOUS / EDUCATION IN SCIENCE, NATURAL HISTORY
The Anderson School of Natural History, on Penikese Island (Massachusetts), was founded by Louis Agassiz (1807-1873). In June and July of that year, Agassiz gave the first marine biology course.

1873. GEOLOGY
Raphael Pumpelly (1837-1923) made what may have been the first use in the United States of thin sections in a petrographic investigation. The work related to his study of copper ores in Michigan.

1873. ORGANIZATIONS—ACADEMIC
A scientific school was organized at Princeton University. This was part of a trend in higher education at the time, one observer reporting that the number of scientific schools appended to academic institutions increased from seventeen in 1870 to a high of seventy in 1873.

1873. ORGANIZATIONS—SOCIETIES AND ASSOCIATIONS
The American Metrological Society was established at Columbia University. It had as a primary goal to determine or promulgate standard weights, measures, and money. It also undertook to promote the metric system.

1873. ZOOLOGY
Addison Emery Verrill (1839-1926) carried out a close study of animal life in Vineyard Sound, the largest such ecological undertaking in the country up to that time. (He was assisted by Sidney I. Smith, 1843-1926, the brother of his wife.) The result was "Report upon the Invertebrate Animals of Vineyard Sound and the Adjacent Waters, with an Account of the Physical Characters of the Region," *Report of United States Commissioner of Fisheries* 1:295-747.

1873 (February 1). GENERAL OR MISCELLANEOUS / METEOROLOGY AND CLIMATOLOGY, OCEANOGRAPHY
Oceanographer and meteorologist Matthew Fontaine Maury (b.1806) died at Lexington, Virginia.

1873 (December 14). GENERAL OR MISCELLANEOUS / GEOLOGY, ZOOLOGY
Geologist and zoologist Louis Agassiz (b.1807) died at Cambridge, Massachusetts.

1873-1876. PHYSICS
Edward Charles Pickering (1846-1919) published *Elements of Physical Manipulation* (Boston), in two volumes. Drawing upon his experience in having set up the first instructional physics laboratory in the United States, at the Massachusetts Institute of Technology, this work was the first American physics laboratory manual.

1874. GOVERNMENT—FEDERAL / METEOROLOGY AND CLIMATOLOGY
The Smithsonian Institution's program for the collection of meteorological data was transferred to the U.S. Army Signal Service.

1874. ORGANIZATIONS—OBSERVATORIES
A trust was established by California philanthropist James Lick (1796-1876) that was the foundation for the Lick Observatory. The first director was Edward Singleton Holden (1846-1914).

1874. ORGANIZATIONS—SOCIETIES AND ASSOCIATIONS
The American Association for the Advancement of Science was chartered in the state of Massachusetts, thus signifying its permanent character. In the same year, the category of Fellow, emphasizing research involvement, was instituted, and eligibility to hold office was restricted to this group. The Association divided into sections for (a) mathematics, physics, and chemistry, and (b) natural history. Further sectioning took place in 1892, 1895, and 1961.

1874. PALEONTOLOGY / ZOOLOGY
Othniel Charles Marsh (1831-1899) published "Fossil Horses in America," *American Naturalist* 8:288-294.

1874. PHILOSOPHY
John Fiske (1842-1901) published *The Outlines of Cosmic Philosophy Based on the Doctrine of Evolution* (London).

1874. RELIGION AND THEOLOGY
John William Draper (1811-1882) published *History of the Conflict Between Religion and Science* (New York).

1875. EXPLORATION AND SURVEYING
John Wesley Powell (1834-1902) published *Report on the Exploration of the Colorado River of the West and Its Tributaries* (Washington, D.C.).

1875. GOVERNMENT—STATE / AGRICULTURE
The state of Connecticut contributed to the establishment of the country's first agricultural experiment station. First located at Wesleyan University in Middletown under the headship of Wilbur Olin Atwater (1844-1907), it was moved to New Haven in 1877 where Samuel W. Johnson (1830-1909) took charge. At that time it became fully state funded.

1875. ORGANIZATIONS—ACADEMIC / PHYSICS
Henry Augustus Rowland (1848-1901) was made professor of physics in the new Johns Hopkins University and went to Europe for a year to examine laboratories and to procure apparatus.

1875. ORGANIZATIONS—SOCIETIES AND ASSOCIATIONS / NATURAL HISTORY
The Agassiz Association, for the promotion of nature study, was established. The chief organizer was Massachusetts schoolteacher Harlan H. Ballard (1853-1934). By the 1890s, there were some 1,200 "chapters" (often quite small) with a membership that consisted largely of children.

1875. PALEONTOLOGY
Edward Drinker Cope (1840-1897) published *The Vertebrata of the Cretaceous Formations of the West*, Report of the U.S. Geological Survey of the Territories (Hayden Survey), II.

1875. PERIODICALS AND PUBLISHING / BOTANY
The *Botanical Gazette* (for a short time, *Botanical Bulletin*) began publication. It was established by John Merle Coulter (1851-1928) and initially had the intention to serve the Midwest in a way similar to the New York-based *Bulletin of Torrey Botanical Club*. Instead, it stressed

professional interests and concerns, unlike the amateur sympathies of the *Bulletin*.

1875. RELIGION AND THEOLOGY / MEDICINE
Mary Baker Eddy (1821-1910), founder of the Christian Science Church, published *Science and Health with Key to the Scripture* (Boston).

1875. ZOOLOGY / PHYSIOLOGY
John Call Dalton (1825-1889) published *Experimentation on Animals, as a Means of Knowledge in Physiology, Pathology, and Practical Medicine* (New York).

1875. ZOOLOGY / PHYSIOLOGY
Russell Henry Chittenden (1856-1943) published in Liebig's *Annalen* his discovery of the natural occurrence of free glycine. Working on scallops, Chittenden's work was done while an undergraduate student of Samuel W. Johnson (1830-1909) at the Sheffield Scientific School (Yale University).

1876. ASTRONOMY
Charles A. Young (1834-1908) measured the rotation rate of the sun, the first effective gauge of this phenomenon.

1876. EVOLUTION / RELIGION AND THEOLOGY
With the assistance of clergyman George Frederick Wright (1838-1921), Asa Gray (1810-1888) collected his essays on evolution and on related religious concerns in *Darwiniana* (New York).

1876. GENERAL OR MISCELLANEOUS / ENGINEERING AND APPLIED SCIENCE
The first national conference on engineering education took place at the Franklin Institute. It was organized by Alexander Holley (1832-1882).

1876. GENERAL OR MISCELLANEOUS / EVOLUTION
Darwin's promoter, English biologist Thomas Henry Huxley (1825-1895), visited the United States where he lectured on evolution. He was orator at the inauguration of Daniel Coit Gilman (1831-1908) as first president of Johns Hopkins University.

1876. GENERAL OR MISCELLANEOUS / TECHNOLOGY AND INVENTION
Thomas Alva Edison (1847-1931) moved his private laboratory to Menlo Park, New Jersey, and there set up the first establishment in the world for the production of inventions, a harbinger of the industrial research laboratory. Eleven years later it was moved to West Orange, New Jersey.

1876. ORGANIZATIONS—ACADEMIC
The Johns Hopkins University was opened. Of the six departments in the new institution, half were in the sciences: chemistry and physics; mathematics; and natural science, encompassing geology, mineralogy, botany, and zoology. Among the professorial appointments were Ira Remsen (1846-1927) in chemistry, who helped to promote the incorporation of laboratory research with teaching, and British mathematician James Joseph Sylvester (1814-1897).

1876. ORGANIZATIONS—FAIRS AND EXPOSITIONS /
ORGANIZATIONS--MUSEUMS
The Centennial Exhibition in Philadelphia was notable especially for its displays of technology. Many collections from the Exposition afterward went to the Smithsonian Institution's National Museum where a new building was provided for them by Congress.

1876. ORGANIZATIONS—SOCIETIES AND ASSOCIATIONS / CHEMISTRY
The American Chemical Society was established. Initial interest was for a local organization in New York City, with Charles F. Chandler (1836-1925) taking the lead, but the outcome was a broader-based group. John William Draper (1811-1882) was made the first president. In spite of its name, however, the Society met in New York until 1891, when the annual meeting was held at Newport, Rhode Island, and thereafter a more truly national structure for the Society evolved.

1876. ZOOLOGY
James Orton (1830-1877) published *Comparative Zoology, Structural and Systematic, for Use in Schools and Colleges* (New York).

1876. ZOOLOGY
David Starr Jordan (1851-1931) published *Manual of the Vertebrates of the Northern United States* (Chicago). A thirteenth edition appeared in 1929.

1876. ZOOLOGY / EMBRYOLOGY
Alpheus Spring Packard, Jr. (1839-1905) published a pioneering American work in comparative embryology, *Life Histories of Animals, Including Man; or, Outlines of Comparative Embryology* (New York, 1876). It also appeared as *Outlines of Comparative Embryology* (New York, 1878).

1876. ZOOLOGY / ENTOMOLOGY
Alpheus Spring Packard, Jr. (1839-1905) published *A Monograph of the Geometrid Moths or Phalaenidae of the United States* (Washington, D.C.), as one of the reports of the U.S. Geological Survey of the Territories (F. V. Hayden, in charge).

1876 (March). TECHNOLOGY AND INVENTION / ELECTRICITY AND ELECTRONICS
Alexander Graham Bell (1847-1922) patented the telephone.

1876-1878. CHEMISTRY / PHYSICS
Josiah Willard Gibbs (1839-1903) published his major scientific contribution, "On the Equilibrium of Heterogeneous Substances," *Transactions of Connecticut Academy of Sciences* 3. The memoir, consisting of two papers, was some 300 pages in length. Copies were sent to scientists throughout the world by Gibbs. The paper included Gibbs's important phase rule and made fundamental contributions to various topics, becoming a foundation for physical chemistry. In 1878, he published an abstract in *American Journal of Science*, 3rd series 16.

1876-1880. BOTANY
Sereno Watson (1826-1892) published the two-volume *Botany of California* (Cambridge, Mass., 1876, 1880).

1877. ANTHROPOLOGY AND ETHNOLOGY
Ethnologist Lewis Henry Morgan (1818-1881) published *Ancient Society, or Researches in the Lines of Human Progress from Savagery through Barbarism to Civilization* (New York).

1877. ANTHROPOLOGY AND ETHNOLOGY
John Wesley Powell (1834-1902) published *Introduction to the Study of Indian Languages* (Washington, D.C.).

1877. ARCHAEOLOGY
Ephraim George Squier (1821-1888) published *Peru: Incidents of Travel and Exploration in the Land of the Incas* (New York and London). He related his discovery of Latin American civilizations that predated the Inca.

1877. ASTRONOMY
The two satellites of Mars were discovered by Asaph Hall (1829-1907). He named them Deimos and Phobos.

1877. GENERAL OR MISCELLANEOUS / BOTANY
Charles Darwin's *The Different Forms of Flowers on Plants of the Same Species* was dedicated to Asa Gray (1810-1888).

1877. GOVERNMENT—FEDERAL / ENTOMOLOGY
Outbreak of locusts led to the appointment of a U.S. Entomological Commission (in the Department of Interior), headed by Charles V. Riley (1843-1895). In 1880, the Commission was incorporated in the Department of Agriculture and became its Division of Entomology. (The Division became the Bureau of Entomology in 1906.)

1877. ORGANIZATIONS—INDUSTRY / ELECTRICITY AND ELECTRONICS
Bell Telephone Company (after 1899, American Telephone and Telegraph Company, AT&T) was founded.

1877. PALEONTOLOGY
The first three quarries of large dinosaurs in the country were discovered in Colorado and Wyoming. (Among the early workers in all three was Samuel Wendell Williston, 1851-1918, who was collecting for Othniel Charles Marsh, 1831-1899.)

1877. CHEMISTRY
Ira Remsen (1846-1927) published the text, *Principles of Theoretical Chemistry: With Special Reference to the Constitution of Chemical Compounds* (Philadelphia).

1877-1880. GENERAL OR MISCELLANEOUS / ZOOLOGY
Edward Sylvester Morse (1838-1925) was professor of zoology in Tokyo University.

1878. ASTRONOMY
Charles Sanders Peirce (1839-1914) published "Photometric Researches," *Annals of Harvard College Observatory*. The work related to the shape of the Milky Way galaxy.

1878. BOTANY / PATHOLOGY
Thomas J. Burrill (1839-1916) established that bacteria were responsible for "fire blight," a disease of apple and pear orchards. His work led to a new conceptualization of plant disease.

1878. GEOLOGY
John Wesley Powell (1834-1902) published *Report on the Lands of the Arid Region of the United States, with a More Detailed Account of the Lands of Utah* (45th Congress, 2nd Session, HR Executive Document 73; Washington, D.C.).

1878. GEOLOGY
Joseph LeConte (1823-1901) published *Elements of Geology: A Textbook for Colleges and for the General Reader* (New York). A fifth, revised edition appeared in 1903, partly redone by Herman LeRoy Fairchild (1850-1943).

1878. GOVERNMENT—FEDERAL / GEOLOGY, NATURAL HISTORY
At the request of the government, initiated by Representative Abram S. Hewitt (1822-1903) of New York, the National Academy of Sciences undertook a review of the several geological and natural history surveys at work in the West. James Dwight Dana (1813-1895) chaired the committee. The committee report was adopted by the Academy in November 1878. Though the Congress failed to endorse the Academy's

recommendations, the report was a prelude to the abolition of the various existing surveys and formation of the U.S. Geological Survey in 1879.

1878. GOVERNMENT—FEDERAL / GEOPHYSICS AND GEODESY
The United States Coast Survey was changed to the Coast and Geodetic Survey.

1878. ORGANIZATIONS—ACADEMIC / PHYSICS
The second college-level teaching laboratory in physics in the country, at Wellesley College, was established by Sarah Frances Whiting (1847-1927). Her preparation for this undertaking had involved study at the Massachusetts Institute of Technology, where laboratory physics had been established.

1878. ORGANIZATIONS—SOCIETIES AND ASSOCIATIONS / MICROBIOLOGY AND MICROSCOPY
The American Microscopical Society was established.

1878. PERIODICALS AND PUBLISHING / MATHEMATICS
The *American Journal of Mathematics* was founded by James Joseph Sylvester (1814-1897) at Johns Hopkins University. It was the earliest mathematical research journal in the United States. The first issue of the *Journal* contained the article, "Researches in the Lunar Theory" by George William Hill (1838-1914), an innovative approach to the study of three mutually attracting bodies that became a foundational piece for celestial mechanics.

1878. PHYSICS
Alfred Marshall Mayer (1836-1897) reported the results of his experiments on magnetized needles floating in corks on water in a magnetic field. The stable patterns that were demonstrated came to be cited in relation to later interests in illustrating atomic structures. Mayer published in: *American Journal of Science*, 3rd series 15 and 16, and in *Scientific American*, supplement 5, all in 1878.

1878. PHYSICS
Albert Abraham Michelson (1852-1931) carried out his first measurement of the speed of light, a subject he continued to study for the rest of his life.

1878 (May 13). GENERAL OR MISCELLANEOUS / PHYSICS
Physicist and head of the Smithsonian Institution Joseph Henry (b.1797) died at Washington, D.C.

1878 (October 15). ORGANIZATIONS—INDUSTRY / ELECTRICITY AND ELECTRONICS
Thomas A. Edison (1847-1931) established the Edison Electric Light Company in New York, although the achievement of a viable incandescent lightbulb did not come until 1879. With the financial input of J.P. Morgan, in 1892 the company combined with Thomson-Houston to form the General Electric Company.

ca.1878. MEDICINE / PATHOLOGY
William Henry Welch (1850-1934) established the first teaching laboratory in pathology in the United States, at the Bellevue Hospital medical school in New York.

1878-1906. ORGANIZATIONS—RESEARCH INSTITUTIONS / ZOOLOGY
The Chesapeake Zoological Laboratory convened each summer at various marine locations along the Atlantic coast and in the West Indies.

1879. CHEMISTRY
Saccharin was developed at Johns Hopkins University and was subsequently patented by Constantin Fahlberg (1850-1910), a research fellow of Professor Ira Remsen (1846-1927). The two debated the fact of who was originator. Fahlberg devised means to manufacture the substance and produced it commercially in Germany.

1879. CHEMISTRY
Arthur Michael (1853-1942) was the first to achieve synthesis of a natural glucoside. The work appeared in "On the Synthesis of Helicin and Phenolglucoside," *American Chemical Journal* 1:305-312.

1879. GENERAL OR MISCELLANEOUS / INFORMATION ACCESS
Samuel Hubbard Scudder (1837-1911) published *Catalogue of Scientific Serials of All Countries ... 1633-1878* (Cambridge, Mass.).

1879. GENERAL OR MISCELLANEOUS / ZOOLOGY
Charles Otis Whitman (1842-1910) went as professor of zoology at the Imperial University of Tokyo, where he remained for two years. His influence was lasting and he has been referred to as the "father of zoology in Japan."

1879. INFORMATION ACCESS / MEDICINE
John Shaw Billings (1838-1913) established *Index Medicus*, a classified index to current medical literature of the world. In 1880, Billings also began the *Index Catalogue of the Surgeon General's Office, U.S. Army (Army Medical Library)*; Billings' continued his involvement with the publication until 1895.

1879. MICROBIOLOGY AND MICROSCOPY
Joseph Leidy (1823-1891) identified parasitic amoebae.

1879. PERIODICALS AND PUBLISHING / CHEMISTRY
The *American Chemical Journal* began publication at Johns Hopkins University under the direction of Ira Remsen (1846-1927). Its emphasis was on pure research. The same year, the *Journal of American Chemical Society* began publication. Remsen served as editor of the *American Chemical Journal*. In 1913, it was merged with the *Journal of American Chemical Society*.

1879. PHYSICS / ELECTRICITY AND ELECTRONICS
Edwin Herbert Hall (1855-1938) discovered the "Hall effect" while a graduate student at Johns Hopkins University. The Hall effect showed that a current in a conductor that was in a magnetic field produced an electromotive force, at right angles to both the current and the field.

1879. TECHNOLOGY AND INVENTION / ELECTRICITY AND ELECTRONICS
Thomas Alva Edison (1847-1931) succeeded in operating a practical incandescent electric lamp using a carbon-thread filament. In the same year, Joseph Swan in England achieved a similar goal.

1879. TECHNOLOGY AND INVENTION / PHOTOGRAPHY
George Eastman (1854-1932) devised means of producing a photographic dry plate and went on to develop the Eastman Kodak Company (so named in 1893). By 1884, Eastman had perfected the initial version of

his Kodak camera, including rolled film that took the place of the glass plate. The first Kodak was manufactured in 1888.

1879 (March 3). GOVERNMENT—FEDERAL / ANTHROPOLOGY AND ETHNOLOGY
The Bureau of American Ethnology (until 1894, the Bureau of Ethnology) was established at the Smithsonian Institution. This event was part of the government reorganization that consolidated the geological surveys into the U.S. Geological Survey, while transferring the anthropological work carried out by those surveys to the Smithsonian. John Wesley Powell (1834-1902) was put in charge of the Bureau, which was discontinued in 1964 when it was merged with the Smithsonian's Department of Anthropology.

1879 (March 3). GOVERNMENT—FEDERAL / GEOLOGY
Legislation establishing the U.S. Geological Survey became effective. The new organization consolidated the existing federal surveys in the West that had been directed by Ferdinand V. Hayden (1829-1887), Clarence R. King (1842-1901), John Wesley Powell (1834-1902), and George M. Wheeler (1842-1905). It had been suggested by a committee of the National Academy of Sciences. Subsequently, King was appointed to head the Survey, situated in the Department of Interior; Powell succeeded him as director in 1881. One outcome of the reorganization was that Edward Drinker Cope (1840-1897) lost to Othniel Charles Marsh (1831-1899) access to paleontological patronage of the federal surveys.

1879 (March 3). GOVERNMENT—FEDERAL / PUBLIC HEALTH
The National Board of Health was established by Congress to prepare a national public health plan. The head of the Board was James L. Cabell who was president of the American Public Health Association; the representative of the U.S. Army, John Shaw Billings (1838-1913), was vice-chair and its most influential member. Grants were made to individual scientists for particular studies and a yellow fever commission was sent to Cuba. In June 1879, Congress gave the Board power and funds to direct quarantine relating to the yellow fever epidemic but its efforts met with difficulties and opposition. The quarantine power terminated in 1883. The Board continued in name until 1893, but it was long an ineffective body whose funds and mission were transferred to the

Marine Hospital Service. Though of minimal immediate impact, the Board helped to establish a precedent for federal assistance for university research.

1879-1881. PHYSICS / INSTRUMENTS AND INSTRUMENTATION
Samuel Pierpont Langley (1834-1906) developed the temperature-sensitive bolometer to measure energy of radiation as related to wavelength. In 1881, he made an excursion to Mt. Whitney to conduct solar energy and atmospheric absorption studies with the instrument. The important outcome was given in "Researches on Solar Heat and Its Absorption by the Earth's Atmosphere, A Report of the Mount Whitney Expedition," *Professional Papers of the Signal Service*, no. 15 (1884).

1879-1884. PALEONTOLOGY / BOTANY
Leo Lesquereux (1806-1889) published his three-volume *Description of the Coal Flora of the Carboniferous Formation in Pennsylvania and Throughout the U.S.* (Second Geological Survey of Pennsylvania: Report of Progress, P) (Harrisburg, Penn.).

1880. BOTANY
Charles Edwin Bessey (1845-1915) published *Botany for High Schools and Colleges* (New York), an influential textbook that gave new direction to botanical teaching. Among its features was the emphasis given to laboratory study in botany, and the inclusion of cryptogamic botany and physiological anatomy. The book was an American version of Julius von Sachs' *Lehrbuch der Botanik*.

1880. BOTANY
Charles Sprague Sargent (1841-1927) conducted a survey for the U.S. Census, which was presented in the publication *Report on the Forests of North America (Exclusive of Mexico)* (Washington, D.C., 1884).

1880. EVOLUTION
Alpheus Hyatt (1838-1902) wrote on his influential idea that species' changes have the features of youth, maturity, and age with death as do individual organisms. (The best representation of his ideas appeared this year in "The Genesis of the Tertiary Species of Planorbis at Steinheim," *Anniversary Memoirs of Boston Society of Natural History*.)

1880. GENERAL OR MISCELLANEOUS / EDUCATION IN SCIENCE, PHYSICS
In this year, there were only eleven secondary schools with laboratory-based teaching in physics.

1880. GEOLOGY
Clarence Edward Dutton (1841-1912) published *Report on the Geology of the High Plateaus of Utah* (U.S. Geographical and Geological Survey of the Rocky Mountain Region, volume 32) (Washington, D.C.).

1880. ORGANIZATIONS—SOCIETIES AND ASSOCIATIONS / ENGINEERING AND APPLIED SCIENCE
The American Society of Mechanical Engineers was founded. Robert Henry Thurston (1839-1903) was the first president.

1880. PALEONTOLOGY
Othniel Charles Marsh (1831-1899) published *Odontornithes: A Monograph on the Extinct Toothed Birds of North America* (U.S. Geological Exploration of the 40th Parallel, vol. 7) (Washington, D.C.). Charles Darwin considered Marsh's work on the extinct birds to be the best evidence then known in support of evolution.

1880. ASTRONOMY / PHOTOGRAPHY
By about this date, Henry Draper (1837-1882) had succeeded in establishing photography as the most effective means of carrying out astronomical studies.

1880 (August 28). GENERAL OR MISCELLANEOUS / CHEMISTRY, GEOLOGY
Chemist and geologist Charles Thomas Jackson (b.1805) died at Somerville, Massachusetts.

ca.1880. GENERAL OR MISCELLANEOUS / ENGINEERING AND APPLIED SCIENCE
By this period, most engineers in the United States were trained in college.

1880s. ZOOLOGY / ORNITHOLOGY
During this time, Foster E.L. Bcal (1840-1916) in Iowa and Stephen A. Forbes (1844-1930) in Illinois began the study of the contents of the

stomachs of birds to determine their dietary practices. Part of the motivation was to understand the relations of the birds to the insect population and the effects on agriculture.

1881. INSTRUMENTS AND INSTRUMENTATION / ORGANIZATIONS—INDUSTRY
After this date, John A. Brashear (1840-1920) was able to turn his full attention to the production of scientific instruments and his company subsequently became a major provider of astrophysical apparatus worldwide.

1881. PERIODICALS AND PUBLISHING / ASTRONOMY
Seth Carlo Chandler (1846-1913), at the Harvard College Observatory, with John Ritchie (1853-1939), established Science Observer Code for the telegraphic dissemination of information on new comets.

1881. PERIODICALS AND PUBLISHING / NATURAL HISTORY
The *Bulletin* of the American Museum of Natural History was established, chiefly through the efforts of Robert Parr Whitfield (1828-1910), a curator in geology and paleontology at the Museum.

1881. ZOOLOGY—HUMAN
Henry Newell Martin (1848-1896) published the textbook, *The Human Body* (New York). It had several editions.

1882. GEOLOGY
Clarence Edward Dutton (1841-1912) published *The Tertiary History of the Grand Canyon District* (U.S. Geological Survey Monograph, no. 2) (Washington, D.C.).

1882. ORGANIZATIONS—SOCIETIES AND ASSOCIATIONS
The American Association for the Advancement of Science organized nine sections. Prior to this, there were two sections under the direction of a vice-president with chairs of subdivisions as required.

1882. ORGANIZATIONS—SOCIETIES AND ASSOCIATIONS / BOTANY
The American Forestry Association was established.

1882. PERIODICALS AND PUBLISHING / ASTRONOMY
Simon Newcomb (1835-1909) established the *Astronomical Papers Prepared for the Use of the American Ephemeris and National Almanac*, published by the Naval Observatory. Most of his own contributions appeared in this source.

1882. PHYSICS
Johann Bernhard Stallo (1823-1900) published *Concepts and Theories of Modern Physics* (New York). The book was a critique of the foundations of "corpuscular-kinetic" or mechanistic views of matter, which were then prevalent. Stallo's ideas were largely dismissed by leading American physicists of the time. The book was published in a German translation in 1901.

1882. PHYSICS / INSTRUMENTS AND INSTRUMENTATION
A machine for producing precision diffraction gratings used in spectroscopy was devised by Henry Augustus Rowland (1848-1901). With his apparatus Rowland produced ruled lines on a surface that was spherically curved and capable of focusing all of the spectrum on a circle.

1882. TECHNOLOGY AND INVENTION / ELECTRICITY AND ELECTRONICS
Electric lighting was made feasible through Thomas A. Edison's (1847-1931) steam-powered Pearl Street central power station in New York. It began operation on 4 September.

1882. ZOOLOGY / ENTOMOLOGY
Augustus Radcliffe Grote (1841-1903) published *An Illustrated Essay on the Noctuidae of North America* (London), a semipopular entomological work.

1882 (May 30). GENERAL OR MISCELLANEOUS / GEOLOGY
Geologist and science educator William Barton Rogers (b.1804) died in Boston, Massachusetts.

1882-1888. GEOLOGY
Thomas Chrowder Chamberlin (1843-1928) and his colleagues published a series of notable papers in the *Report of the United States Geological*

Survey. During this period (and until 1907), Chamberlin was head of the Survey's glacial division.

1883. GENERAL OR MISCELLANEOUS
Henry Augustus Rowland (1848-1901), vice-president of the American Association for the Advancement of Science, gave a notable address called "A Plea for Pure Science."

1883. GENERAL OR MISCELLANEOUS
Four time zones were established in the United States. The decision drew upon an 1879 report by Cleveland Abbe (1838-1916) and previously designated zones used by the railroads.

1883. ORGANIZATIONS—ACADEMIC / PSYCHOLOGY
Granville Stanley Hall (1846-1924) established the first American psychological laboratory at Johns Hopkins University.

1883. ORGANIZATIONS—SOCIETIES AND ASSOCIATIONS / BOTANY
The American Botanical Club was founded by botanists in the American Association for the Advancement of Science.

1883. ORGANIZATIONS—SOCIETIES AND ASSOCIATIONS / NATURAL HISTORY
The American Society of Naturalists was founded. It was initially known as the Society of Naturalists of the Eastern United States.

1883. ORGANIZATIONS—SOCIETIES AND ASSOCIATIONS / ORNITHOLOGY
The American Ornithologists' Union was founded by Joel Asaph Allen (1838-1921), William Brewster (1851-1919), and Elliott Coues (1842-1899). Allen served as first president, from 1883-1890.

1883. PALEONTOLOGY
Edward Drinker Cope (1840-1897) published *The Vertebrata of the Tertiary Formations of the West, Book I*, Report of the U.S. Geological Survey of the Territories, III [Hayden Survey]. It became known as "Cope's Bible."

1883. PERIODICALS AND PUBLISHING
The journal *Science* began publication. Growing out of an idea of Alexander Graham Bell (1847-1922) in the previous year, during its first eight years Bell and his father-in-law, Gardiner G. Hubbard (1822-1897), contributed approximately $100,000 for the publication.

1883. PHYSICS / ELECTRICITY AND ELECTRONICS
Thomas Alva Edison (1847-1931) discovered what is known as the Edison effect, the outflow of electricity from heated metal. The phenomenon later was used (by others) in the vacuum tube used in radio and television.

1883. PSYCHOLOGY
Granville Stanley Hall (1846-1924) published his paper, "The Contents of Children's Minds," *Princeton Review* 11:249-273, which helped to promote studies in child development.

1883. ZOOLOGY / ENTOMOLOGY
George Henry Horn (1840-1897) and John Lawrence LeConte (1825-1883) published "Classification of the Coleoptera of North America," *Smithsonian Miscellaneous Collections*, 26:1-567.

1883 (May 24). ENGINEERING AND APPLIED SCIENCE
The Brooklyn Bridge was opened. Begun in 1869, it had a span of 1595.5 feet, by a wide margin the largest suspension bridge built up to that time.

1884. BOTANY
Leo Lesquereux (1806-1889) and Thomas Potts James (1803-1882) published *Manual of the Mosses of North America* (Boston). Lesquereux had begun work on it with William S. Sullivant, who died in 1873.

1884. FUNDS AND FUNDING
The Elizabeth Thompson Science Fund was established by New Yorker Thompson (1821-1899) with an endowment of $25,000. The fund was managed by a group of Harvard scientists, including Charles S. Minot (1852-1914) (embryology) and Edward C. Pickering (1846-1919) (astronomy).

1884. GOVERNMENT—FEDERAL / AGRICULTURE, ZOOLOGY
The federal government formed the Bureau of Animal Industry in the Department of Agriculture. The new Bureau had as part of its mission the regulation of activities in order to control disease among farm animals. Daniel E. Salmon (1850-1914) was put in charge.

1884. ORGANIZATIONS—SOCIETIES AND ASSOCIATIONS / AGRICULTURE, CHEMISTRY
The Association of Official Agricultural Chemists was established.

1884. ORGANIZATIONS—SOCIETIES AND ASSOCIATIONS / ENGINEERING AND APPLIED SCIENCE
The American Institute of Electrical Engineers was established.

1884. RELIGION AND THEOLOGY
Arnold Henri Guyot (1807-1884) published *Creation, or the Biblical Cosmogony in the Light of Modern Science* (New York). Guyot died at Princeton, New Jersey on 8 February.

1884. TECHNOLOGY AND INVENTION / ELECTRICITY AND ELECTRONICS
Nikola Tesla (1856-1943) invented the electric alternator. This also was the year he emigrated to the United States.

1884 (August 26). TECHNOLOGY AND INVENTION
Ottmar Mergenthaler (1854-1899) patented the linotype machine for the setting of type.

1884 (October). GENERAL OR MISCELLANEOUS / ASTRONOMY
At an international conference at Washington, the prime meridian was established as running through Greenwich, England.

1884-1886. GOVERNMENT—FEDERAL
In 1884, Congress established a Joint Commission to Consider the Present Organization of the Signal Service, Geological Survey, Coast and Geodetic Survey, and the Hydrographic Office of the Navy Department, with a View to Secure Greater Efficiency and Economy of Administration of the Public Service. Chaired by Senator William B. Allison (Iowa), it came to be known as the Allison Commission. The Commission submitted its report in 1886, which left the government

science establishment essentially intact and by implication endorsed the role of the federal government in science as represented in the agencies involved in the review. The Commission did not endorse the idea of a central department of science which had been suggested to it by a committee of the National Academy of Sciences.

1884-1887. PHARMACOLOGY AND PHARMACY
John Uri Lloyd (1849-1936) and his brother Curtis G. Lloyd (1859-1926) published the two-volume work, *Drugs and Medicines of North America* (Cincinnati).

1885. GEOLOGY
Israel Charles White (1848-1927) published "The Geology of Natural Gas," *Science* 5:521-552. This paper included White's anticlinal theory regarding accumulations of oil and natural gas. In his work, he related such accumulations to specific gravities and geological structures. Although not an idea specific to White, he helped to formulate and promote it for commercial benefit.

1885. MEDICINE / PATHOLOGY
Francis Delafield (1841-1915) and Theophil Mitchell Prudden (1849-1924) published *A Handbook of Pathological Anatomy and Histology* (New York). Originally published by Delafield in 1872 as *A Handbook of Post-Mortem Examinations and of Morbid Anatomy*, a seventh edition appeared in 1904.

1885. ORGANIZATIONS—ACADEMIC
Georgia School of Technology was founded. In 1948, its name was changed to Georgia Institute of Technology.

1885. ORGANIZATIONS—SOCIETIES AND ASSOCIATIONS / PARAPSYCHOLOGY
The American Society for Psychical Research was established.

1885. ZOOLOGY / PHYSIOLOGY
Henry Pickering Bowditch (1840-1911) published experimental results of studies showing the fatiguelessness of the nerve trunk. The paper, "Note on the Nature of Nerve-Force," appeared in the *Journal of Physiology* 6 (1885): 133-135.

1885 (December). GENERAL OR MISCELLANEOUS
Geologist Grove Karl Gilbert (1843-1918) presented a presidential address to the Society of American Naturalists, "The Inculcation of Scientific Method," in which he promoted the value of devising and testing multiple hypotheses.

1886. BIOLOGY—GENERAL
William T. Sedgwick (1855-1921) and Edmund Beecher Wilson (1856-1939) published *General Biology* (New York). As a textbook, it helped to reorient academic teaching away from the taxonomic and phylogenetic.

1886. GOVERNMENT—FEDERAL / ZOOLOGY
A Division of Economic Ornithology and Mammalogy (after 1891, the word Economic was deleted) was established in the U.S. Department of Agriculture. Clinton Hart Merriam (1855-1942) became its head.

1886. ORGANIZATIONS—INDUSTRY / CHEMISTRY
Arthur D. Little Company was founded with a partnership of Arthur D. Little (1863-1935) and Roger B. Griffin (?-1893) known as Griffin & Little, Chemical Engineers.

1886. ORGANIZATIONS—SOCIETIES AND ASSOCIATIONS
The intercollegiate scientific honor society, Sigma Xi, was founded at Cornell University. Professor of paleontology Henry Shaler Williams (1847-1918) was organizer of the society.

1886. PHYSICS / EDUCATION IN SCIENCE
Edwin Herbert Hall (1855-1938) first published what came to be known as the Harvard *Descriptive List of Elementary Physical Experiments*, in a provisional form. This publication, intended to serve the purposes of examination for college admission, had an important influence on the teaching of physics in secondary schools.

1886. TECHNOLOGY AND INVENTION / ORGANIZATIONS—INDUSTRY
An economically feasible means of producing aluminum was invented by American Charles Martin Hall (1863-1914). During the same year, Paul-Louis-Toussaint Heroult in France independently devised a similarly practical means and the procedure came to be known as the

Hall-Heroult process. Hall procured a patent in 1889 and the company he founded came to be known as the Aluminum Company of America.

1886. ZOOLOGY / INSTRUMENTS AND INSTRUMENTATION
Charles Sedgwick Minot (1852-1914), at the Harvard Medical School, invented the automatic rotary microtome.

1886. ZOOLOGY / ORNITHOLOGY
The American Ornithologists' Union prepared a Code of Nomenclature that later served as a foundation for the International Code of Zoological Nomenclature.

1886 (June 30). GOVERNMENT—FEDERAL / BOTANY
The Division of Forestry in the U.S. Department of Agriculture was established. Its head was Bernhard E. Fernow (1851-1923).

1887. ASTRONOMY
Samuel P. Langley's (1834-1906) *The New Astronomy* (Boston) was an introduction to astrophysics for the general reader. It appeared originally in *Century Magazine*, 1884-1887.

1887. GOVERNMENT—FEDERAL / PUBLIC HEALTH
The U.S. Marine Hospital Service began the establishment of laboratories and the promotion of research relating to infectious diseases. In this year, the Service established a small-scale diagnostic laboratory in New York.

1887. ORGANIZATIONS—OBSERVATORIES / INSTRUMENTS AND INSTRUMENTATION
The Lick 36-inch refracting telescope, the largest in the world at the time, was finished and mounted on Mt. Hamilton near San Francisco (December 31). Warner and Swasey Company of Cleveland, manufacturer of machine tools, had been given the contract to build the mounting in 1886; the lens was made by Alvan Clark and Sons. In 1888, the Lick Observatory was formally transferred to the University of California. The 36-inch refractor was first used in June 1888.

1887. ORGANIZATIONS—SOCIETIES AND ASSOCIATIONS / AGRICULTURE
The Association of American Agricultural Colleges and Experiment Stations was established.

1887. ORGANIZATIONS—SOCIETIES AND ASSOCIATIONS / PHYSIOLOGY
The American Physiological Society was founded. (Of the 24 founders, one-quarter were students of Johns Hopkins University professor Henry Newell Martin, 1848-1896.)

1887. PERIODICALS AND PUBLISHING / CHEMISTRY
The *Journal of Analytical and Applied Chemistry* was founded and continued until 1893, when it merged with the *Journal of American Chemical Society*. It was edited by Edward Hart (1854-1931).

1887. PERIODICALS AND PUBLISHING / PSYCHOLOGY
The *American Journal of Psychology* was established by Granville Stanley Hall (1846-1924); it was the first American journal devoted to psychology. Hall edited the publication for many years.

1887. PERIODICALS AND PUBLISHING / ZOOLOGY
The *Journal of Morphology* was founded by Charles Otis Whitman (1842-1910), then director of the Allis Lake Laboratory in Milwaukee, Wisconsin. This was the earliest zoology and anatomy journal in the country.

1887. PHYSICS
Albert Abraham Michelson (1852-1931) and Edward W. Morley (1838-1923) undertook experiments with their light beam interferometer on ether drift. Michelson had begun the experiment in 1881. The Michelson-Morley experiment found no difference in the course of two beams of light split from a single source, traveling parallel and perpendicular to the direction of the motion of the earth (i.e., they were unable to detect any evidence of earth's motion relative to a supposed stationary ether). The negative result had consequences in the later abandonment of the ether concept.

1887. ZOOLOGY / ORNITHOLOGY
Robert Ridgway (1850-1929) published *Manual of North American Birds* (Philadelphia).

1887. ZOOLOGY / PHYSIOLOGY
Isaac Ott (1847-1916) discovered the brain's heat-regulating center.

1887 (February 25). BIOLOGY—GENERAL / ECOLOGY
Stephen Alfred Forbes (1844-1930) presented early views on ecological community in a lecture, "The Lake as a Microcosm," to the Scientific Association of Peoria, Illinois. It was published in the Association's *Bulletin* 1 (1887): 77-87 [and reprinted in *Bulletin of Illinois State Natural History Survey* 15 (1925): 537-550].

1887 (March 2). GOVERNMENT—FEDERAL / AGRICULTURE
The Hatch Act was passed by Congress to provide support for state agricultural experiment stations; an Office of Experiment Stations was created in the Department of Agriculture. In 1906, the Adams Act restricted funding to original research.

1887 (August 19). GENERAL OR MISCELLANEOUS / ZOOLOGY
Zoologist Spencer Fullerton Baird (b.1823) died at Woods Hole, Massachusetts.

1888. ASTRONOMY
Charles Augustus Young (1834-1908) published his frequently used *A Textbook of General Astronomy for Colleges and Scientific Schools* (Boston).

1888. GENERAL OR MISCELLANEOUS / CHEMISTRY
Arthur Amos Noyes (1866-1936) was one of the first Americans to study physical chemistry with Wilhelm Ostwald at Leipzig University. Noyes later taught at MIT and California Institute of Technology.

1888. GOVERNMENT—FEDERAL / GEOLOGY
John Wesley Powell (1834-1902) was put in charge of a survey of regions where agriculture depended on irrigation, in anticipation of possible dam projects. An immediate result was that sale of public land in these areas was suspended temporarily. The survey and its regulatory features encountered considerable political opposition and resulted in severe cuts in funding for government science in the early 1890s.

1888. MEDICINE / CHEMISTRY
Frederick George Novy (1864-1957) and Victor Clarence Vaughan (1851-1929) published *Ptomaines and Leucomaines, or the Putrefactive and Physiological Alkaloids* (Philadelphia). A much revised fourth edition appeared in 1902 as *Cellular Toxins, or the Chemical Factors in the Causation of Disease* (Philadelphia and New York).

1888. ORGANIZATIONS—ACADEMIC / BOTANY
Cornell University appointed Liberty Hyde Bailey, Jr. (1858-1954) to the first American professorship in practical and experimental horticulture.

1888. ORGANIZATIONS—ACADEMIC / PSYCHOLOGY
James McKeen Cattell (1860-1944) was appointed professor of psychology at the University of Pennsylvania, the first such professorship anywhere in the world.

1888. ORGANIZATIONS—INDUSTRY / ELECTRICITY AND ELECTRONICS
George Westinghouse (1846-1914) bought various patents of Nikola Tesla (1856-1943) relating to alternating current (AC) dynamos, motors, and related devices that made it possible to transmit electrical current over long distance. In the resulting conflict with Thomas A. Edison (1847-1931), who favored direct current, alternating current won out and was the foundation for the emergent electric power industry. (The Westinghouse Electric Company was founded in 1886.)

1888. ORGANIZATIONS—RESEARCH INSTITUTIONS / ZOOLOGY
The Marine Biological Laboratory was opened in Woods Hole, Massachusetts. Among the founders was Charles Otis Whitman (1842-1910), who served as first director from 1893 to 1908. The first student to receive a problem in the laboratory was Cornelia M. Clapp, 1849-1934, teacher at Mt. Holyoke Seminary.

1888. ORGANIZATIONS—SOCIETIES AND ASSOCIATIONS / ANATOMY
The American Association of Anatomists was founded.

1888. ORGANIZATIONS—SOCIETIES AND ASSOCIATIONS / GEOGRAPHY AND CARTOGRAPHY
The National Geographic Society was established for the promotion of popular interest in the subject. Its first president was Gardiner G.

Hubbard (1822-1897). The Society's *National Geographic Magazine* began publication the same year.

1888. ORGANIZATIONS—SOCIETIES AND ASSOCIATIONS / GEOLOGY
The Geological Society of America was founded (it was known until 1889 as the American Geological Society). (Among the major founders was Alexander Winchell, 1824-1891, who has been called "father" of the organization.) James Hall, Jr. (1811-1898) was the first president.

1888. PERIODICALS AND PUBLISHING / ANTHROPOLOGY AND ETHNOLOGY
The *American Anthropologist* was established.

1888. PERIODICALS AND PUBLISHING / GEOLOGY
The *American Geologist* was established and was published in Minnesota until 1905. (Among the founders were Alexander Winchell, 1824-1891, and his brother Newton Horace Winchell, 1839-1914.) In 1905, it was superseded by *Economic Geology*.

1888. PERIODICALS AND PUBLISHING / MATHEMATICS
Publications of American Statistical Association was established. In 1922, the name was changed to *Journal of American Statistical Association*.

1888. PERIODICALS AND PUBLISHING / PHILOSOPHY
The philosophy of science journal, *Monist*, was founded, of which Paul Carus (1852-1919) was editor. Carus also was editor of the previously established *The Open Court*.

1888. ZOOLOGY
Alpheus Spring Packard, Jr. (1839-1905) published *The Cave Fauna of North America, With Remarks on the Anatomy of the Brain and Origin of the Blind Species* (Memoirs of National Academy of Sciences, vol. 4) (Washington, D.C.). The work's significance included its integration of taxonomic, anatomical, and evolutionary approaches.

1888. ZOOLOGY / ENTOMOLOGY
Paleontologist and entomologist Samuel Wendell Williston (1851-1918) published the first edition of his *The Manual of North American Diptera*

(New Haven), though under a somewhat different title. A third and much expanded edition appeared in 1908 with the title as given here.

1888. ZOOLOGY / ENTOMOLOGY
John Henry Comstock (1849-1931) published *An Introduction to Entomology* with illustrations by his wife, Anna Botsford Comstock (1854-1930).

1888 (January 30). GENERAL OR MISCELLANEOUS / BOTANY
Botanist Asa Gray (b.1810) died at Cambridge, Massachusetts.

1888 (November 24). ORGANIZATIONS—SOCIETIES AND ASSOCIATIONS / MATHEMATICS
The New York Mathematical Society had its beginnings in a meeting of the interested faculty of Columbia University. A *Bulletin* began publication in 1890. In 1894, the organization became the American Mathematical Society, which began issuing its *Transactions* in 1900.

1888-1889. ZOOLOGY / ENTOMOLOGY
Samuel Hubbard Scudder (1837-1911) published *Butterflies of the Eastern United States and Canada* (Cambridge, Mass.), in three volumes.

1888-1891. EVOLUTION
John Thomas Gulick (1832-1923) published two papers that presented his views of the importance of isolation as a means of supplementing natural selection in the establishment of species. His ideas were developed through studies of animal forms in the Hawaiian Islands. The papers were: "Divergent Evolution Through Cumulative Segregation," *Journal of the Linnean Society, Zoology* 20 (1888): 189-274, and "Intensive Segregation [or Divergence Through Independent Transformation]," *ibid*. 23 (1891): 312-380.

1889. CHEMISTRY / BOTANY
Thomas Burr Osborne (1859-1929), at the Connecticut Agricultural Experiment Station, began his lifetime of work on the proteins of plant seeds. In his investigations, he developed procedures that came to be standard for the separation of proteins from plant seed.

1889. GEOLOGY
William Morris Davis (1850-1934) published "The Rivers and Valleys of Pennsylvania," *National Geographic Magazine* 1:183-253, which was the starting point for development of his influential work in landscape analysis. The next year he published "The Rivers of Northern New Jersey With Notes on the Classification of Rivers in General," *National Geographic Magazine* 2:81-110. These works helped to introduce the concept of the cycle of erosion, Davis's most significant contribution.

1889. ORGANIZATIONS—SOCIETIES AND ASSOCIATIONS / ENTOMOLOGY
The American Association of Economic Entomologists was established.

1889. ORGANIZATIONS—SOCIETIES AND ASSOCIATIONS / GEOPHYSICS AND GEODESY
The United States took up membership in the International Geodetic Association.

1889 (February). ORGANIZATIONS—SOCIETIES AND ASSOCIATIONS / ASTRONOMY
The Astronomical Society of the Pacific was established. Edward Singleton Holden (1846-1914) was particularly prominent in the founding and early affairs of the Society, which was characterized by its inclusion of professionals and amateurs.

1889 (February 9). GOVERNMENT—FEDERAL / AGRICULTURE
The U.S. Department of Agriculture was elevated to cabinet status.

1890. ASTRONOMY
The *Draper Catalogue of Stellar Spectra*, with 10,351 stars classified by Williamina Paton Stevens Fleming (1857-1911), was published (*Annals of Harvard College Observatory*, volume 27). The classification scheme of Fleming later was replaced by that of Annie Jump Cannon (1863-1941).

1890. AWARDS AND PRIZES / PHYSICS
Henry Augustus Rowland (1848-1901) won a gold medal and a grand prize at the Paris Exposition for his concave spectral grating and his map of the solar spectrum. Rowland is best remembered for his invention of ruling of the spectral grating.

1890. BIOLOGY—GENERAL
Clinton Hart Merriam (1855-1942), who came to be noted especially for his work on life zones of faunal distribution, reported on the subject in his "Results of a Biological Survey of the San Francisco Mountain Region and Desert of the Little Colorado, Arizona," *North American Fauna* 3:119-136.

1890. GENERAL OR MISCELLANEOUS / FUNDS AND FUNDING
Prior to the era of large foundations for support, the National Academy of Sciences in this year had an endowment for support of research that amounted to only $94,000. These funds yielded an annual income of $4,000.

1890. GOVERNMENT—FEDERAL / GEOLOGY
Beginning in this year, the U.S. Geological Survey became increasingly interested in Alaska, prompted by the discovery of gold in the territory and in northwestern Canada during the 1880s. Alfred Hulse Brooks (1871-1924) was put in charge of the new Alaska branch of the Survey in 1903 and retained the post (with the exception of a period during World War I) until his death.

1890. GOVERNMENT—FEDERAL / METEOROLOGY AND CLIMATOLOGY
The U.S. Weather Bureau was established by Congress in the Department of Agriculture, transferring the weather service from the Army Signal Corps. Cleveland Abbe (1838-1916), associated with the enterprise since 1871, accompanied the move to the new bureau. In 1940, it was transferred to the Department of Commerce and in 1965 to the newly established Environmental Science Services Administration.

1890. ORGANIZATIONS—OBSERVATORIES
The Smithsonian Astrophysical Observatory was established by the Institution's secretary, Samuel Pierpont Langley (1834-1906). The establishment had the support of both private and governmental funds.

1890. ORGANIZATIONS—RESEARCH INSTITUTIONS / BIOLOGY--GENERAL
Cold Spring Harbor Laboratory (Cold Spring Harbor, New York) was established.

1890. ORGANIZATIONS—SOCIETIES AND ASSOCIATIONS / ZOOLOGY
The American Society of Zoologists was founded. Initially known as the American Morphological Society, the name was changed in 1903.

1890. PSYCHOLOGY
William James (1842-1910) published *The Principles of Psychology* (New York), in two volumes.

1890. TECHNOLOGY AND INVENTION / CHEMISTRY
Stephen Moulton Babcock (1843-1931) reported his test for butterfat in milk and cream (*Bulletin of the University of Wisconsin Experiment Station*, no. 24). Not requiring scientific training, the test became widespread and helped in the development of the dairy industry.

1890. TECHNOLOGY AND INVENTION / COMPUTERS AND INFORMATION SCIENCE
An electrically operated punched card calculating system of Herman Hollerith (1860-1929) was used on the U.S. census.

1890 (April 30). GOVERNMENT—FEDERAL / ZOOLOGY
The National Zoological Park in Washington, D.C., was established by an act of Congress.

1890 (September 25). GOVERNMENT—FEDERAL / ENVIRONMENT AND CONSERVATION
Yosemite National Park was established by Congress. John Muir (1838-1914), who was first to show the glacial origin of the Yosemite Valley, and Robert U. Johnson (1853-1937) were instrumental in promoting the project.

1890-1903. ZOOLOGY / CONCHOLOGY AND MALACOLOGY
William Healey Dall (1845-1927) published his chief work in malacology, "Contributions to the Tertiary Fauna of Florida," *Transactions of Wagner Free Institute of Science of Philadelphia* 3, parts 1-6.

1891. ASTRONOMY
Hubert Anson Newton (1830-1896) published "On the Capture of Comets by Planets, Especially Their Capture by Jupiter," *Memoirs of National Academy of Sciences* 6:7-23.

1891. ASTRONOMY
Seth Carlo Chandler (1846-1913) published early reports on his astronomical work on the discovery of variation of latitude. The investigations on which he reported had begun at least as early as 1884-1885. About the same time, similar conclusions were reached by Friedrich Kustner in Berlin. Chandler's work appeared as "On the Variation of Latitude" in *Astronomical Journal* 11 (1891), 12 (1892), 13 (1893).

1891. ASTRONOMY / INSTRUMENTS AND INSTRUMENTATION
George Ellery Hale (1868-1938) first used his spectroheliograph to take a picture of the sun.

1891. ENVIRONMENT AND CONSERVATION
Nathaniel Southgate Shaler (1841-1906) published *Nature and Man in America* (New York), promoting the idea of the relations of environment and history.

1891. GENERAL OR MISCELLANEOUS / AERONAUTICS
Samuel P. Langley (1834-1906) published *Experiments in Aerodynamics* (Washington, D.C.).

1891. GENERAL OR MISCELLANEOUS / GEOLOGY
Geologist George Perkins Merrill (1854-1929) published *Stones for Building and Decoration* (New York). Subsequent editions appeared in 1897 and 1903.

1891. ORGANIZATIONS—BOTANICAL GARDENS
The New York Botanical Garden was incorporated by the New York state legislature. Promoted by Nathaniel Lord Britton (1859-1934), Elizabeth Knight Britton (1858-1934), and the Torrey Botanical Club, in 1895 the city of New York made available 250 acres in Bronx Park. N.L. Britton was made director in 1896 and the public opening took place in 1900.

1891. PERIODICALS AND PUBLISHING / MATHEMATICS
The *Bulletin of New York Mathematical Society* was established. In 1894, it became *Bulletin of American Mathematical Society*.

1891. PERIODICALS AND PUBLISHING / PSYCHOLOGY, ZOOLOGY
The *Journal of Comparative Neurology* was founded by Clarence L. Herrick (1858-1904), at the University of Cincinnati. In 1904, Robert M. Yerkes (1876-1956) became a staff member and the name was changed to *Journal of Comparative Neurology and Psychology*.

1891. TECHNOLOGY AND INVENTION / ELECTRICITY AND ELECTRONICS
The Tesla coil, a transformer that produced high voltage at high frequency, was invented by Nikola Tesla (1856-1943).

1891. ORGANIZATIONS—OBSERVATORIES
Harvard University set up its Boyden astronomical station at Arequipa, Peru, which continued in operation until 1927. Solon Irving Bailey (1854-1931) and William Henry Pickering (1858-1938), younger brother of Harvard observatory director Edward Charles Pickering (1846-1919), both were involved in its establishment. Bailey for many years exercised general supervision over the facility.

1891. PERIODICALS AND PUBLISHING / ASTRONOMY
George Ellery Hale (1868-1938) and William W. Payne (1837-1928) founded the journal, *Astronomy and Astro-Physics*, formerly the *Sidereal Messenger*.

1891. ZOOLOGY—HUMAN / NEUROBIOLOGY
During his years on the faculty of Clark University, 1889-1892, neurologist Henry Herbert Donaldson (1857-1938) studied the brain of Laura Bridgman, a blind deaf-mute. In 1891, he published an important work on the subject and in 1895 he published *The Growth of the Brain: A Study of the Nervous System in Relation to Education* (London and New York).

1891 (March 3). GOVERNMENT—FEDERAL / ENVIRONMENT AND CONSERVATION
Congress passed the Forest Reserve Act, which resulted in the president setting aside 13,000,000 acres of public land as forest reserves.

1892. ASTRONOMY
Edward Emerson Barnard (1857-1923) discovered the fifth satellite of Jupiter, using the Lick Observatory's 36-inch refractor. In the same year, Barnard's observations of a cloud of gas given off by a nova as it brightens was evidence that they are exploding stars.

1892. CHEMISTRY
Arthur Amos Noyes (1866-1936) published the text, *Notes on Qualitative Analysis* (Boston). Under various titles, it had ten editions, the last published posthumously in 1942. Noyes helped to disseminate electrolytic dissociation theory of Wilhelm Ostwald in this work, directing analysis to the level of ions rather than molecules.

1892. GENERAL OR MISCELLANEOUS / ANTHROPOLOGY AND ETHNOLOGY
Alexander F. Chamberlain (1865-1914) received the first American Ph.D. in anthropology. Conferred by Clark University, the work was supervised by Franz Boas (1858-1942).

1892. GEOLOGY
George Frederick Wright (1838-1921) published *Man and the Glacial Period* (New York). Wright's contention that humans lived in North America during the Pleistocene era resulted in a widespread controversy in the scientific community.

1892. ORGANIZATIONS—OBSERVATORIES
The University of Chicago's Yerkes Observatory was established.

1892. ORGANIZATIONS—RESEARCH INSTITUTIONS / BIOLOGY—GENERAL
The Wistar Institute of Anatomy and Biology was founded in Philadelphia.

1892. ORGANIZATIONS—SOCIETIES AND ASSOCIATIONS / ENVIRONMENT AND CONSERVATION
The Sierra Club was established.

1892. ORGANIZATIONS—SOCIETIES AND ASSOCIATIONS / PSYCHOLOGY
The American Psychological Association was established. Granville Stanley Hall (1846-1924), who had convened the organizational meeting at Clark University, was made the first president.

1892. PSYCHOLOGY
Logician and psychologist Christine Ladd-Franklin (1847-1930) developed a color theory that was controversial.

1892. ZOOLOGY—HUMAN / EMBRYOLOGY
Charles Sedgwick Minot (1852-1914) published *Human Embryology* (New York).

1892. MEDICINE
William Osler (1849-1919) published the important textbook, *The Principles and Practice of Medicine* (New York), while holding the position of professor of medicine at Johns Hopkins University.

1892 (April 19). TECHNOLOGY AND INVENTION
Charles E. Duryea (1861-1938) and his brother J. Frank Duryea (1869-1967), in Springfield, Mass., achieved the first operational American automobile. It was publicly demonstrated on September 21, 1893.

1893. GEOLOGY
Bailey Willis (1857-1949) published results of laboratory research in geology, "The Mechanics of Appalachian Structure," *Report of United States Geological Survey*, 13, part 2:211-281. The work was especially noted in Europe.

1893. MEDICINE / AGRICULTURE
Theobald Smith (1859-1934), with Fred Lucius Kilborne (1858-1936), published *Investigations Into the Nature, Causation, and Prevention of Texas or Southern Cattle Fever* (Washington, D.C.). The research demonstrated that the disease was caused by a micro-organism transmitted by a tick, the first time such a carrying agent was shown. The work was done by Smith, Kilborne, and Curtice Cooper in the U.S. Bureau of Animal Industry.

1893. MEDICINE / DISEASE
William Hallock Park (1863-1939) discovered the role of healthy humans in the transmission of diphtheria.

1893. ORGANIZATIONS—ACADEMIC / MEDICINE
The Johns Hopkins Medical School was opened.

1893. PERIODICALS AND PUBLISHING / ASTRONOMY
Popular Astronomy was established. It was published by Goodsell Observatory, Carleton College, Northfield, Minnesota.

1893. PERIODICALS AND PUBLISHING / GEOLOGY
The *Journal of Geology* was established by Thomas Chrowder Chamberlin (1843-1928) at the newly established University of Chicago.

1893 (July). PERIODICALS AND PUBLISHING / PHYSICS
The Physical Review began publication. It was produced by Edward L. Nichols (1854-1937) at Cornell University, with a subsidy from the University and was American physicists' first research journal. In 1913, it was taken over by the American Physical Society.

1894. BOTANY / EVOLUTION
Charles Edwin Bessey (1845-1915) published the first of his two classic papers on plant phylogeny, "Evolution and Classification," *Proceedings of American Association for the Advancement of Science* 42:237-251. The second paper, "Phylogenetic Taxonomy of Flowering Plants" appeared in *Annals of Missouri Botanical Garden* 2 (1915): 109-164.

1894. EVOLUTION
Alpheus Hyatt (1838-1902) published the important neo-Lamarckian study, "Phylogeny of an Acquired Characteristic," *Proceedings of American Philosophical Society* 32:349-647.

1894. METEOROLOGY AND CLIMATOLOGY
William Morris Davis (1850-1934) published *Elementary Meteorology* (Boston), a textbook.

1894. ORGANIZATIONS—OBSERVATORIES
Percival Lowell (1855-1916) established an observatory at Flagstaff, Arizona.

1894. ORGANIZATIONS—SOCIETIES AND ASSOCIATIONS / BOTANY
The Botanical Society was formally established, as an association for professional botanists (defined in terms of serious interest in research and publication). In 1892, a botanical section (G) had been established in the American Association for the Advancement of Science and in 1893 the national botanical organization was formulated with the election of charter members at the AAAS meeting.

1894. ORGANIZATIONS—SOCIETIES AND ASSOCIATIONS / MATHEMATICS
The American Mathematical Society was formed from the existing New York Mathematical Society.

1894. PERIODICALS AND PUBLISHING / MATHEMATICS
American Mathematical Monthly was established; it was published by Kidder Institute in Missouri. In 1916, it became the official publication of the Mathematical Association of America.

1894. PERIODICALS AND PUBLISHING / PSYCHOLOGY
James McKeen Cattell (1860-1944) and James Mark Baldwin (1861-1934) founded *Psychological Review*. Cattell served as editor until 1903.

1894-1896. PHYSICS / INSTRUMENTS AND INSTRUMENTATION
During these years, Ernest Fox Nichols (1869-1924) was a student in Berlin. Assisted by Ernst Pringsheim, he constructed an improved radiometer (the Nichols radiometer). It was particularly effective in infrared measurements and the use of the instrument constituted his chief scientific work.

1894-1927. GOVERNMENT—FEDERAL / ENTOMOLOGY
During these years, when the division of insects was under the leadership of Leland Ossian Howard (1857-1950), the U.S. Department of Agriculture's budget for work in entomology increased from an annual $30,000 to $3,000,000.

1895. ASTRONOMY
Simon Newcomb (1835-1909) published *The Elements of the Four Inner Planets and the Fundamental Constants of Astronomy* (Washington, D.C.). A Paris conference the next year concluded that, beginning in 1901, the constants (with modest change) should become the standard for ephemerides of all nations.

1895. ASTRONOMY
James Edward Keeler (1857-1900) at the Allegheny Observatory confirmed James Clerk Maxwell's contention that the rings of Saturn are particulate in nature. The work was published as "Spectroscopic Proof of the Meteoritic Constitution of Saturn's Rings," *Astrophysical Journal* 1 (1895): 416-427.

1895. INFORMATION ACCESS / CHEMISTRY
Arthur Amos Noyes (1866-1936) established "Review of American Chemical Research," which began publication in *Technology Quarterly* and later appeared in *Journal of American Chemical Society*. In 1907, it became *Chemical Abstracts*. The first editor of *Chemical Abstracts* (1907-1910) was William Albert Noyes (1857-1941), distant cousin to Arthur Noyes.

1895. ORGANIZATIONS—SOCIETIES AND ASSOCIATIONS / EDUCATION IN SCIENCE
The National Science Teachers Association was founded.

1895. PERIODICALS AND PUBLISHING
James McKeen Cattell (1860-1944) acquired ownership and became editor of *Science*, a relationship he retained for the remainder of his life.

1895. PERIODICALS AND PUBLISHING / ASTRONOMY
The *Astrophysical Journal* was established under the editorship of George Ellery Hale (1868-1938) and James Edward Keeler (1857-1900). (Edwin Brant Frost, 1866-1935, served as an assistant editor from this date and as editor 1902-1935.)

1895. PHYSICS / ACOUSTICS
Wallace Clement Ware Sabine (1868-1919) began his work in architectural acoustics, answering a request of Harvard University

President Charles W. Eliot (1834-1926) to determine the nature of acoustical problems in a University lecture hall. This was the initiation of Sabine's contributions that turned architectural acoustics from a strictly trial-and-error activity to a quantitatively based applied science.

1895. ZOOLOGY / ENTOMOLOGY
John Henry Comstock (1849-1931) published *A Manual for the Study of Insects* with illustrations by his wife, Anna Botsford Comstock (1854-1930).

1895. ZOOLOGY / PHYSIOLOGY
George Neil Stewart (1860-1930) published *Manual of Physiology with Practical Exercises* (New York). There was an eighth edition in 1918. The work was notable for the inclusion of the student exercises and experiments involving mammals.

1895-1914. ZOOLOGY / ENTOMOLOGY
Alpheus Spring Packard, Jr. (1839-1905) published *Monograph of the Bombycine Moths of North America* (Memoirs of National Academy of Sciences, vols. 7, 9, 12) (Washington, D.C.).

1896. ASTRONOMY
At the Lick Observatory, William Wallace Campbell (1862-1938) began a program of observations aimed at delineating the path of the sun among the other stars. The plan not only achieved Campbell's intent but resulted in data for other significant stellar studies.

1896. ASTRONOMY
The stellar spectra classification by Antonia Maury (1866-1952) of 681 bright stars in the northern sky, using a scheme that reflected greater degrees of differentiation, was published in *Annals of Harvard College Observatory* vol. 28, part 1. Although praised by some, later work at the Harvard observatory followed the classification scheme perfected by Annie Jump Cannon (1863-1941), who joined the Observatory staff in this year and remained there until her death.

1896. ENGINEERING AND APPLIED SCIENCE / METALLURGY
Albert Sauveur (1863-1939) published "The Microstructure of Steel and the Current Theories of Hardening," *Transactions of American Institute*

of Mining Engineers 26:863-906. The paper was widely noticed and discussion of it at a meeting of the American Institute was published in 1898 in the Institute's transactions.

1896. EVOLUTION
Edward Drinker Cope (1840-1897) published *The Primary Factors of Organic Evolution* (Chicago), in which he defended the idea of acquired characteristics.

1896. GOVERNMENT—FEDERAL / BIOLOGY—GENERAL
The Division of the Biological Survey was formed from the older Division of Economic Ornithology and Mammalogy in the U.S. Department of Agriculture. These events affirmed the agency's research orientation. In 1905, it became the Bureau of Biological Survey. It was incorporated into the newly established Fish and Wildlife Service in 1940.

1896. GOVERNMENT—FEDERAL / ENVIRONMENT AND CONSERVATION
At the invitation of the U.S. Department of Interior, the National Academy of Sciences appointed a committee, chaired by Charles Sprague Sargent (1841-1927), which in 1897 issued a report "On the Inauguration of a Rational Forest Policy for the Forested Lands of the United States." One outcome was that President Grover Cleveland established 21,000,000 acres of forest reserves. Legislation in 1897 essentially confirmed Cleveland's action and established the rudiments of a policy for dealing with the forests. The president of the Academy and important in the formation of the study committee was Oliver Wolcott Gibbs (1822-1908).

1896. MEDICINE / INSTRUMENTS AND INSTRUMENTATION
The first diagnostic X-ray photograph in the United States was produced by Michael I. Pupin (1858-1935). In the same year, Dr. Emil H. Grube was first in the country to use X-rays in the treatment of breast cancer.

1896. ORGANIZATIONS—ACADEMIC / MATHEMATICS
Leonard Eugene Dickson (1874-1954) received the first doctorate in mathematics from the University of Chicago.

1896. ORGANIZATIONS—INDUSTRY / COMPUTERS AND INFORMATION SCIENCE
The Tabulating Machine Company was founded by Herman Hollerith (1860-1929). Through mergers thereafter, it was part of the origins of what became the International Business Machines Company (IBM).

1896. PERIODICALS AND PUBLISHING / CHEMISTRY
The *Journal of Physical Chemistry* began publication as the first journal on the subject published outside of Germany. It was founded by Wilder Dwight Bancroft (1867-1953), at Cornell University, who financed as well as edited (or coedited) the journal until he transferred it to the American Chemical Society in 1932.

1896. PERIODICALS AND PUBLISHING / GEOPHYSICS AND GEODESY
Louis Agricola Bauer (1865-1932) founded the journal *Terrestrial Magnetism and Atmospheric Electricity* (later, *Journal of Geophysical Research*). He served as editor until 1927.

1896. PERIODICALS AND PUBLISHING / MEDICINE
The *Journal of Experimental Medicine* was established.

1896. RELIGION AND THEOLOGY
Andrew Dickson White (1832-1918) published *A History of the Warfare of Science with Theology in Christendom* (London).

1896. TECHNOLOGY AND INVENTION / AERONAUTICS
In Washington, a steam-powered aerial machine of Samuel P. Langley (1834-1906) flew three-quarters of a mile before crashing.

1896. ZOOLOGY / EVOLUTION
Edmund Beecher Wilson (1856-1939) published *The Cell in Development and Inheritance* (New York). In 1925, a third, revised and enlarged edition appeared. Wilson's central idea in doing the book was to trace the relations between cell and evolutionary theory. The book has been labeled as the most influential book on cytology of the twentieth century.

1897. GEOLOGY
George Perkins Merrill (1854-1929) published *A Treatise on Rocks, Rock Weathering, and Soils* (New York). A second edition appeared in 1906.

1897. ORGANIZATIONS—INDUSTRY / CHEMISTRY
Dow Chemical Company was founded by Herbert Henry Dow (1866-1930). The company's initial product was chlorine bleach.

1897. ORGANIZATIONS—OBSERVATORIES / INSTRUMENTS AND INSTRUMENTATION
The University of Chicago's Yerkes Observatory at Williams Bay, Wisconsin, was dedicated. A facility directed by George Ellery Hale (1868-1938), it contained a 40-inch refractor telescope, the largest in the world. The lens had originally been made by Alvan G. Clark (1832-1897) for a southern California observatory, but that group was unable to follow through because of financial reverses.

1897. ORGANIZATIONS—SOCIETIES AND ASSOCIATIONS / BOTANY
The Society for Plant Morphology and Physiology was established.

1897. PALEONTOLOGY
Charles Schuchert (1858-1942) published *Synopsis of American Fossil Brachiopoda* (Bulletin of U.S. Geological Survey, 87) (Washington, D.C.).

1897. PHYSICS / ELECTRICITY AND ELECTRONICS
Ernst J. Berg (1871-1941) and Charles Proteus Steinmetz (1865-1923) published the textbook, *Theory and Calculation of Alternating Current Phenomena* (New York). Steinmetz subsequently wrote a number of other influential texts on electricity.

1897 (April 12). GENERAL OR MISCELLANEOUS / PALEONTOLOGY, ZOOLOGY
Zoologist and paleontologist Edward Drinker Cope (b.1840) died at Philadelphia, Pennsylvania.

1897-1900. ANTHROPOLOGY AND ETHNOLOGY
The Jesup North Pacific Expedition, substantially the effort of Franz Boas (1858-1942), undertook to determine cultural relations between peoples of Siberia and the Northwest Coast.

1898. ORGANIZATIONS—SOCIETIES AND ASSOCIATIONS
The Association for Maintaining the American Women's Table at the Zoological Station (at Naples) was established. In 1917, it became the Association to Aid Scientific Research by Women. The Society last met in 1932.

1898. ORGANIZATIONS—SOCIETIES AND ASSOCIATIONS / BOTANY
The Sullivant Moss Society was founded, chiefly due to the efforts of Elizabeth Knight Britton (1858-1934). After 1948, it was known as the American Bryological Society.

1898. PALEONTOLOGY
Charles Doolittle Walcott (1850-1927) published an important work on fossil jellyfish.

1898. PERIODICALS AND PUBLISHING / METALLURGY
The journal *Metallographist* (from 1904, *Iron and Steel Magazine*) was founded by Albert Sauveur (1863-1939), who also served as editor until the publication's demise in 1906.

1898. PERIODICALS AND PUBLISHING / PHYSIOLOGY
The *American Journal of Physiology* was established.

1899. ASTRONOMY
The ninth satellite of Saturn (Phoebe) was found by William Henry Pickering (1858-1938).

1899. BIOLOGY—GENERAL / STATISTICS
Charles Benedict Davenport (1866-1944) published a manual, *Statistical Methods with Special Reference to Biological Variation* (New York), which helped introduce biometric methodology into the United States (and especially the work of Karl Pearson). A fourth edition appeared in 1936.

1899. EXPLORATION AND SURVEYING
The Harriman Alaska Expedition of this year had biologist Clinton Hart Merriam (1855-1942) as scientific director and report editor.

1899. GEOLOGY / PALEONTOLOGY
Amadeus William Grabau (1870-1946) published an early work that studied the environment of old sedimentary rocks in light of knowledge of the conditions of life among modern organisms, "The Relations of Marine Bionomy to Stratigraphy," *Bulletin of Buffalo Society of Natural Sciences*, 6 no. 4 (September 1899): 319-356. It was a step toward the development of paleoecology.

1899. ORGANIZATIONS—SOCIETIES AND ASSOCIATIONS / ASTRONOMY
The Astronomical and Astrophysical Society of America was founded, with its first meeting at George Ellery Hale's (1868-1938) Yerkes Observatory. In 1914, the name was changed to American Astronomical Society, astrophysics having achieved a secure place in the discipline.

1899. ORGANIZATIONS—SOCIETIES AND ASSOCIATIONS / MICROBIOLOGY AND MICROSCOPY
The Society of American Bacteriologists was organized. (It later became the American Society for Microbiology.)

1899. ORGANIZATIONS—SOCIETIES AND ASSOCIATIONS / PHYSICS
The American Physical Society was established at a meeting at Columbia University. This first professional society devoted to physics in the United States was founded largely due to the efforts of Arthur Gordon Webster (1863-1923). Henry A. Rowland (1848-1901) was the Society's first president.

1899. PERIODICALS AND PUBLISHING / BIOLOGY—GENERAL
Biological Bulletin was established. It was published by the Marine Biological Laboratory, Woods Hole, Mass.

1899. PERIODICALS AND PUBLISHING / ZOOLOGY
The journal *Bird Lore* was founded by Frank Michler Chapman (1864-1945). He served as editor until 1934. It later became *Audubon Magazine*.

1899. ZOOLOGY
William Keith Brooks (1848-1908) published *The Foundations of Zoology* (New York).

1899. ZOOLOGY / EVOLUTION
Hermon Carey Bumpus (1862-1943) published "The Elimination of the Unfit as Illustrated by the Introduced Sparrow, *Passer domesticus*," *Biological Lectures from the Marine Biological Laboratory, Wood's Hole, Mass., 1898*, 4 (Boston, 1899): 209-226. Among his conclusions was that extremely variable individuals were more susceptible to selective elimination. This study of English sparrows as affected by a severe winter storm is notable as an instance of natural selection in operation.

1899. ZOOLOGY / PHYSIOLOGY
Jacques Loeb (1859-1924) reported on his successful experiments in raising sea urchins from unfertilized eggs to the larval stage by altering the concentration of salt in the solution. A result of these widely noted studies was to increase interest in work involving the application of physico-chemical approaches (specifically, osmotic pressures) to biological phenomenon.

1900. ASTRONOMY
Sherburne Wesley Burnham (1838-1921) capped years of astronomical observation with *A General Catalogue of 1290 Double Stars Discovered from 1871 to 1899 by S. W. Burnham, Arranged in Order of Right Ascension with All the Micrometrical Measures of Each Pair* (Publications of the Yerkes Observatory, no. 1) (Chicago).

1900. ASTRONOMY
The spiral character of some nebulae was discovered by James Edward Keeler (b.1857) through photographic evidence. Keeler, director of the Lick Observatory, died at San Francisco on 12 August of this year.

1900. AWARDS AND PRIZES / MATHEMATICS
Mathematician George Abram Miller (1863-1951) was given a prize for his work in group theory by the Academy of Sciences of Cracow, reportedly the Academy's first prize in pure mathematics to an American.

1900. CHEMISTRY
Moses Gomberg (1866-1947) produced the earliest stable free radical, triphenylmethyl. Organic chemists at the time assumed free radicals were not likely to exist. The report of the event appeared in *Berichte der Deutschen chemischer Gesellschaft* 33 (1900): 3150-3163, and *Journal of American Chemical Society* 22 (1900): 757-771.

1900. MEDICINE / DISEASE
William Hallock Park (1863-1939) discovered the role of healthy humans in the transmission of typhoid fever.

1900. MEDICINE / DISEASE
Walter Reed (1851-1902) was appointed head of an army Yellow Fever Board to investigate the causes of an outbreak of the disease among United States forces in Cuba. He was accompanied by James Carroll (1854-1907), Jesse W. Lazear (1866-1900), and Aristides Agramonte y Simoni (1868-1931). Aided by work done from as early as 1881 by Carlos Finlay (1833-1915), the Board concluded that yellow fever was spread by a mosquito of the genus Aedes. The discovery was reported in "The Etiology of Yellow Fever, A Preliminary Note," *Philadelphia Medical Journal* 6 (October 27, 1900): 190-196. In 1901, Reed established that the disease was caused by a filterable virus, the first such human disease shown to be caused by such an agent. Their studies of the course of the infection helped to lead to its control. Lazear died as a result of his involvement in the study.

1900. ORGANIZATIONS—HOSPITALS / BIOCHEMISTRY AND MOLECULAR BIOLOGY
The first biochemical research laboratory in a hospital was established at McLean Hospital for the Insane (Waverley, Massachusetts), with Otto K.O. Folin (1867-1934) in charge.

1900. ORGANIZATIONS—INDUSTRY / ELECTRICITY AND ELECTRONICS
The General Electric Company's research laboratory was established at Schenectady, New York. It was perhaps the first such industrial research facility in the United States.

1900-1901. ORGANIZATIONS—SOCIETIES AND ASSOCIATIONS / ENVIRONMENT AND CONSERVATION
State Audubon societies affiliated under the National Committee of Audubon Societies of America. In 1905, it became the National Association of Audubon Societies for the Protection of Wild Birds and Animals, and in 1940 the National Audubon Society. (The first Audubon Society was organized in 1886.)

1900. PERIODICALS AND PUBLISHING / MATHEMATICS
The *Transactions of American Mathematical Society* began publication as a research outlet.

1900. PERIODICALS AND PUBLISHING / NATURAL HISTORY
American Museum Journal was established. It was published by the American Museum of Natural History. In 1919, it became *Natural History*.

1900. PERIODICALS AND PUBLISHING / ORGANIZATIONS—SOCIETIES AND ASSOCIATIONS
Science, owned and edited by James McKeen Cattell (1860-1944), became the official journal of the American Association for the Advancement of Science. Upon his death, ownership was vested in the Association.

1900. PHYSICS / ACOUSTICS
Wallace Clement Ware Sabine (1868-1919) published "Reverberation," *The American Architect* (republished in his *Collected Papers on Acoustics*, Cambridge, Mass., 1922), that delineated the basic components of an acoustically effective auditorium. Sabine's proposed acoustical law related the reverberation time to the volume of the room and amount of absorbent material it contained.

1901. BIOCHEMISTRY AND MOLECULAR BIOLOGY
Jokichi Takamine (1854-1922) isolated a substance from the adrenal glands that later was known as adrenaline or epinephrine. Although eventually credited with having been the first to isolate a pure hormone, the concept of hormones was not known at the time.

1901. GENERAL OR MISCELLANEOUS
Theodore Roosevelt (1858-1919) became president, the first since John Quincy Adams (1767-1848) with a personal interest and knowledge in science.

1901. GENERAL OR MISCELLANEOUS / CHEMISTRY
The American Chemical Society conducted a census. In this survey, there were more than 22,000 enrollees in college chemistry courses, of which physical chemistry accounted for only 2.5%. Inorganic chemistry courses had 59% of the total, analytical chemistry 24%. Organic chemistry enrollment accounted for 12% of the total.

1901. GENETICS
Thomas Harrison Montgomery, Jr. (1873-1912) presented research results that contained the kernel of the idea of inheritance from two parents by way of the chromosomes. A report of this work appeared in "A Study of the Chromosomes of Germ Cells of Metazoa," *Transactions of American Philosophical Society*, new series 20:154-236.

1901. GOVERNMENT—FEDERAL / PUBLIC HEALTH
Funds were appropriated for construction of the Hygienic Laboratory in Washington, D.C. The laboratory, associated with the Marine Hospital Service, had started on a small scale on Staten Island in 1887 and had moved to Washington in the early 1890s.

1901. MATHEMATICS
Edwin Bidwell Wilson (1879-1964) published *Vector Analysis* (New York). The textbook was based on the lectures of Josiah Willard Gibbs (1839-1903).

1901. MICROBIOLOGY AND MICROSCOPY / ZOOLOGY
Gary Nathan Calkins (1869-1943) published *The Protozoa* (New York), an early English-language work on that subject.

1901. ORGANIZATIONS—ACADEMIC / CHEMISTRY
Theodore William Richards (1868-1928) was offered a professorship in physical chemistry at the University of Goettingen, which he did not accept. In order not to lose him, Harvard University promoted him, reduced his teaching, and built a new laboratory (Wolcott Gibbs

Laboratory). It was a distinct honor to be offered such a foreign professorship, however, and greatly promoted his position among American chemists.

1901. ORGANIZATIONS—INDUSTRY / CHEMISTRY
The Monsanto corporation was established by John Francisco Queeny (1859-1933) in St. Louis, Missouri.

1901. ORGANIZATIONS—RESEARCH INSTITUTIONS / ZOOLOGY—HUMAN
The Rockefeller Institute for Medical Research was founded by John D. Rockefeller. In 1965, it became Rockefeller University.

1901. PERIODICALS AND PUBLISHING / ANATOMY
The American Journal of Anatomy was established by Franklin Paine Mall (1862-1917), Charles S. Minot (1852-1914), and George S. Huntington (1861-1927).

1901. PERIODICALS AND PUBLISHING / BOTANY
American Botanist was established.

1901. TECHNOLOGY AND INVENTION
The mercury vapor arc lamp was invented by Peter C. Hewitt (1861-1921). An improved and commercially viable version appeared in 1903.

1901. ZOOLOGY / NEUROBIOLOGY
Florence Rena Sabin (1871-1953) published *An Atlas of the Medulla and Midbrain* (Baltimore).

1901. ZOOLOGY / ORNITHOLOGY
For what was probably the first time, the use of leg bands for the study of bird migrations was suggested. It appeared in Leon Jacob Cole (1877-1948), "Suggestions for a Method of Studying the Migration of Birds," *Michigan Academy of Science Report* 3 (1902): 67-70.

1901 (March 3). GOVERNMENT—FEDERAL / PHYSICAL SCIENCES—GENERAL
The U.S. Congress voted to establish the National Bureau of Standards (after 1903, in the new Department of Commerce and Labor). Much of

the promotion for the new agency had been done by Samuel Wesley Stratton (1861-1931) who, in 1899, had been made head of the federal government's work on weights and measures. Stratton was made first director of the Bureau, and had modeled his concept of a national laboratory on the German Physikalisch-Technische Reichsanstalt. Among the new Bureau's functions was to conduct research. In 1988, it was renamed the National Institute of Standards and Technology.

1901 (April 16). GENERAL OR MISCELLANEOUS / PHYSICS
Physicist Henry Augustus Rowland (b.1848) died in Baltimore, Maryland.

1901 and 1902. GENETICS
In these years, Clarence Erwin McClung (1870-1946) published work on the accessory or X chromosome in insects and its possible linkage to sex determination, and was among the first to give substantive evidence for the association of particular characteristics and a specific chromosome. The results were presented in: "Notes on the Accessory Chromosome," *Anatomischer Anzeiger* 20 (1901): 220-226; and "The Accessory Chromosome: Sex Determinant?," *Biological Bulletin* 3 (1902): 43-84.

1901-1905. PSYCHOLOGY
Edward Bradford Titchener (1867-1927) published *Experimental Psychology* (New York), in four volumes. It consisted of two student and two teacher manuals.

1901-1913. GENERAL OR MISCELLANEOUS / CHEMISTRY
Ira Remsen (1846-1927) served as second president of Johns Hopkins University and continued as professor of chemistry.

1901-1919. ZOOLOGY / ORNITHOLOGY
Robert Ridgway (1850-1929) published "The Birds of North and Middle America," *Bulletin of United States National Museum* no. 50, in eight parts (1901, 1902, 1904, 1907, 1911, 1914, 1916, 1919).

1902. ASTRONOMY
Charles Augustus Young (1834-1908) published *Manual of Astronomy* (Boston), a well-known textbook.

1902. CHEMISTRY
Arthur Amos Noyes (1866-1936) published *The General Principles of Physical Science* (New York). Later editions appeared, with the collaboration of Miles Sherrill (1877-1965), under the eventual title of *A Course of Study in Chemical Principles*. The work was of great importance in the teaching of physical chemistry.

1902. INFORMATION ACCESS / ZOOLOGY
Albert Hassall (1862-1942) and Charles Wardell Stiles (1867-1941) began publication of *Index-catalogue of Medical and Veterinary Zoology* (Washington, D.C.). Publication went on over a period of years.

1902. ORGANIZATIONS
The Nantucket Maria Mitchell Association was founded. It came to include an observatory, science library, and natural science museum.

1902. ORGANIZATIONS—FOUNDATIONS
The General Education Board (Rockefeller) was founded. It was active in science especially during the 1920s.

1902. ORGANIZATIONS—RESEARCH INSTITUTIONS / FUNDS AND FUNDING
The Carnegie Institution of Washington was founded with a ten-million-dollar endowment from Andrew Carnegie (between 1902 and 1911, Carnegie gave an additional 12 million dollars). At first there were grants for individual researchers, but in time the Institution came to concentrate its efforts in support of its own research facilities. As of 1902, when the Institution was chartered, the total American endowment for the support of scientific research was less than three million dollars. About 80% was at the Smithsonian Institution and Harvard University.

1902. ORGANIZATIONS—SOCIETIES AND ASSOCIATIONS / ANTHROPOLOGY AND ETHNOLOGY
The American Anthropological Association was founded.

1902. ORGANIZATIONS—SOCIETIES AND ASSOCIATIONS / BOTANY
The Wild Flower Preservation Society of America was founded. (Among those primarily responsible for the establishment was Elizabeth Knight Britton, 1858-1934.)

1902. PHYSICS
Josiah Willard Gibbs (1839-1903) published *Elementary Principles in Statistical Mechanics Developed with Special Reference to the Rational Foundation of Thermodynamics* (New Haven).

1902. ZOOLOGY—HUMAN / PHYSIOLOGY
Russell Henry Chittenden (1856-1943) began his studies of protein requirements of men, the result of which was a lower daily requirement than was thought to be true at the time. A related work was his *Physiological Economy in Nutrition with Special Reference to the Minimal Proteid Requirements of the Healthy Man, An Experimental Study* (New York, 1904).

1902. GOVERNMENT—FEDERAL / PUBLIC HEALTH
The Marine Hospital Service was renamed the Public Health and Marine Hospital Service. In 1912, the name was changed to Public Health Service.

1902. PHYSICS / METEOROLOGY AND CLIMATOLOGY
Arthur Edwin Kennelly (1861-1939) postulated that a discontinuity in the upper atmosphere could reflect radio waves and thus account for the trans-Atlantic radiotelegraph transmission of Guglielmo Marconi. In England, Oliver Heaviside made a similar suggestion and it came to be called the Kennelly-Heaviside layer. The layer of electrically charged particles later was designated the ionosphere.

1902 (September 23). GENERAL OR MISCELLANEOUS / GEOLOGY
Geologist and ethnologist John Wesley Powell (b.1834) died in Haven, Maine.

1902-1903. GENETICS
Graduate student Walter Stanborough Sutton (1877-1916) published two papers in this period that put forth and supported his idea of the pairing of chromosomes as the basis of Mendel's concepts. The papers were: "Morphology of the Chromosome Group in Brachystola magna," *Biological Bulletin* 4 (1902): 24-39; and "The Chromosomes in Heredity," *ibid.* 4 (1903): 231-251.

Chronology

1902-1904. PHYSICS
Edward W. Morley (1838-1923) and Dayton Clarence Miller (1866-1941) carried out repetitions of the 1887 Michelson-Morley experiments to detect the stationary luminiferous ether.

1903. ASTRONOMY
Edward Charles Pickering (1846-1919) issued the first-ever *Photographic Map of the Entire Sky*, based on work of the Harvard College Observatory in Cambridge and in Arequipa, Peru.

1903. GEOLOGY / CHEMISTRY
Henry Stephens Washington (1867-1934) published *Chemical Analyses of Igneous Rocks Published from 1884 to 1900* (Washington, D.C.); an enlarged edition appeared in 1917.

1903. GEOLOGY / CHEMISTRY
C(harles) Whitman Cross (1854-1949), Joseph Paxson Iddings (1857-1920), Louis V. Pirsson (1860-1919), and Henry S. Washington (1867-1934) published *Quantitative Classification of Igneous Rocks, Based on Chemical and Mineral Characters, With Systematic Nomenclature* (Chicago). The classification of rocks by quantitative chemical composition became known as the C.I.P.W. system, for the four authors.

1903. GOVERNMENT—FEDERAL
President Theodore Roosevelt (1858-1919) appointed a Committee of Organization of Government Scientific Work. Charged with the examination of duplicate efforts in government science, they concluded that there was little. The Committee reports were not published and little resulted from its work.

1903. ORGANIZATIONS—RESEARCH INSTITUTIONS / BOTANY
The Carnegie Institution of Washington's Department of Plant Biology was founded.

1903. ORGANIZATIONS—SOCIETIES AND ASSOCIATIONS / BIOLOGY—GENERAL, MEDICINE
The Society for Experimental Biology and Medicine was established.

1903. ORGANIZATIONS—SOCIETIES AND ASSOCIATIONS / BOTANY
The American Mycological Society was established. In 1906, it combined with other organizations to form the Botanical Society of America.

1903. PHYSICS
Albert Abraham Michelson (1852-1931) published *Light Waves and Their Uses* (Chicago). The book was based on his 1899 Lowell lectures.

1903. ORGANIZATIONS—SOCIETIES AND ASSOCIATIONS / BOTANY
The American Society for Horticultural Science was founded. Liberty Hyde Bailey, Jr. (1858-1954) of Cornell University was the first president.

1903 (April 28). GENERAL OR MISCELLANEOUS / PHYSICS
Mathematician and theoretical physicist Josiah Willard Gibbs (b.1839) died in New Haven, Connecticut.

1903 (December 17). TECHNOLOGY AND INVENTION / AERONAUTICS
Orville Wright (1871-1948) and Wilbur Wright (1867-1912) conducted four flights from level ground near Kill Devil Hills, south of Kitty Hawk, North Carolina. The longest flight was more than a half mile airborne. These were the first successful flights with a powered heavier-than-air device.

ca.1903. ORGANIZATIONS—INDUSTRY / CHEMISTRY
DuPont chemical company, as part of a general reorganization, created a small department for research on explosives, the company's primary business.

1904. BIOLOGY—GENERAL
Edmund Duncan Montgomery (1835-1911) published *The Vitality and Organization of Protoplasm* (Austin, Texas).

1904. CHEMISTRY / PHYSICS
Bertram Borden Boltwood (1870-1927) began his research in radiochemistry, work which eventually won him high rank among early investigators of the chemical identity and sequencing of the radioelements that followed the discovery of radioactivity by Henri

Becquerel in 1896. About this time, Boltwood, as well as Herbert McCoy, and Robert Strutt in England, confirmed that radium was a product of uranium decay.

1904. GENERAL OR MISCELLANEOUS / DISEASE
Army surgeon William Gorgas (1854-1920) was given charge of disease control during the building of the Panama Canal. Through the control of mosquitoes, he essentially eradicated malaria and yellow fever in the canal region.

1904. GEOLOGY / CHEMISTRY
Henry Stephens Washington (1867-1934) published the textbook, *Manual of the Chemical Analysis of Rocks* (Washington, D.C.). A fourth edition appeared in 1930.

1904. ORGANIZATIONS—FAIRS AND EXPOSITIONS
An International Congress of Arts and Sciences was held at the St. Louis World's Fair.

1904. ORGANIZATIONS—RESEARCH INSTITUTIONS
During this year, several research facilities supported by the Carnegie Institution of Washington came into being. George Ellery Hale (1868-1938) achieved funding for the Mount Wilson Observatory at Pasadena, California. Louis Agricola Bauer (1865-1932) established and became head of the Carnegie Institution's Department of Terrestrial Magnetism. (Bauer's most substantial achievement related to the mapping of magnetic fields upon the oceans.) Charles Benedict Davenport (1866-1944) became director of the Station for Experimental Evolution at Cold Spring Harbor, New York, having arranged for support of the facility by the Carnegie Institution.

1904. ORGANIZATIONS—SOCIETIES AND ASSOCIATIONS / GEOGRAPHY AND CARTOGRAPHY
The Association of American Geographers was founded. In 1948, it absorbed the American Society of Professional Geographers.

1904. ORGANIZATIONS—SOCIETIES AND ASSOCIATIONS / PSYCHOLOGY
A group of psychologists formed "The Experimentalists" under the direction of Edward B. Titchener (1867-1927). In 1929, it became the Society of Experimental Psychologists.

1904. PERIODICALS AND PUBLISHING / MEDICINE
The *Journal of Infectious Diseases* was established. (The editor from its founding until 1941 was Ludvig Hektoen, 1863-1951.)

1904. PERIODICALS AND PUBLISHING / ZOOLOGY
The *Journal of Experimental Zoology* was founded. From this year until 1946, Ross Granville Harrison (1870-1959) served as managing editor.

1904. PHYSICS
Arthur Gordon Webster (1863-1923) published the textbook, *The Dynamics of Particles and of Rigid, Elastic, and Fluid Bodies* (Leipzig).

1904. PHYSICS
DeWitt Bristol Brace (1859-1905) experimentally tested the Lorentz-Fitzgerald contraction hypothesis, using optical methods. Contrary to Brace's interpretation, it was later demonstrated that this result did not disprove the contraction hypothesis. Relevant is his paper, "Double Refraction in Matter Moving Through the Ether," *Philosophical Magazine*, new series 7 (1904): 317-328.

1904. PSYCHOLOGY
Granville Stanley Hall (1846-1924) published *Adolescence: Its Psychology, and Its Relations to Physiology, Anthropology, Sociology, Sex, Crime, Religion, and Education* (New York). He presented there the idea that maturation progressed through the historical stages of the human race itself.

1904 and 1905. ASTRONOMY
Charles Dillon Perrine (1867-1951) discovered the sixth and seventh satellites of Jupiter.

1904-1908. MEDICINE / SURGERY
Work carried out by Alexis Carrel (1873-1944) during these years laid the basis for future progress in heart and blood vessel surgery and in organ transplant. He came to the United States in 1904.

1905. ASTRONOMY
Percival Lowell (1855-1916) predicted the existence of a ninth planet, beyond Neptune, and began a search for the object from his observatory at Flagstaff, Arizona.

1905. CHEMISTRY / NUTRITION
Lafayette Benedict Mendel (1872-1935), of Yale University and Thomas B. Osborne (1859-1929), of the Connecticut Agricultural Experiment Station at New Haven, began a long period of collaboration on studies in nutrition from which in excess of 100 papers resulted. In 1909, they began their studies of the nutritional effects of proteins, which over the years received financial support from the Carnegie Institution of Washington.

1905. EVOLUTION
John Thomas Gulick (1832-1923) published *Evolution, Racial and Habitudinal* (Washington, D.C.).

1905. GENETICS
Edmund Beecher Wilson (1856-1939) and Nettie Maria Stevens (1861-1912), working separately, published on the likelihood of a chromosomal determination of sex (i.e., XX chromosomes for female and XY for male). Their work also was the first research support for the idea that hereditary characteristics are associated with particular chromosomes. Wilson's work was in "The Chromosomes in Relation to the Determination of Sex in Insects," *Science* 22:500-502; that of Stevens was "Studies in Spermatogenesis with Especial Reference to the 'Accessory Chromosome'," *Publications of Carnegie Institution of Washington*, no. 36.

1905. GEOLOGY
Thomas Chrowder Chamberlin (1843-1928) published "Fundamental Problems of Geology," *Yearbook [1904] of Carnegie Institution of Washington*, no. 3. The work contained the important planetesimal

hypothesis, on the origins of the earth, which superseded the then-current theory of Laplace. Chamberlin's report represented work on which Forest Ray Moulton (1872-1952) was a collaborator. In 1916, Chamberlin published on the planetesimal ideas in his *The Origin of the Earth* (Chicago).

1905. GEOLOGY
The meteor origin of Great Barringer Meteor Crater in Arizona was proposed by Daniel Barringer (1860-1929). His contention argued against volcanic origin of the site.

1905. GOVERNMENT—FEDERAL / BOTANY
The U.S. Forest Service was first referred to by that designation at the time of transfer of the federal forest reserves from the Department of the Interior to the Department of Agriculture.

1905. MEDICINE
The first direct transfusion of blood was carried out by George Washington Crile (1864-1943).

1905. ORGANIZATIONS—SOCIETIES AND ASSOCIATIONS / SOCIOLOGY
The American Sociological Association was founded.

1905. PERIODICALS AND PUBLISHING / BIOCHEMISTRY AND MOLECULAR BIOLOGY
The *Journal of Biological Chemistry* was established by John Jacob Abel (1857-1938) and Christian A. Herter (1865-1910).

1905. PERIODICALS AND PUBLISHING / GEOLOGY
The journal *Economic Geology* was established.

1905. PHYSICS
Robert Williams Wood (1868-1955) published *Physical Optics* (New York and London), particularly notable for its experimental features. There were three editions of the work.

Chronology

1905. PUBLIC HEALTH
The American Public Health Association published *Report of the Committee on Standard Methods of Water Analysis* (Chicago). The committee was under the leadership of W.H. Welch.

1905-1914. BOTANY / PATHOLOGY
Erwin Frink Smith (1854-1927) published *Bacteria in Relation to Plant Diseases* (Washington, D.C.), in three volumes.

1906. ASTRONOMY
Simon Newcomb (1835-1909) published *A Compendium of Spherical Astronomy with Its Applications to the Determination and Reduction of Positions of the Fixed Stars* (New York).

1906. ASTRONOMY
Percival Lowell (1855-1916) published *Mars and Its Canals* (New York). In 1908, he published *Mars as the Abode of Life* (New York).

1906. CHEMISTRY / NUTRITION
Graham Lusk (1866-1932) published *Elements of the Science of Nutrition* (Philadelphia). A fourth edition appeared in 1928.

1906. GENERAL OR MISCELLANEOUS
The first edition of *American Men of Science* was published by James McKeen Cattell (1860-1944) and with support from the Carnegie Institution of Washington.

1906. GEOLOGY
Eugene Woldemar Hilgard (1833-1916) published *Soils, Their Formation, Properties, Composition, and Relations to Climate and Plant Growth in the Humid and Arid Regions* (New York).

1906. GOVERNMENT—FEDERAL / AGRICULTURE
The U.S. Department of Agriculture first established a biophysical research laboratory. It was organized and directed by Lyman James Briggs (1874-1963), who remained with the Department until 1917.

1906. GOVERNMENT—FEDERAL / CHEMISTRY
The U.S. Food and Drug Act was passed and became effective 1 January 1907, prohibiting the interstate sale of adulterated or unlabeled products. The U.S. Department of Agriculture's Bureau of Chemistry, which in 1902 had begun studies of the human effects of food additives, was given charge of enforcement. Harvey Washington Wiley (1844-1930) was head of the Bureau, having held that position since 1883. The Food and Drug Administration, under that name, was provided for in 1930.

1906. MEDICINE / DISEASE
Howard Taylor Ricketts (1871-1910) published *Infection, Immunity, and Serum Therapy* (Chicago).

1906. ORGANIZATIONS—RESEARCH INSTITUTIONS / ASTRONOMY
The Carnegie Institution of Washington established a meridian astrometry department with Dudley Observatory director Lewis Boss (1846-1912) in charge. The project had the object to refine knowledge of positions and proper motions of stars.

1906. ORGANIZATIONS—RESEARCH INSTITUTIONS / GEOPHYSICS AND GEODESY
The Carnegie Institution of Washington's Geophysical Laboratory was founded.

1906. ORGANIZATIONS—SOCIETIES AND ASSOCIATIONS / BIOCHEMISTRY AND MOLECULAR BIOLOGY
The American Society of Biological Chemists was founded. In 1987, it became the American Society for Biochemistry and Molecular Biology.

1906. ORGANIZATIONS—SOCIETIES AND ASSOCIATIONS / BOTANY
The Botanical Society, Society for Plant Morphology and Physiology, and American Mycological Society combined to form the Botanical Society of America.

1906. ORGANIZATIONS—SOCIETIES AND ASSOCIATIONS / ENTOMOLOGY
The Entomological Society of America was established.

1906. ORGANIZATIONS—SOCIETIES AND ASSOCIATIONS / SEISMOLOGY
The Seismological Society of America was founded.

1906. TECHNOLOGY AND INVENTION / ELECTRICITY AND ELECTRONICS
The National Electric Signalling Company, based on inventions of Reginald Aubrey Fessenden (1866-1932), sent the first long-distance voice (radio) message from a station at Brant Rock, Massachusetts.

1906. TECHNOLOGY AND INVENTION / ELECTRICITY AND ELECTRONICS
Ernst Frederik Werner Alexanderson (1878-1975) invented a high-frequency alternator that had a significant impact on radio communications.

1906. ZOOLOGY
Neurologist Henry Herbert Donaldson (1857-1938) published an early paper that used the white rat as experimental animal (rather than the frog). From that point, Donaldson, as director of the Wistar Institute of Anatomy and Biology (Philadelphia), developed a Wistar Institute strain of rat that was widely used in experimentation.

1906. ZOOLOGY
Herbert Spencer Jennings (1868-1947) published *Behavior of the Lower Organisms* (New York). Among the implications of the book was an effective challenge to the idea of physicochemical tropisms in animals promoted especially by Jacques Loeb (1859-1924).

1906 (February 27). GENERAL OR MISCELLANEOUS / ASTRONOMY, PHYSICS
Astronomer and physicist Samuel Pierpont Langley (b.1834) died in Aiken, South Carolina.

1907. AWARDS AND PRIZES / PHYSICS
Albert Abraham Michelson (1852-1931) became the first American to win a Nobel Prize in science. In receiving the physics prize, he was honored for "his optical precision instruments and the spectroscopic and metrological investigations carried out with their aid."

1907. BIOLOGY—GENERAL
Ross Granville Harrison (1870-1959) first reported on his pioneering *in vitro* experimental work on developing nerve fiber in embryos, presenting an innovative technique in tissue culture (hanging drop

method) that had applications in a number of other areas of biological research.

1907. CHEMISTRY / GEOLOGY
The use of the decay of uranium as a means of gauging the age of rocks was discovered by Bertram Borden Boltwood (1870-1927). Boltwood's radioactive dating of rocks, based on lead, became the established procedure by the 1930s.

1907. MATHEMATICS
Mathematician William Fogg Osgood (1864-1943) published *Lehrbuch der Funktiontheorie* (Leipzig and Berlin).

1907. MEDICINE
The seven-volume work, *Modern Medicine: Its Theory and Practice* (Philadelphia and New York) was published. William Osler (1849-1919) was editor of the work.

1907. ORGANIZATIONS—RESEARCH INSTITUTIONS / NUTRITION
Francis Gano Benedict (1870-1957) was made director of the Boston Nutrition Laboratory, a facility established for him by the Carnegie Institution. He remained at its head until 1937.

1907. ORGANIZATIONS—SOCIETIES AND ASSOCIATIONS / AGRICULTURE
The American Society of Agronomy was established.

1907. ORGANIZATIONS—SOCIETIES AND ASSOCIATIONS / EDUCATION IN SCIENCE
The American Federation of the Mathematical and the Natural Sciences was established and included a number of organizations of teachers of mathematics and science.

1907. ORGANIZATIONS—SOCIETIES AND ASSOCIATIONS / MEDICINE
The American Association for Cancer Research was established.

1907. PHILOSOPHY
William James (1842-1910) published *Pragmatism: A New Name for Some Old Ways of Thinking* (New York).

1907. TECHNOLOGY AND INVENTION / CHEMISTRY
Leo H. Baekeland (1863-1944) produced bakelite, the first plastic that solidified upon heating. Produced from a chemical reaction between phenol and formaldehyde, the introduction of the substance in 1909 gave rise to the modern plastics industry. Baekeland published the first report on his achievement in "The Synthesis, Constitution, and Uses of Bakelite," *Journal of Industrial and Engineering Chemistry* 1 (March 1909): 149-161.

1907. TECHNOLOGY AND INVENTION / ELECTRICITY AND ELECTRONICS
Lee deForest (1873-1961) patented the triode or three-element vacuum tube. Originally called the audion, it was a basic component of radio technology and is described in U.S. Patent Office, "841,387. A Device for Amplifying Feeble Electrical Currents. Lee de Forest, New York, N.Y.," *Official Gazette* 126 (January 15, 1907): 908.

1907. ZOOLOGY / CHEMISTRY, PATHOLOGY
Harry Gideon Wells (1875-1943) published *Chemical Pathology* (Philadelphia), which achieved a fifth edition in 1925.

1907-1910. ANTHROPOLOGY AND ETHNOLOGY
Frederick Webb Hodge (1864-1956) published *Handbook of American Indians North of Mexico* (Washington, D.C.). The work drew upon the efforts of the Bureau of American Ethnology as it had been formulated by John Wesley Powell (1834-1902).

1908. ASTRONOMY
The "Revised Harvard Photometry" was published as volumes 50 and 54 of the *Annals of Harvard College Observatory*, a summation of Observatory director Edward C. Pickering's (1846-1919) photometric studies. The catalogue included the magnitudes of more than 45,000 stars brighter than the seventh magnitude.

1908. ASTRONOMY
In "1777 Variables in the Magellanic Clouds," *Annals of Harvard College Observatory*, 60 no. 4, Henrietta Swan Leavitt (1868-1921) made an observation relating periods and luminosity of stars. This related to her discovery of the period-luminosity relation of Cepheid (pulsating) variable stars, and the discovery was further confirmed in

work she did in 1912 on another group of stars. Her work made it possible to calculate the distance of the Cepheids.

1908. ASTRONOMY
George Ellery Hale (1868-1938) discovered magnetic fields in sunspots, the first extraterrestrial magnetic fields known.

1908. ASTRONOMY / INSTRUMENTS AND INSTRUMENTATION
A 60-inch reflecting telescope, then the largest in the world, was installed on Mt. Wilson. The disk was given by the father of George Ellery Hale (1868-1938).

1908. GENETICS
Vernon Lyman Kellogg (1867-1937) published *Inheritance in Silkworms* (Stanford University Publications, series 1), an early American work on genetics.

1908. GEOLOGY / CHEMISTRY
The work of chief chemist Frank Wigglesworth Clarke (1847-1931) at the U.S. Geological Survey was published in *The Data of Geochemistry* (Survey Bulletin no. 330). A fifth edition appeared in 1924, the year of Clarke's retirement.

1908. GOVERNMENT—FEDERAL / PHYSICS
A naval radiotelegraphy laboratory was established at the National Bureau of Standards; Louis Winslow Austin (1867-1932) was at its head. In 1923, it became part of the newly founded Naval Research Laboratory of the U.S. Navy (Austin remained with the NBS).

1908. ORGANIZATIONS—INDUSTRY
General Motors Company was established in Flint, Michigan by William Crapo Durant (1861-1947).

1908. ORGANIZATIONS—SOCIETIES AND ASSOCIATIONS / BOTANY
The American Phytopathological Society was established.

1908. ORGANIZATIONS—SOCIETIES AND ASSOCIATIONS / CHEMISTRY, ENGINEERING AND APPLIED SCIENCE
The American Institute of Chemical Engineers was founded.

1908. ORGANIZATIONS—SOCIETIES AND ASSOCIATIONS / MEDICINE
In the aftermath of antivivisection attacks on the Rockefeller Institute, a Defense Committee in Support of Medical Research was organized by the American Medical Association. Walter Bradford Cannon (1871-1945) was appointed to chair the Committee.

1908. ORGANIZATIONS—SOCIETIES AND ASSOCIATIONS / NATURAL HISTORY
The American Nature Study Society was established.

1908. ORGANIZATIONS—SOCIETIES AND ASSOCIATIONS / PALEONTOLOGY
The Paleontological Society was established.

1908. ORGANIZATIONS—SOCIETIES AND ASSOCIATIONS / PHARMACOLOGY AND PHARMACY
The American Society for Pharmacology and Experimental Therapeutics was founded.

1908. PHYSICS / INSTRUMENTS AND INSTRUMENTATION
Dayton Clarence Miller (1866-1941) invented a mechanical means of recording patterns of sound photographically. He called his device the phonodeik.

1908. TECHNOLOGY AND INVENTION
Henry Ford (1863-1947) introduced the Model T Ford.

1908. ZOOLOGY / EMBRYOLOGY
Frank Rattray Lillie (1870-1947) published *Development of the Chick, An Introduction to Embryology* (New York). There was a second edition in 1919, and a third (by Howard L. Hamilton, 1916-) in 1952.

1908 (December 9). GENERAL OR MISCELLANEOUS / CHEMISTRY
Chemist Oliver Wolcott Gibbs (b.1822) died at Newport, Rhode Island, the last survivor of the original members of the National Academy of Sciences.

1908-1910. GEOLOGY / SEISMOLOGY
The State Earthquake Investigation Commission published a two-volume report, *The California Earthquake of April 18, 1906* (Carnegie Institution

of Washington, Publication no. 87). Preparation of the report was supervised by Andrew Cowper Lawson (1861-1952).

1909. ASTRONOMY
Vesto Melvin Slipher (1875-1969) discovered dust and gas in interstellar space.

1909. BIOCHEMISTRY AND MOLECULAR BIOLOGY
Edward Tyson Reichert (1855-1931) contended that chemical differences in "vital substances" could account for all taxonomic and individual differences of organisms. Reichert drew on the differing forms of hemoglobin crystals of more than 100 animals. With mineralogist Amos Peaslee Brown (1864-1917), Reichert published *The Differentiation and Specificity of Corresponding Proteins and Other Vital Substances in Relation to Biological Classification and Organic Evolution: The Crystallography of Hemoglobins* (Washington, D.C., 1909).

1909. BIOCHEMISTRY AND MOLECULAR BIOLOGY
Phoebus Aaron Levene (1869-1940) concluded that a sugar, ribose, was contained in nucleic acids. Two decades later, he discovered that there was a second sugar, deoxyribose, that was present in other nucleic acids. (These two nucleic acids were what came to be known as ribonucleic acid or RNA, and deoxyribonucleic acid or DNA.)

1909. BOTANY
Frederic Edward Clements (1874-1945) published *Genera of Fungi* (Minneapolis).

1909. BOTANY / PATHOLOGY
Benjamin Minge Duggar (1872-1956) published *Fungus Diseases of Plants* (New York), the first such work on plant pathology published anywhere.

1909. GENETICS
William Ernest Castle (1867-1962), working with John C. Phillips (1876-1938), proved through the transplant of ovaries in a guinea pig that the differentiation between germ and somatic cells proposed by August Weismann to be true. "A Successful Ovarian Transplantation in the Guinea Pig, and Its Bearing on Problems of Genetics," with Phillips,

Science 30 (1909): 312-313 is a presentation of research results on the topic.

1909. GEOLOGY / PALEONTOLOGY
Charles Doolittle Walcott (1850-1927) discovered the Middle Cambrian Burgess Shale deposit, described by some as the most important of all fossil discoveries and notable especially because of its preservation of parts of soft-bodied specimens.

1909. GEOPHYSICS AND GEODESY
John Fillmore Hayford (1868-1925) published *The Figure of the Earth and Isostasy from Measurements in the United States* (Washington, D.C.).

1909. MEDICINE / PATHOLOGY
(Francis) Peyton Rous (1879-1970) conducted experiments on cancer in chickens. In injecting a filtered solution of cancer tissue into healthy chickens, the chickens developed cancer, and Rous concluded that the introduced disease was caused by a virus. The initial research was reported in 1910 and more definitive results in 1911, but it was not until 1966 that Rous's contribution was sufficiently recognized that he was given a Nobel Prize.

1909. MINERALOGY AND CRYSTALLOGRAPHY
Newton Horace Winchell (1839-1914) and his son Alexander Newton Winchell (1874-1958) published *Elements of Optical Mineralogy* (New York). Under the authorship of A.N. Winchell, several subsequent editions were published.

1909. ORGANIZATIONS / PUBLIC HEALTH
John D. Rockefeller established the Rockefeller Sanitary Commission for the Eradication of Hookworm Disease with a one-million-dollar grant. Concentrated in the southern states, the commission continued for five years; Charles Wardell Stiles (1867-1941) served as medical director.

1909. PERIODICALS AND PUBLISHING / NATURAL HISTORY
The *American Midland Naturalist* was founded by Julius Arthur Nieuwland (1878-1936). He served as editor for much of the remainder of his life.

1909. PERIODICALS AND PUBLISHING / PHARMACOLOGY AND PHARMACY
John Jacob Abel (1857-1938) founded the *Journal of Pharmacology and Experimental Therapeutics*. He served as editor until 1932.

1909. PHYSICS
The first American publication on special relativity was published by Gilbert N. Lewis (1875-1946) and Richard Chace Tolman (1881-1948), "The Principle of Relativity and Non-Newtonian Mechanics," *Proceedings of American Academy of Arts and Sciences* 44:711-726.

1909 (April 6). EXPLORATION AND SURVEYING
Robert E. Peary (1856-1920) and Matthew A. Henson (1866-1955) were the first to reach the North Pole.

1909 (July 11). GENERAL OR MISCELLANEOUS / ASTRONOMY
Astronomer Simon Newcomb (b.1835) died in Washington, D.C.

1910. GENERAL OR MISCELLANEOUS / MEDICINE
With support from the Carnegie Foundation, Abraham Flexner (1866-1959) issued an influential critical report on medical education that led to widespread change, including the closing of a number of substandard schools.

1910. GENETICS
Edward Murray East (1879-1938) published the paper "A Mendelian Interpretation of Variation That Is Apparently Continuous," *American Naturalist* 44:65-82. This paper was important in the United States in helping to point out that Mendelian inheritance applied to all inherited characters. (Similar work also was done about the same time by H. Nilsson-Ehle in Sweden.)

1910. GENETICS
Thomas Hunt Morgan (1866-1945) found among the *Drosophila* with which he was working an individual with white eyes. His subsequent research for the first time found conclusive evidence of hereditary characteristics that are linked to specific chromosomes. The white eye characteristic was associated with the X or accessory chromosome and therefore sex-linked.

1910. GEOLOGY
Frank Bursley Taylor (1860-1938), independently of others, first published details of his theoretical ideas on continental drift in "Bearing of the Tertiary Mountain Belt on the Origin of the Earth's Plan," *Bulletin of Geological Society of America* 21:179-226. Taylor's proposal included the idea that South America and Africa once were joined.

1910. GEOLOGY
Thomas Augustus Jaggar, Jr. (1871-1953) devised a volcano classification based on viscosity. His paper, "Japanese Volcanoes and Volcano Classification," *M.I.T. Bulletin of the Society of Arts* (February 1910), related to this achievement.

1910. GOVERNMENT—FEDERAL / ENTOMOLOGY
The Federal Insecticide Act was passed as protective regulatory legislation. Among those who contributed to its formulation was Ezra Dwight Sanderson (1878-1944).

1910. GOVERNMENT—FEDERAL / MINING
The Bureau of Mines was established by Congress in the Department of Interior. Its focus was on mining safety. After about 1915, a system of regional experiment stations developed in the West, generally located on university campuses.

1910. MICROBIOLOGY AND MICROSCOPY
Hans Zinsser (1878-1940) and Philip H. Hiss, Jr. (1868-1913) published *A Textbook of Bacteriology* (New York and London) which eventually had eight editions in Zinsser's lifetime (Hiss died in 1913) and, in the hands of others, fourteen by 1968.

1910. MICROBIOLOGY AND MICROSCOPY / AGRICULTURE
S. Henry Ayers (1884-?), of the U.S. Department of Agriculture, issued a report that was convincing evidence that pasteurized was safer than raw milk.

1910. ORGANIZATIONS—ACADEMIC / GENETICS
The University of Wisconsin recruited Leon Jacob Cole (1877-1948) to establish a department of experimental breeding (beginning in 1918, named genetics) which was the earliest of its kind in the United States.

1910. ORGANIZATIONS—INDUSTRY / ELECTRICITY AND ELECTRONICS
William David Coolidge (1873-1975), at the General Electric Company, achieved success with his development of a means of making tungsten wire, changing the metal's grains from cubic to fiber form. Extensively used thereafter in incandescent light bulbs, the process was one of the earliest notable commercial successes from an American industrial research laboratory. Initially produced in 1908, it was described in Coolidge's "Ductile Tungsten," *Transactions of American Institute of Electrical Engineers* 29, part 2 (1910): 961-965.

1910. ORGANIZATIONS—OBSERVATORIES
The Lick Observatory established a southern station in Chile. This was done under director William Wallace Campbell (1862-1938).

1910. ORGANIZATIONS—RESEARCH INSTITUTIONS / EUGENICS
The Eugenics Records Office was established at Cold Spring Harbor, New York by Charles Benedict Davenport (1866-1944). Initially supported by Mary Averell (Mrs. E.H.) Harriman (1851-1932), after 1918 the facility was funded by the Carnegie Institution.

1910. ORGANIZATIONS—SOCIETIES AND ASSOCIATIONS / GEOLOGY
The Petrologists' Club, in Washington, D.C., was founded by Charles Whitman Cross (1854-1949) and others. The organizational as well as many other meetings thereafter took place in Cross's home. (Cross was with the U.S. Geological Survey.)

1910. PALEONTOLOGY
Charles Schuchert (1858-1942) published the classic "Paleogeography of North America," *Bulletin of Geological Society of America*, vol. 20:427-606.

1910. PHYSICS
Robert A. Millikan (1868-1953) devised an oil drop method that allowed him to measure an electronic charge and to prove the existence of electrons.

1910. PHYSICS / ELECTRICITY AND ELECTRONICS
George Washington Pierce (1872-1956) published *Principles of Wireless Telegraphy* (New York).

Chronology 177

1910. ZOOLOGY / ENTOMOLOGY
William Morton Wheeler (1865-1937) published *Ants: Their Structure, Development and Behavior* (Columbia University Biological Series, 9) (New York).

1910. PSYCHOLOGY
Edward Bradford Titchener (1867-1927) published *A Text-Book of Psychology* (New York).

1910 (August 26). GENERAL OR MISCELLANEOUS / PHILOSOPHY, PSYCHOLOGY
Psychologist and philosopher William James (b.1842) died in Chocorua, New Hampshire.

1910-1912. ZOOLOGY—HUMAN / EMBRYOLOGY
The two-volume *Handbook of Human Embryology* (Philadelphia) was published. Edited by Franklin Paine Mall (1862-1917) at Johns Hopkins University and Franz Keibel at Freiburg, it included contributions by American and German authors.

1910 and 1918. MATHEMATICS
Oswald Veblen (1880-1960) and John Wesley Young (1879-1932) published *Projective Geometry* I (New York, 1910) and II (Boston, 1918).

1911. ANTHROPOLOGY AND ETHNOLOGY
Alice C. Fletcher (1838-1923) published *The Omaha Tribe* (Washington, D.C.).

1911. ANTHROPOLOGY AND ETHNOLOGY
Franz Boas (1858-1942) published *The Mind of Primitive Man* (New York).

1911. BIOLOGY—GENERAL
Edward Asahel Birge (1851-1950) and Chancey Juday (1871-1944), both at the University of Wisconsin, published "The Inland Lakes of Wisconsin: The Dissolved Gases," *Bulletin of Wisconsin Geological and Natural History Survey*, science series 7, no. 22. The work addressed the biological relations of the gases.

1911. BIOLOGY—GENERAL
Samuel Ottmar Mast (1871-1947) published *Light and the Behavior of Organisms* (New York).

1911. EXPLORATION AND SURVEYING / ZOOLOGY
Frank Michler Chapman (1864-1945), a curator in the American Museum of Natural History, made the first of a succession of expeditions to South America in support of his zoogeographic interests.

1911. GENETICS
Alfred Henry Sturtevant (1891-1970), at the time an undergraduate working with Thomas Hunt Morgan (1866-1945), first produced a genetic map for sex-linked genes of *Drosophila*. The map reflected the strength of linkage of factors as a means of reflecting the spatial relations of genes on a chromosome. He published results of his work on gene mapping in 1913.

1911. GENETICS / EUGENICS
Charles Benedict Davenport (1866-1944) published *Heredity in Relation to Eugenics* (New York).

1911. GEOGRAPHY AND CARTOGRAPHY
Ellen Churchill Semple (1863-1932) published *Influences of Geographic Environment* (New York).

1911. GEOGRAPHY AND CARTOGRAPHY / BOTANY
The first inclusive view of the topography, climate, soils, and vegetation of the United States was presented in Isaiah Bowman's (1878-1950) *Forest Physiography* (New York).

1911. GEOLOGY / SEISMOLOGY
Harry Fielding Reid (1859-1944) published "The Elastic-Rebound Theory of Earthquakes," *Publications of the University of California, Bulletin of the Department of Geology* 6:413-444. Reid's theory of elastic-rebound as the cause of earthquakes grew (in part) out of his work with the California state earthquake commission that followed in the aftermath of the earthquake of 18 April 1906.

Chronology

1911. INSTRUMENTS AND INSTRUMENTATION
A gyrocompass invented by Elmer Ambrose Sperry (1860-1930) was installed on the battleship *Delaware*.

1911. MATHEMATICS
Louis Charles Karpinski (1878-1956) and David Eugene Smith (1860-1944) published *The Hindu-Arabic Numerals* (Boston).

1911. MATHEMATICS
The Princeton University thesis of Robert D. Carmichael (1879-1967) may have been the first noteworthy American work on difference equations. His thesis adviser was George D. Birkhoff (1884-1944).

1911. MINERALOGY AND CRYSTALLOGRAPHY
Frederick Eugene Wright (1877-1953) published *The Methods of Petrographic-Microscope Research: Their Relative Accuracy and Range of Application* (Publications of Carnegie Institution of Washington, no. 158). It was significant in the furtherance of a quantitative approach to the measurement of optical aspects of crystals.

1911. NATURAL HISTORY
Anna Botsford Comstock (1854-1930) published *The Handbook of Nature-Study for Teachers and Parents* (Ithaca, N.Y.). It was translated into eight languages and with 24 editions between 1911 and 1939.

1911. ORGANIZATIONS—FOUNDATIONS
The Carnegie Corporation of New York was founded. It was active in science from 1919 to about 1930.

1911. ORGANIZATIONS—INDUSTRY
General Motors engaged the Arthur D. Little Company to develop its first research laboratory.

1911. ORGANIZATIONS—INDUSTRY / COMPUTERS AND INFORMATION SCIENCE
The Computer-Tabulating-Recording Company was incorporated. Thomas J. Watson (1874-1956) took charge in 1914 and in 1924 it took the name International Business Machines Company (IBM). The

company's roots included Herman Hollerith's (1860-1929) Tabulating Machine Company of 1896.

1911. ORGANIZATIONS—RESEARCH INSTITUTIONS / GEOLOGY
Thomas Augustus Jaggar, Jr. (1871-1953) established the Hawaii Volcano Observatory, in which he was financially aided by the Massachusetts Institute of Technology and the Volcano Research Association of Honolulu. Jaggar directed the facility until 1940.

1911. ORGANIZATIONS—SOCIETIES AND ASSOCIATIONS / ASTRONOMY
The American Association of Variable Star Observers was established.

1911. PALEONTOLOGY
Samuel Wendell Williston (1851-1918) published *American Permian Vertebrates* (Chicago).

1911. PERIODICALS AND PUBLISHING / ZOOLOGY
The *Journal of Animal Behavior* was founded, with Robert Mearns Yerkes (1876-1956) as editor.

1911. TECHNOLOGY AND INVENTION
A functional electric self-starter for automobiles was invented by Charles F. Kettering (1876-1958). It was installed in the Cadillac in 1912.

ca.1911. ZOOLOGY / PHYSIOLOGY
After about this time the gradient theory of Charles Manning Child (1869-1954), concerning the regeneration of an organism through physiological stages, became a significant idea in studies in morphogenesis.

1911-1916. CHEMISTRY / BOTANY
During these years, Thomas Burr Osborne (1859-1929) and H. Gideon Wells (1875-1943) demonstrated means to differentiate virtually all proteins from different plant sources.

1911-1941. ANTHROPOLOGY AND ETHNOLOGY
The *Handbook of American Indian Languages* (Washington, D.C. [etc.]) was published in four volumes. Franz Boas (1858-1942) was contributing editor.

1912. ASTRONOMY
John Adelbert Parkhurst (1861-1925) published "Yerkes Actinometry," *Astrophysical Journal* 36:169-227. Reporting on his work at the Yerkes Observatory, the paper related to characteristics of certain stars in terms of magnitudes (both visual and photographic), color indexes, and spectral class.

1912. ASTRONOMY
The first systematic study of the light from eclipsing binary stars was given by Henry Norris Russell (1877-1957). The work was published in a series of four papers in *Astrophysical Journal*: "On the Determination of the Orbital Elements of Eclipsing Variable Stars," vol. 35:315-340 and 36:54-74; and, written with Harlow Shapley (1885-1972), "On the Darkening at the Limb in Eclipsing Variables," vol. 36:239-254 and 385-408.

1912. ASTRONOMY
Vesto Melvin Slipher's (1875-1969) study of the spectrum of Andromeda resulted in the first reading of the velocity of a nebula in relation to earth.

1912. AWARDS AND PRIZES / MEDICINE, PHYSIOLOGY
Alexis Carrel (1873-1944) won the Nobel Prize in medicine or physiology. He was chosen "in recognition of his work on vascular suture and the transplantation of blood-vessels and organs."

1912. CHEMISTRY / METALLURGY
Albert Sauveur (1863-1939) published *The Metallography of Iron and Steel* (Cambridge, Mass.). Three subsequent editions appeared under the title *The Metallography and Heat Treatment of Iron and Steel*.

1912. ENGINEERING AND APPLIED SCIENCE / CHEMISTRY
William Merriam Burton (1865-1954), at Standard Oil of Indiana, patented a thermal cracking process for the separation of gasoline from crude oil. The chemically based procedures significantly increased the amount of gasoline produced.

1912. MATHEMATICS
Edwin Bidwell Wilson (1879-1964) published *Advanced Calculus* (Boston) which was a standard text for a number of years.

1912. MEDICINE / PHYSIOLOGY
Harvey Williams Cushing (1869-1939) published *The Pituitary Body and Its Disorders* (Philadelphia).

1912. ORGANIZATIONS—RESEARCH INSTITUTIONS / FUNDS AND FUNDING
Frederick Gardner Cottrell (1877-1948) founded the nonprofit Research Corporation, to which he dedicated his patents (especially relating to electrical precipitation of particles for dust and smoke control). The organization distributed income from patents as research grants for the furtherance of research and its applications for public benefit. In turn, other patents were assigned to the Corporation for support of its work.

1912. ORGANIZATIONS—RESEARCH INSTITUTIONS / OCEANOGRAPHY
A marine research station previously established at La Jolla, California by William Emerson Ritter (1856-1944) was taken over by the University of California to become the Scripps Institution for Biological Research (later, Oceanography).

1912. ORGANIZATIONS—SOCIETIES AND ASSOCIATIONS / ENGINEERING AND APPLIED SCIENCE
The Institute of Radio Engineers was established.

1912. ORGANIZATIONS—SOCIETIES AND ASSOCIATIONS / MEDICINE
The Harriet Newell Lowell Society for Dental Research was established in Boston.

1912. PALEONTOLOGY
Charles Doolittle Walcott (1850-1927) published *Cambrian Brachiopoda* (Washington, D.C.) in two volumes.

1912. PHYSICS
Owen W. Richardson (1879-1959) and his Princeton University student, Karl T. Compton (1887-1954), performed the first preliminary confirmation of Einstein's quantum equation of the photoelectric effect.

It was subsequently proven definitively by Robert A. Millikan (1868-1953).

1912. PHYSIOLOGY
Jacques Loeb (1859-1924) published *The Mechanistic Conception of Life* (Chicago). The origins of life were explained in terms of chemistry and physics.

1912. ZOOLOGY / PHYSIOLOGY
Henry Gray Barbour (1885/1886-1943) published his initial and foundational work on the relation of the brain to temperature regulation of the body in *Archiv fur experimentelle Pathologie und Pharmakologie* 70:1-26. His research in this area continued throughout his career.

1912. ORGANIZATIONS—ACADEMIC / MATHEMATICS
George David Birkhoff (1884-1944) was appointed assistant professor of mathematics at Harvard University and professor in 1919. A Ph.D. graduate of the University of Chicago (1907), he was commonly considered to be the leading American mathematician of his time.

1913. ASTRONOMY
Henry Norris Russell (1877-1957) promulgated a theory relating to the evolution of stars. In 1905, the idea was independently put forth by Danish astronomer Ejnar Hertzsprung, and it later became known as the Hertzsprung-Russell diagram, a plotting of star magnitudes against their spectral classes.

1913. CHEMISTRY / NUTRITION
Elmer Verner McCollum (1879-1967) discovered what later became known as vitamin A, a fat soluble vitamin as compared to the water soluble vitamin B. About the same time, Lafayette B. Mendel (1872-1935) and Thomas Burr Osborne (1859-1929) reached a similar conclusion but published somewhat later.

1913. GENERAL OR MISCELLANEOUS / ASTRONOMY
The International Solar Union adopted the Harvard Classification of stars.

1913. GENERAL OR MISCELLANEOUS / GEOLOGY
Oscar Edward Meinzer (1876-1948) became head of the ground-water division of the U.S. Geological Survey and retained the position until 1946. During this period, he largely established ground-water hydrology as a scientific field.

1913. GENERAL OR MISCELLANEOUS / TECHNOLOGY AND INVENTION
The first real assembly line was begun by Henry Ford (1863-1947), in which vehicles moved along a conveyor belt.

1913. MATHEMATICS
George David Birkhoff (1884-1944) published his most notable work, "Proof of Poincare's Geometric Theorem," *Transactions of American Mathematical Society* (1913): 14-22, which Poincare had been unable to demonstrate. Birkhoff's solution was a major mathematical occasion at the time.

1913. MEDICINE / INSTRUMENTS AND INSTRUMENTATION
The first artificial kidney was devised by John Jacob Abel (1857-1938). About the same time, Cecil Drinker (1887-1956) and Alfred Newton Richards (1876-1976) devised apparatus to supply blood to the kidneys, which allowed control of the procedure.

1913. ORGANIZATIONS—FOUNDATIONS
The Rockefeller Foundation was founded with funding from John D. Rockefeller. It was active in science from 1919.

1913. ORGANIZATIONS—INDUSTRY / ELECTRICITY AND ELECTRONICS
Research of Irving Langmuir (1881-1957) led to General Electric's application for patent on a nitrogen-filled light bulb which proved to be a great financial success for the company.

1913. ORGANIZATIONS—RESEARCH INSTITUTIONS
The Mellon Institute was established as a nonprofit research institution. In 1967, it was merged with Carnegie Institute of Technology to form Carnegie Mellon University and became a research section of the university.

1913. ORGANIZATIONS—SOCIETIES AND ASSOCIATIONS /
BIOLOGY—GENERAL
The Federation of American Societies for Experimental Biology was founded.

1913. ORGANIZATIONS—SOCIETIES AND ASSOCIATIONS / GENETICS
The American Genetic Association was established, based on the American Breeders Association which had been founded in 1903. (William Ernest Castle, 1867-1962, had a role in both events.)

1913. ORGANIZATIONS—SOCIETIES AND ASSOCIATIONS / MEDICINE
The American Society for Experimental Pathology was established.

1913. ORGANIZATIONS—SOCIETIES AND ASSOCIATIONS / MEDICINE
The American Society for Control of Cancer was established. In 1929, it became the American Cancer Society.

1913. PERIODICALS AND PUBLISHING / GENETICS
The *Journal of Heredity* was begun, a successor to the *American Breeders' Magazine*.

1913. PERIODICALS AND PUBLISHING / PHYSICS
Physical Review became the official journal of the American Physical Society.

1913. PHYSICS / ELECTRICITY AND ELECTRONICS
The hot-cathode X-ray tube was devised by William D. Coolidge (1873-1975).

1913. PSYCHOLOGY
John Broadus Watson (1878-1958) published the influential paper, "Psychology as a Behaviorist Views It," *Psychological Review* 20:158-177. This was a first presentation of the behaviorist views, which rejected introspection as a part of psychology. In 1914, he published *Behavior: An Introduction to Comparative Psychology* (New York).

1913. ZOOLOGY / ECOLOGY
Victor Ernest Shelford (1877-1968) published *Animal Communities in Temperate America* (Chicago). Important for its promotion of succession

studies in American ecology, it also included early generalized ideas that were fundamental to animal ecology.

1913. ZOOLOGY / PHYSIOLOGY
Physiologist Lawrence J. Henderson (1878-1942) published *The Fitness of the Environment* (New York).

1913 (December). GENERAL OR MISCELLANEOUS / FUNDS AND FUNDING
The American Association for the Advancement of Science, at the instigation of James McKeen Cattell (1860-1944), formed the Committee of 100 on Research. Among the aims of the committee was fund-raising to support individual research but in this they had little success. Although it continued into the 1920s, the committee's activities were mainly manpower surveys of scientists.

1914. ASTRONOMY
Seth Barnes Nicholson (1891-1963), at the Lick Observatory, made the first of his four discoveries of moons of Jupiter (Jupiter IX). Later, on the staff of the Mt. Wilson Observatory, he found Jupiter X and XI (1938) and XII (1951).

1914. AWARDS AND PRIZES / CHEMISTRY
Theodore William Richards (1868-1928) won the Nobel Prize in chemistry. He was chosen "in recognition of his accurate determinations of the atomic weight of a large number of chemical elements." This was the first such prize in chemistry to be awarded to an American.

1914. BIOCHEMISTRY AND MOLECULAR BIOLOGY
Edward Calvin Kendall (1886-1972) isolated for the first time the crystalline thyroxine, but he was unable to achieve status as the first to uncover its chemical structure (discovered in England in 1926). Kendall's achievement made it possible to use pure thyroxine in the treatment of thyroid deficiency.

1914. BIOCHEMISTRY AND MOLECULAR BIOLOGY
Walter Jennings Jones (1865-1935) published *Nucleic Acids: Their Chemical Properties and Physiological Conduct* (London). There was a second edition in 1920.

1914. CHEMISTRY / NUTRITION
By this date, Robert Runnels Williams (1886-1965) had established that a nitrogenous base was the element of rice polishings that was a cure for beriberi. The substance, which was to be established as vitamin B1 (thiamine), was isolated in small amounts by Dutch chemists, B.C.P. Jansen and W.F. Donath, in 1926; by 1933, Williams had pure crystals and he and his collaborators subsequently developed means to increase the output beyond that of other researchers.

1914. GENETICS
In *Piebald Rats and Selection: An Experimental Test of the Effectiveness of Selection and of the Theory of Gametic Purity in Mendelian Crosses* (Carnegie Institution of Washington Publication no. 195), William Ernest Castle (1867-1962) and John C. Phillips (1876-1938) made the point that genes were permanently changed in selection by bringing them into association with other germ cells. In a 1919 paper, based on further experiments, Castle rescinded this controversial conclusion.

1914. HISTORY OF SCIENCE—FIELD OF STUDY / CHEMISTRY
Edgar Fahs Smith (1854-1928) published *Chemistry in America: Chapters From the History of the Science in the United States* (New York).

1914. MEDICINE / DISEASE
Hans Zinsser (1878-1940) published *Infection and Resistance* New York). In 1939, it appeared, with John Franklin Enders (1897-1985) and LeRoy D. Fothergill (1901-1967), as *Immunity: Principles and Application in Medicine and Public Health*.

1914. MICROBIOLOGY AND MICROSCOPY / GEOLOGY
At this time, Joseph Augustine Cushman (1881-1949) undertook his early attempts to use Foraminifera to determine the age of well samples. He became known especially for his use of the Protozoan order as an aid to petroleum location.

1914. ORGANIZATIONS / ENGINEERING AND APPLIED SCIENCE
The Engineering Foundation was established in New York with a grant from instrument maker and machine tool manufacturer Ambrose Swasey

(1846-1937). The Foundation had the purpose of encouraging engineering research.

1914. ORGANIZATIONS—RESEARCH INSTITUTIONS / ZOOLOGY
The Carnegie Institution of Washington's Department of Embryology was founded.

1914. PHYSICS
Theodore Lyman (1874-1954) published *The Spectroscopy of the Extreme Ultra-Violet* (New York). A revised edition appeared in 1928.

1914. CHEMISTRY
Theodore William Richards (1868-1928) discovered that lead from different mineral sources can have different atomic masses, a finding that supported the idea of isotopes recently proposed by Frederick Soddy.

ca.1914. ZOOLOGY
Clarence Cook Little (1888-1971), at Harvard University as an administrator and graduate student, began the development of an inbred strain of mice (later known as DBA) that came to be widely used by biological and medical researchers. He later also developed a second strain, C57BL.

1915. ASTRONOMY
Walter S. Adams (1876-1956), at the Mount Wilson Observatory, discovered the first White Dwarf star, Sirius B, using spectroscopic methods. During this time period, Adams produced a means, known as spectroscopic parallax, that made it possible to estimate the distance of a star from its spectrum.

1915. GENERAL OR MISCELLANEOUS / CHEMISTRY
Agnes Fay Morgan (1884-1968) was appointed to the faculty at the University of California (Berkeley) and while there she was the first to make chemistry a constituent aspect of teaching in home economics.

Chronology 189

1915. GENERAL OR MISCELLANEOUS / HISTORY OF SCIENCE—FIELD OF STUDY
Belgian-born historian of science George Alfred Leon Sarton (1884-1956) came to the United States. In 1916, he received a temporary appointment at Harvard University. In 1918, he was given an appointment as research associate in the history of science by the Carnegie Institution of Washington although he remained physically at Harvard.

1915. GENETICS
Thomas Hunt Morgan (1866-1945), Calvin B. Bridges (1889-1938), Alfred H. Sturtevant (1891-1970), and Hermann J. Muller (1890-1967) published *The Mechanism of Mendelian Heredity* (New York), a summation of their work on *Drosophila*.

1915. GEOLOGY
Louis Valentine Pirsson (1860-1919) and Charles Schuchert (1858-1942) published *A Text-Book of Geology* (New York) which was a standard American text for several decades.

1915. GEOLOGY
(Charles) David White (1862-1935) published "Some Relations in Origin Between Coal and Petroleum," *Journal of Washington Academy of Sciences* 5:189-212. In this paper, he first presented his theory of carbon ratio that related the increase of fixed carbon in coal to a corresponding elevation in the grade of the accompanying oil.

1915. GOVERNMENT—FEDERAL / AERONAUTICS
The National Advisory Committee for Aeronautics was established by Congress. The originator of the idea and first head was Charles Doolittle Walcott (1850-1927), secretary of the Smithsonian Institution; NACA was a replacement for an advisory committee on aeronautics that had been formed under the Smithsonian. The members of the new Committee were appointed by the President, a majority of whom were government scientists. The Committee had the function to supervise and conduct scientific studies relating to flight and continued until 1958 when it was succeeded by the National Aeronautics and Space Administration.

1915. ORGANIZATIONS—SOCIETIES AND ASSOCIATIONS / ECOLOGY
The Ecological Society of America was established.

1915. ORGANIZATIONS—SOCIETIES AND ASSOCIATIONS / GEOGRAPHY AND CARTOGRAPHY
Isaiah Bowman (1878-1950) became director of the American Geographical Society. He worked to move the Society from an amateur organization to one devoted to research.

1915. ORGANIZATIONS—SOCIETIES AND ASSOCIATIONS / MATHEMATICS
The Mathematical Association of America was established at a convention in Columbus, Ohio, as an organization particularly interested in mathematical education.

1915. PERIODICALS AND PUBLISHING
James McKeen Cattell (1860-1944) founded *Scientific Monthly*, which he edited until his death. From 1900-1915, Cattell had edited *Popular Science Monthly*.

1915. PERIODICALS AND PUBLISHING
The *Proceedings of National Academy of Sciences* began publication.

1915. TECHNOLOGY AND INVENTION / ELECTRICITY AND ELECTRONICS
Alexander Graham Bell (1847-1922) in New York and Thomas A. Watson (1854-1934) in San Francisco conducted the first North American coast-to-coast telephone call. The first transatlantic radiotelephone call also took place this year (between Paris, France, and Arlington, Virginia).

1915. GENERAL OR MISCELLANEOUS / CHEMISTRY, PHYSICS
By about this date, the United States was the leading producer of radium. The country's foremost authority on radioactivity, Yale University professor Bertram Borden Boltwood (1870-1927), through assistance and advice to mining and other interests, helped to achieve this status.

1915. GOVERNMENT—FEDERAL / TECHNOLOGY AND INVENTION
The U.S. Navy established the Naval Consulting Board, chaired by Thomas A. Edison (1847-1931). The membership consisted chiefly of individuals chosen by the nation's engineering societies. The Board was

charged with the review of inventions and ideas that might have military usefulness. During World War I, the Board received some 110,000 suggestions of which only one ever went into production.

1916. BOTANY / ECOLOGY
Frederic Edward Clements (1874-1945) published *Plant Succession* (Washington, D.C.), the botanist and ecologist's most important work.

1916. CHEMISTRY / NUTRITION
Elmer Verner McCollum (1879-1967) initiated the practice of labeling vitamins with letters of the alphabet when he demonstrated that the normal growth of rats required fat-soluble A and water-soluble B factors.

1916. CHEMISTRY / PHYSICS
Gilbert Newton Lewis (1875-1946) published "The Atom and the Molecule," *Journal of American Chemical Society* 38:762-785. This paper contained his important idea of this year that chemical bonding involved a pair of electrons shared by two atoms.

1916. GENERAL OR MISCELLANEOUS / MEDICINE
Margaret Sanger (1879-1966) established the first American birth control clinic.

1916. GEOLOGY / GEOGRAPHY AND CARTOGRAPHY
Nevin Melancthon Fenneman (1865-1945) published "Physiographic Divisions of the United States," *Annals of Association of American Geographers* 6:19-98, and accompanying map. This was the foundation for similar maps and was adopted by the U.S. Geological Survey and other bodies.

1916. GOVERNMENT—FEDERAL / ENVIRONMENT AND CONSERVATION
The National Park Service was established by Congress in the Department of Interior. Stephen Tyng Mather (1867-1930) was the first director.

1916. MEDICINE / PATHOLOGY
By this date, Joseph Goldberger (1874-1929) had reached the well-supported conclusion that pellagra was a disease caused by dietary deficiency.

1916. MICROBIOLOGY AND MICROSCOPY / PATHOLOGY
The disease microorganisms in arthropods that can be passed on to some vertebrates, studied by Howard Taylor Ricketts (1871-1910), were named Rickettsia by Brazilian physician Henrique da Rocha-Lima.

1916. ORGANIZATIONS—ACADEMIC / CHEMISTRY
Roger Adams (1889-1971) was appointed assistant professor of chemistry at the University of Illinois. In 1926, he became department head; during his long tenure (nearly 30 years in charge of the program) Illinois became the major center for production of graduate chemists for American industry.

1916. ORGANIZATIONS—SOCIETIES AND ASSOCIATIONS / PHYSICS
The Optical Society of America was founded.

1916. PERIODICALS AND PUBLISHING / GENETICS
The journal *Genetics* was founded by ten individuals.

1916. PHYSICS
Robert Andrews Millikan (1868-1953) confirmed Einstein's 1905 equation that explained the photoelectric effect and, working with the photoelectric effect, he proved Planck's constant.

1916. PSYCHOLOGY
Lewis Madison Terman (1877-1956) devised the English-language version of the French Binet-Simon intelligence test; it was known as the Stanford-Binet test.

1916. ZOOLOGY / PHYSIOLOGY
Edmund Newton Harvey (1887-1959) published "The Mechanism of Light Production in Animals," *Science* 44:208-209.

1916. GOVERNMENT—FEDERAL
In a July 25 letter from President Woodrow Wilson to William H. Welch (1850-1934), president of the National Academy of Sciences, the establishment of a National Research Council for the promotion of research related to national security was approved. The idea for the Council had been promoted by George Ellery Hale (1868-1938) and he became chair at an organizing meeting in September. The Council represented academic, government, and industrial interests and both science and engineering. In 1917, Robert A. Millikan (1868-1953) became the NRC executive officer. The expenses of the Council were covered both by the government and private sources (especially the Carnegie Corporation and the Rockefeller Foundation).

1916 (October 28). GENERAL OR MISCELLANEOUS / ASTRONOMY, METEOROLOGY AND CLIMATOLOGY
Astronomer and meteorologist Cleveland Abbe (b.1838) died in Chevy Chase, Maryland.

1916 (November 12). GENERAL OR MISCELLANEOUS / ASTRONOMY
Astronomer Percival Lowell (b.1855) died in Flagstaff, Arizona.

1916-1923. INFORMATION ACCESS / ICHTHYOLOGY AND PISCICULTURE
Bashford Dean (1867-1928) published his three-volume *Bibliography of Fishes* (New York).

1917. MATHEMATICS
Edward V. Huntington (1874-1952) published *The Continuum, and Other Types of Serial Order, With an Introduction to Cantor's Transfinite Numbers* (Cambridge, Massachusetts), originally published in *Annals of Mathematics* in 1905.

1917. MILITARY AND WAR / PHYSICS
When the United States became involved in World War I, the need for a means of combatting the submarine danger was the most pressing. In February, the Naval Consulting Board appointed a Special Problems Committee and General Electric's Willis R. Whitney (1868-1958) directed efforts relating to underwater sound detection. The National Research Council also became involved with the problem of submarine detection. During the War, these various activities were carried out at

facilities at Nahant, Massachusetts and at New London, Connecticut, the latter becoming the more important. In all, however, little real progress was made during the War.

1917. ORGANIZATIONS—ACADEMIC / BIOLOGY, PHYSICS
William Duane (1872-1935) was made professor of biophysics at Harvard University, apparently the first such appointment in the country.

1917. ORGANIZATIONS—ACADEMIC / ZOOLOGY
Florence Rena Sabin (1871-1953) was promoted to the title of professor of histology, the first woman to serve as professor in the Johns Hopkins Medical School. She had been appointed to the faculty in 1902, also the first woman in such a position in the School.

1917. ORGANIZATIONS—OBSERVATORIES / INSTRUMENTS AND INSTRUMENTATION
A 100-inch refracting telescope was installed at the Mount Wilson Observatory, the largest in the world at the time and remained so until 1948. The lens was funded by John D. Hooker and the Carnegie Institution of Washington provided a grant for the telescope mounting.

1917. ORGANIZATIONS—SOCIETIES AND ASSOCIATIONS / ENVIRONMENT AND CONSERVATION
The Save-the-Redwoods League was founded by Ralph Works Chaney (1890-1971), Henry Fairfield Osborn (1857-1935), and Madison Grant (1865-1937).

1917. PERIODICALS AND PUBLISHING / ANTHROPOLOGY AND ETHNOLOGY
Franz Boas (1858-1942) founded the *International Journal of American Linguistics*, which he supported financially and served as editor until his death.

1917. PERIODICALS AND PUBLISHING / PSYCHOLOGY
The *Journal of Applied Psychology* was established.

1917. PHYSICS
Richard Chace Tolman (1881-1948) published *The Theory of the Relativity of Motion* (Berkeley).

1917. ZOOLOGY
Thomas Barbour (1884-1946) and Leonhard Hess Stejneger (1851-1943) published *Check-list of North American Amphibians and Reptiles* (Cambridge, Mass.). The work reached a fifth edition during Stejneger's lifetime.

1917. ZOOLOGY / CONCHOLOGY AND MALACOLOGY
Henry Edward Crampton (1875-1956) published the first of his studies on Partula, a genus of land snail, *Studies on the Variation, Distribution, and Evolution of the Genus Partula I: The Species Inhabiting Tahiti* (Carnegie Institution of Washington). The second part, *The Species of the Mariana Islands, Guam, and Saipan* appeared in 1925 and part three, *The Species Inhabiting Moorea*, in 1932.

1917. ZOOLOGY / EMBRYOLOGY
Frank Rattray Lillie (1870-1947) published "The Free-Martin; A Study of the Action of Sex Hormones in the Foetal Life of Cattle," *Journal of Experimental Zoology* 23:371-452.

1917. ZOOLOGY / PHYSIOLOGY
Horatio Hackett Newman (1875-1957) published *The Biology of Twins* (Chicago). In 1923, he published *The Physiology of Twinning* (Chicago).

1917. ZOOLOGY / PHYSIOLOGY
Physiologist Lawrence J. Henderson (1878-1942) published *The Order of Nature* (Cambridge, Mass.).

1917-1919. MILITARY AND WAR / PSYCHOLOGY
Robert Mearns Yerkes (1876-1956) was attached to the Office of Surgeon General and was effectively head of efforts for psychological testing for the U.S. Army during the War.

1917-1925. ZOOLOGY / ICHTHYOLOGY AND PISCICULTURE
Ichthyologist Carl H. Eigenmann (1863-1927) published the five-part "American Characidae" in *Memoirs of Museum of Comparative Zoology of Harvard College*, volume 43.

1918. ASTRONOMY
Robert G. Aitken (1864-1951) published *The Binary Stars* (New York).

1918. ENGINEERING AND APPLIED SCIENCE / ELECTRICITY AND ELECTRONICS
The superheterodyne radio receiver was invented by Edwin H. Armstrong (1890-1954).

1918. GENETICS / AGRICULTURE
Ernest Brown Babcock (1877-1954) and Roy E. Clausen (1891-1956) published the textbook, *Genetics in Relation to Agriculture* (New York), the first such work to address both fields of concern. A second edition appeared in 1927.

1918. GEOLOGY
Everette Lee DeGolyer (1886-1956) published on "The Theory of Volcanic Origin of Salt Domes," *Bulletin of American Institute of Mining and Metallurgical Engineers* 137:987-1000, and in the years thereafter developed means of detecting salt domes by the methods of geophysics. In doing so, he made major contributions to methods of petroleum detection.

1918. MATHEMATICS
Griffith Conrad Evans (1887-1973) published *Functionals and Their Applications* (Providence, R.I.). The work was based on lectures sponsored by the American Mathematical Society.

1918. ORGANIZATIONS—FOUNDATIONS
The Laura Spelman Rockefeller Memorial was founded. It was active in science in the mid-1920s.

1918. ORGANIZATIONS—SOCIETIES AND ASSOCIATIONS / EUGENICS
The Eugenics Research Association was established.

1918. PERIODICALS AND PUBLISHING / PHYSIOLOGY
The *Journal of General Physiology* was established by Jacques Loeb (1859-1924) and Winthrop John Vanleuven Osterhout (1871-1964).

1918. ZOOLOGY / ENTOMOLOGY
John Henry Comstock (1849-1931) published *Wings of Insects* (Ithaca, N.Y.), which represented the most significant development in entomology during that time. His work on the veins of wings was

important for the revelation of phylogenetic relationships of insect groups.

1918. ZOOLOGY—HUMAN / ANATOMY
George Linius Streeter (1873-1948) published a monograph study on development of the labyrinth of the human ear.

1918. ASTRONOMY
The distance to the Andromeda galaxy was calculated by Heber Doust Curtis (1872-1942) as one million light years.

1918. GENERAL OR MISCELLANEOUS / ORGANIZATIONS—INDUSTRY
One of the most significant outcomes for science produced by World War I was the boost it gave to research in industry. In important respects, wartime activities gave birth to the industrial sector of American science. During the 1920s, industrial research organizations and personnel multiplied by three and by 1930 there were some 1600 industrial laboratories. The Chemical Foundation, established in the aftermath of the War to administer seized German patents, was one agency that facilitated the growth of industry.

1918 (February). INSTRUMENTS AND INSTRUMENTATION / AERONAUTICS
David Locke Webster (1888-1976), physicist and member of the Army Signal Corps Air Service Reserve's Science and Research Division, was named as the first scientist pilot with the mission to participate in the design and testing of instrumentation.

1918 (May 11). GOVERNMENT—FEDERAL / FUNDS AND FUNDING
The National Research Council of the National Academy of Sciences, originally formed in 1916, was made permanent by Executive Order of President Woodrow Wilson. Membership on the NRC was not limited to members of the Academy and tended to represent various scientific and engineering societies. It was supported especially by private foundations. In 1919, the Carnegie Corporation gave $5,000,000 as an endowment and for a new building for the NRC and Academy which was completed in 1924. In April 1919, the Council received a grant of $500,000 from the Rockefeller Foundation for postdoctoral fellowships

in chemistry and physics. The first such fellowship program of its kind in the country, the funds were to be distributed over a five-year period.

1918 (July). MILITARY AND WAR / CHEMISTRY
The Chemical Warfare Service was established by the U.S. Army. Earlier poison gas research and related activities had been carried out especially in the Bureau of Mines whose work was taken over by the new agency. This included an experiment station at American University. At the termination of the War the Chemical Warfare Service was given permanent status by Congress but reduced in scale.

1918-1924. ASTRONOMY
The Henry Draper Catalogue of classified stellar spectra, prepared by Annie Jump Cannon (1863-1941), was published as volumes 91-99 of the *Annals of Harvard College Observatory*. In all, some 225,300 stars were included.

1919. ARCHAEOLOGY
William Henry Holmes (1846-1933) published *Handbook of Aboriginal American Antiquities* (Washington, D.C.).

1919. ASTRONOMY
The size of the Milky Way galaxy was accurately estimated by Harlow Shapley (1885-1972), who placed the solar system near the outer edge of the galaxy. Shapley reported on this work in his "Studies Based on the Colors and Magnitudes in Stellar Clusters; Twelfth Paper, Remarks on the Arrangement of the Sidereal Universe," *Astrophysical Journal* 49 (June 1919): 311-336. In arguing against the idea that the sun is near the center of the galaxy, Shapley's most important tool was the period-luminosity relations of the Cepheid variable stars.

1919. ASTRONOMY
Edward Emerson Barnard (1857-1923) published his catalog of dark nebulas.

1919. ASTRONOMY
Tables of the Motion of the Moon (New Haven), published by Ernest W. Brown (1866-1938), drew on thirty years of work and on his theory of lunar motion.

1919. BIOCHEMISTRY AND MOLECULAR BIOLOGY
Otto K.O. Folin (1867-1934) and Hsien Wu (1893-1959) published the summarizing paper on Folin's work, "A System of Blood Analysis," *Journal of Biological Chemistry* 38:81-110.

1919. CHEMISTRY / PHYSICS
A theory of chemical bonding of atoms was brought forth by Irving Langmuir (1881-1957), in which electrovalence and covalence were used as terms to designate the two types of bonding.

1919. GENETICS
Helen Dean King (1869-1955) published results of breeding experiments with brother and sister rats that showed, when carefully chosen, there is no deterioration of the stock in such inbreeding. This contradicted assertions of Darwin. These results appeared originally in a series of articles in *Journal of Experimental Zoology* (1918-1919), which were reprinted in *Studies on Inbreeding* (Philadelphia).

1919. GENETICS
Edward Murray East (1879-1938), with Donald F. Jones (1890-1963), published *Inbreeding and Outbreeding: Their Genetic and Sociological Significance* (Philadelphia and London), in which he argued that inbreeding led to increasing homozygosity.

1919. GENETICS
Thomas Hunt Morgan (1866-1945) published *The Physical Basis of Heredity* (Philadelphia).

1919. GOVERNMENT—FEDERAL / CHEMISTRY
The U.S. government Fixed Nitrogen Research Laboratory was established. Arthur Amos Noyes (1866-1936) (who was chair of the Committee on Nitrate Supply during the war) played an instrumental role in the laboratory's founding. In 1921, it was transferred from the War Department to the Department of Agriculture.

1919. MEDICINE / DISEASE
Walter Abraham Jacobs (1883-1967) and his assistant at the Rockefeller Institute for Medical Research, Michael Heidelberger (1888-1991),

developed Tryparsamide that was greatly useful against sleeping sickness.

1919. MEDICINE / PATHOLOGY
James Ewing (1866-1943) published *Neoplastic Diseases: A Textbook on Tumors* (Philadelphia and London). It was a foundational publication for the study of oncology. A fourth edition appeared in 1940.

1919. ORGANIZATIONS—RESEARCH INSTITUTIONS / PHYSICS
The Wallace Clement Sabine Acoustics Laboratory, or Riverbank Laboratory, was opened in Geneva, Illinois. Intended for W.C. Sabine, he died on 10 January, before he could begin its use. As a consequence, his cousin, Paul Earls Sabine (1879-1958), was appointed director of acoustical research.

1919. ORGANIZATIONS—SOCIETIES AND ASSOCIATIONS
The International Research Council was established at a meeting in Brussels. The National Academy of Sciences played a significant role in the formation of this body during organizational meetings the previous year in London and Paris. In 1931, it was succeeded by the International Council of Scientific Unions.

1919. ORGANIZATIONS—SOCIETIES AND ASSOCIATIONS
The American Society for the Dissemination of Science was established, dedicated to the advance of scientific journalism.

1919. ORGANIZATIONS—SOCIETIES AND ASSOCIATIONS / GEOPHYSICS AND GEODESY
The American Geophysical Union was founded.

1919. ORGANIZATIONS—SOCIETIES AND ASSOCIATIONS / HISTORY OF SCIENCE—FIELD OF STUDY
A section on the history of science was formed in the American Historical Association.

1919. ORGANIZATIONS—SOCIETIES AND ASSOCIATIONS / METEOROLOGY AND CLIMATOLOGY
The American Meteorological Society was established.

1919. ORGANIZATIONS—SOCIETIES AND ASSOCIATIONS / MINERALOGY AND CRYSTALLOGRAPHY
The Mineralogical Society of America was established.

1919. ORGANIZATIONS—SOCIETIES AND ASSOCIATIONS / ZOOLOGY
The American Society of Mammalogists was established.

1919. PHYSICS / ELECTRICITY AND ELECTRONICS
George Washington Pierce (1872-1956) published *Electric Oscillators and Electric Waves* (New York).

1919. PSYCHOLOGY
John Broadus Watson (1878-1958) published *Psychology from the Standpoint of a Behaviorist* (Philadelphia). This was his major work on behaviorism and shortly thereafter he left academia and began working in advertising.

1919. SPACE SCIENCE, TECHNOLOGY, AND EXPLORATION
Robert Hutchings Goddard (1882-1945) published *A Method of Reaching Extreme Altitudes* (Smithsonian Miscellaneous Collections, vol. 71, no. 2) (Washington: Smithsonian Institution, 1919), a significant milestone in the development of rocket technology.

1919. TECHNOLOGY AND INVENTION / ELECTRICITY AND ELECTRONICS
(Louis) Alan Hazeltine (1886-1964) invented the neutrodyne that eliminated most noise from radios of the time.

1919 (February 3). GENERAL OR MISCELLANEOUS / ASTRONOMY
Astronomer Edward Charles Pickering (b.1846) died in Cambridge, Massachusetts.

1919 (March). PERIODICALS AND PUBLISHING / GEOLOGY
Alan Mara Bateman (1889-1971) succeeded John Duer Irving (1874-1918) as editor of the journal *Economic Geology* (founded in 1905) and continued in the position until July 1969.

1919-1923. MATHEMATICS
Leonard Eugene Dickson (1874-1954) published *History of the Theory of Numbers* (Washington), in three volumes.

1919-1925. BOTANY
William Albert Setchell (1864-1943) published "Marine Algae of the Pacific Coast of North America," *University of California Publications in Botany*, vol. 8, parts 1-3.

1920. ASTRONOMY
Employing a stellar interferometer, Albert A. Michelson (1852-1931) established the diameter of Betelgeuse, the first measurement of a star other than the sun.

1920. BOTANY / PATHOLOGY
Erwin F. Smith (1854-1927) published a summation of his work in *An Introduction to Bacterial Diseases of Plants* (Philadelphia).

1920. CHEMISTRY
William Mansfield Clark (1884-1964) published *The Determination of Hydrogen Ions* (Baltimore, Md.), a lasting contribution to studies of acidity.

1920. GENERAL OR MISCELLANEOUS / ELECTRICITY AND ELECTRONICS
The first regularly broadcasting radio station in the United States, KDKA, was established at the Westinghouse plant in Pittsburgh. (Frank Conrad, 1874-1941, an engineer with Westinghouse, was important in this event.)

1920. GENERAL OR MISCELLANEOUS / STATISTICS
Raymond Pearl (1879-1940) proposed a mathematical equation for determination of population until the year 2100.

1920. ORGANIZATIONS—ACADEMIC
The California Institute of Technology was named. Founded in 1891 as Throop Polytechnic Institute, it had been renamed Throop College of Technology in 1913.

1920. ORGANIZATIONS—ACADEMIC
Roscoe Gilkey Dickinson (1894-1945) received the first Ph.D. from the California Institute of Technology (in chemistry).

1920. ORGANIZATIONS—ACADEMIC / GEOGRAPHY AND CARTOGRAPHY
Wallace Walter Atwood (1872-1949) was appointed president of Clark University. One of his central missions, as a geologist and geographer, was to develop the University's Graduate School of Geography.

1920. ORGANIZATIONS—SOCIETIES AND ASSOCIATIONS / GEOLOGY
The Society of Economic Geologists was established.

1920. ORGANIZATIONS—SOCIETIES AND ASSOCIATIONS / MATHEMATICS
The National Council of Teachers of Mathematics was established.

1920. PALEONTOLOGY
Percy Edward Raymond (1879-1952) published *The Appendages, Anatomy and Relationships of Trilobites* (Connecticut Academy of Arts and Science Memoir no. 7) (New Haven).

1920. ZOOLOGY / INSTRUMENTS AND INSTRUMENTATION
The first reported use of electronic amplification in a physiological experiment was made by Alexander Forbes (1882-1965). Results of this work were given in his paper (with Catharine Thacher), "Amplification of Action Currents with the Electron Tube in Recording with the String Galvanometer," *American Journal of Physiology* 52:409-471.

1920. ZOOLOGY / PHYSIOLOGY
Philip Edward Smith (1884-1970) published the results of research on the effects of the removal of the pituitary gland in amphibians in "The Pigmentary, Growth and Endocrine Disturbances Induced in the Anuran Tadpole by the Early Ablation of the Pars Buccalis of the Hypophysis," *American Anatomical Memoirs* 11:1-151.

1920. ZOOLOGY—HUMAN / EMBRYOLOGY
George Linius Streeter (1873-1948) published a notable study on the weight and sizes of embryos and fetuses of humans at various stages of their development. This paper appeared in *Contributions to Embryology* 2:143-170, with a separately published chart.

1920. GENERAL OR MISCELLANEOUS
A group of leading scientists put together the work, *The New World of Science: Its Development During the War* (New York). It was edited by Robert M. Yerkes (1876-1956).

1920 (April 26). ASTRONOMY
Heber Doust Curtis (1872-1942) and Harlow Shapley (1885-1972), at the National Academy of Sciences, debated on the size of the universe, with inconclusive results. A published version of the debate, the result of collaboration between Curtis and Shapley and considerable editing, appeared in *National Research Council Bulletin* 2, part 3 (May 1921): 171-217.

1920 (December). ORGANIZATIONS—SOCIETIES AND ASSOCIATIONS / ENGINEERING AND APPLIED SCIENCE
Under the instigation of Herbert Hoover (1874-1964), who called for an organization that could represent the perspective of engineers in public affairs, the Federated American Engineering Societies was founded, in 1924 changing its name to American Engineering Council. It was followed by the Engineers Joint Council in 1945 and the American Association of Engineering Societies in 1979.

1921. ASTRONOMY
Through photographic evidence, John Charles Duncan (1882-1967) was able to show expansion in the Crab nebula. In 1938, he presented further proof of expansion. The 1921 work appeared in "Changes Observed in the Crab Nebula in Taurus," *Proceedings of National Academy of Sciences* 7:179-180.

1921. GENERAL OR MISCELLANEOUS
The Science Service was established as an agency for the improvement of scientific journalism, the outcome of an idea put forth in 1919 by newspaper owner Edward W. Scripps (1854-1926). The trustees came from among journalists, the American Association for the Advancement of Science, and the National Academy of Sciences-National Research Council. Edwin E. Slosson (1865-1929) was made editor and for many years the Service was situated in the National Academy of Sciences building in Washington.

1921. GENERAL OR MISCELLANEOUS / PHYSICS
Marie Curie first visited the United States. The occasion was the presentation of a fund donated by American women to purchase radium for Curie's research.

1921. MINERALOGY AND CRYSTALLOGRAPHY
Esper Signius Larsen, Jr. (1879-1961) published *The Microscopic Determination of the Nonopaque Minerals* (Bulletin of United States Geological Survey, no. 679). A second edition, with Harry M. Berman (1902-1944), appeared in 1934 (Bulletin no. 848).

1921. ORGANIZATIONS / PSYCHOLOGY
The Psychological Corporation, a nonprofit organization, was founded by James McKeen Cattell (1860-1944) as an agency of applied psychology.

1921. ORGANIZATIONS—SOCIETIES AND ASSOCIATIONS / BIOLOGY—GENERAL
The Union of American Biological Societies was founded.

1921. PHYSICS / INSTRUMENTS AND INSTRUMENTATION
Albert Wallace Hull (1880-1966) published the important paper, "The Effect of a Uniform Magnetic Field on the Motion of Electrons Between Coaxial Cylinders," *Physical Review* 18:31-57. For the set-up, an electron tube producing microwaves, Hull introduced the term magnetron.

1921. ZOOLOGY / CONCHOLOGY AND MALACOLOGY
William Healey Dall (1845-1927) published *Summary of the Marine Shellbearing Mollusks of the Northwest Coast of America from San Diego, California to the Polar Sea* (Bulletin of U.S. National Museum, 112).

1921 (April 2). GENERAL OR MISCELLANEOUS / PHYSICS
Albert Einstein (1879-1955) arrived in New York to lecture at Columbia University on his relativity theory.

1921 (December 9). TECHNOLOGY AND INVENTION / CHEMISTRY
Thomas Midgley, Jr. (1889-1944), working for General Motors Research Corporation (originally Charles Kettering's Dayton Engineering Laboratories), discovered the effectiveness of tetraethyl lead as a gasoline additive in reducing knock in the internal combustion engine.

1921 (December 12). GENERAL OR MISCELLANEOUS / ASTRONOMY
Astronomer Henrietta Swan Leavitt (b.1868) died in Cambridge, Massachusetts.

1922. CHEMISTRY / NUTRITION
Vitamin E was found by Herbert McLean Evans (1882-1971) and Katharine J. Scott [Katharine Scott Bishop] (1889-1975).

1922. CHEMISTRY / NUTRITION
Vitamin D was discovered in cod liver oil by Elmer Verner McCollum (1879-1967). It was used for the treatment of rickets.

1922. GENERAL OR MISCELLANEOUS / ORGANIZATIONS—INDUSTRY, GEOLOGY
David Talbot Day (1859-1925) published the two-volume *Handbook of the Petroleum Industry* (New York).

1922. GEOLOGY
The geological term "pediment" was fixed in the literature as referring to a plain at the foot of mountains that is brought about by normal desert erosion. The occasion was publication of Kirk Bryan's (1888-1950) "Erosion and Sedimentation in the Papago Country, Arizona with a Sketch of the Geology," *Bulletin of United States Geological Survey*, no. 730, pp.19-90.

1922. GEOLOGY
Norman Levi Bowen's (1887-1956) theory of the evolution of igneous rocks was announced. In his work he drew on data from the Geophysical Laboratory of the Carnegie Institution of Washington.

1922. MATHEMATICS
Mathematician Oswald Veblen (1880-1960) published *Analysis Situs* (Colloquium Publications, American Mathematical Society, 5, part 2), an important work on an aspect of geometry.

1922. PERIODICALS AND PUBLISHING / PHYSICAL SCIENCES—GENERAL
Edward Wight Washburn (1881-1934) became editor-in-chief of the *International Critical Tables of Numerical Data, Physics, Chemistry, and Technology*, a project of the International Union of Pure and Applied Chemistry. Seven volumes of the *Tables* appeared, under Washburn's editorship, between 1926 and 1930.

1922. PSYCHOLOGY
Granville Stanley Hall (1846-1924) published *Senescence, the Last Half of Life* (New York), the earliest study of its scope and character.

1922. ZOOLOGY / NEUROBIOLOGY
Alexander Forbes (1882-1965) gave a new research direction in neurophysiology through the union of spinal reflex and nerve conduction study. His presentation appeared in his "The Interpretation of Spinal Reflexes of Present Knowledge of Nerve Conduction," *Physiological Reviews* 2:361-414.

1922. ZOOLOGY—HUMAN / ANATOMY
William King Gregory (1876-1970) published *The Origin and Evolution of the Human Dentition* (Baltimore).

1922 (July 22). GENERAL OR MISCELLANEOUS / CHEMISTRY
Chemist Jokichi Takamine (b.1854) died in New York City.

1922 (August 2). GENERAL OR MISCELLANEOUS / TECHNOLOGY AND INVENTION
Inventor Alexander Graham Bell (b.1847) died in Beinn Bhreagh, Nova Scotia.

1922-1926. BOTANY
Elmer Drew Merrill (1876-1956) published *An Enumeration of Philippine Flowering Plants* (Manila), in four volumes.

1923. ASTRONOMY
Tables for the moon's motion devised by Ernest William Brown (1866-1938) were accepted for the ephemerides of most nations. Henry B. Hedrick (1865-1936) had worked with Brown on the project (*Tables of the Motion of the Moon*, 3 volumes; New Haven, 1919).

1923. AWARDS AND PRIZES / PHYSICS
Robert Andrews Millikan (1868-1953) won the Nobel Prize in physics. He was chosen "for his work on the elementary charge of electricity and the photoelectric effect."

1923. CHEMISTRY / PHYSICS
Gilbert Newton Lewis (1875-1946) published *Valence and the Structure of Atoms and Molecules* (New York).

1923. CHEMISTRY / PHYSICS
Gilbert Newton Lewis (1875-1946) and Merle Randall (1888-1950) published the textbook, *Thermodynamics and the Free Energy of Chemical Substances* (New York).

1923. ENGINEERING AND APPLIED SCIENCE / ELECTRICITY AND ELECTRONICS
Vladimir Zworykin (1889-1982) invented the iconoscope. In 1933, he published "Television with Cathode-Ray Tubes," *Journal of Institution of Electrical Engineers* 73 (October 1933): 437-451, an account of his invention of the iconoscope, an all-electronic camera. Following on earlier mechanical scanners devised by others, Zworykin's contribution made development of television viable.

1923. GEOLOGY / HYDROLOGY
Oscar Edward Meinzer (1876-1948) published *The Occurrence of Ground Water in the United States, with a Discussion of Principles* (U.S. Geological Survey Water Supply Paper 489). During the same year, he also published *Outline of Ground-Water Hydrology, with Definitions* (U.S. Geological Survey Water Supply Paper 494).

1923. GOVERNMENT—FEDERAL / ENGINEERING AND APPLIED SCIENCE
The Naval Research Laboratory was established.

1923. GOVERNMENT—FEDERAL / GEOLOGY
The U.S. Geological Survey carried out the first scientific excursion through the Grand Canyon since the studies by John Wesley Powell (1834-1902) in the late 1860s and early 1870s. Raymond Cecil Moore (1892-1974) was geologist on the new exploration.

1923. GOVERNMENT—FEDERAL / PHYSICAL SCIENCES—GENERAL
In this year the National Bureau of Standards' first director, Samuel Wesley Stratton (1861-1931), who had served since 1901, resigned. By the time of his leaving, the Bureau had became one of the leading centers for research in physics in the country.

1923. MATHEMATICS / EDUCATION IN SCIENCE
The Mathematical Association of America's National Committee on Mathematical Requirements issued the influential *The Reorganization of Mathematics in Secondary Education* (Oberlin, Ohio). John Wesley Young (1879-1932) chaired the committee and edited the report.

1923. MEDICINE / DISEASE
George Dick (1881-1967) and Gladys Dick (1881-1963) discovered that scarlet fever is caused by a streptococci. They developed an antitoxin for the disease. (The discovery by the Dicks was done independently of Alphonse R. Dochez, 1882-1964.)

1923. ORGANIZATIONS—FOUNDATIONS
The International Education Board was founded with Rockefeller funding. It was active in science during its first five years.

1923. ORGANIZATIONS—SOCIETIES AND ASSOCIATIONS / SOCIAL SCIENCES—GENERAL
The Social Science Research Council was founded.

1923. PERIODICALS AND PUBLISHING
James McKeen Cattell (1860-1944) established Science Press at Lancaster, Pennsylvania, to handle his various reference works and periodicals.

1923. PHYSICS
Arthur Holly Compton (1892-1962) proposed a quantum theory of scattering, devised to explain changes in X-ray wavelengths as they scattered at different angles from a target. The phenomenon itself became known as the Compton effect and was presented in his paper, "A Quantum Theory of the Scattering of X-rays by Light Elements," *Physical Review* 21 (1923): 483-502.

1923. ZOOLOGY / PHYSIOLOGY
At the New York Institute for the Study of Malignant Diseases (Buffalo), Carl F. Cori (1896-1984) and Gerty Radnitz Cori (1896-1957) published the first results of their joint research on metabolism of carbohydrates. A series of papers followed thereafter on the relations of the hormones epinephrine and insulin to carbohydrate metabolism.

1923 (February 6). GENERAL OR MISCELLANEOUS / ASTRONOMY
Astronomer Edward Emerson Barnard (b.1857) died in Williams Bay, Wisconsin.

1923 (February 24). GENERAL OR MISCELLANEOUS / CHEMISTRY, PHYSICS
Chemist and physicist Edward Williams Morley (b.1838) died in West Hartford, Connecticut.

1923 (October). ASTRONOMY
Edwin Powell Hubble (1889-1953), using the 100-inch telescope at Mt. Wilson, produced an astronomical plate which led to his discovery of a Cepheid variable star in the nebula of Andromeda. This evidence resolved uncertainty about the status of spiral nebulae and was strong affirmation that the Andromeda nebula was outside our galaxy. In December 1924, announcement was made at a meeting of the American Astronomical Society of the results of Hubble's work relative to Cepheid stars, in which convincing proof was given for the island-universe view of the distribution of matter in space.

1923 (October 26). GENERAL OR MISCELLANEOUS / MATHEMATICS, ELECTRICITY AND ELECTRONICS
Mathematician and electrical engineer Charles Proteus Steinmetz (b.1865) died in Schenectady, New York.

Chronology

1920s. GENERAL OR MISCELLANEOUS / FUNDS AND FUNDING
During this time period, the largest sources of support for science were Rockefeller's General Education Board, International Education Board, and Laura Spelman Rockefeller Memorial. Wickliffe Rose (1862-1931) took charge of the GEB and IEB in 1923 and under his direction, science-related grants to institutions became the most characteristic form of assistance.

1924. CHEMISTRY / ELECTRICITY AND ELECTRONICS
William Francis Giauque (1895-1982) invented cooling by adiabatic demagnetization. This method facilitated studies on electrical and thermal conductivity, superconductors at very low temperatures, and other interests.

1924. ENGINEERING AND APPLIED SCIENCE / STATISTICS
Walter Andrew Shewhart (1891-1967), at Western Electric Company, devised a statistical means of determining when constructive intervention in a production process, to correct for deviation, is warranted. It later came to be called the control chart. He published a general work on the subject in 1931, *Economic Control of Quality of Manufactured Product* (New York).

1924. GEOLOGY
Leason Heberling Adams (1887-1969) published "Temperatures at Moderate Depths Within the Earth," *Journal of Washington Academy of Sciences* 14:459-472.

1924. GEOLOGY
The Division of Geology and Geography of the National Research Council created a Committee on the Measurement of Geologic Time by Atomic Disintegration. Reflecting his interest in such uses of radioactivity, Alfred Church Lane (1863-1948) was appointed to chair the committee. The committee came to take a broader and more eclectic view of measuring methods and dropped "by Atomic Disintegration" from its name. (Lane continued as head of the committee until 1946.)

1924. ORGANIZATIONS—ACADEMIC / PHILOSOPHY
Alfred North Whitehead (1861-1947) came from England to the United States and became professor of philosophy at Harvard University.

1924. ORGANIZATIONS—RESEARCH INSTITUTIONS / PHYSICS
The Franklin Institute's Bartol Research Foundation was established at Swarthmore College. In 1927, William Francis Gray Swann (1884-1962) became director and remained until 1959. Under Swann's direction, the Foundation became a center for cosmic ray studies.

1924. ORGANIZATIONS—SOCIETIES AND ASSOCIATIONS / BOTANY
The American Society of Plant Physiologists was established.

1924. ORGANIZATIONS—SOCIETIES AND ASSOCIATIONS / HISTORY OF SCIENCE—FIELD OF STUDY
The American Society of Medical History was established.

1924. ORGANIZATIONS—SOCIETIES AND ASSOCIATIONS / HISTORY OF SCIENCE—FIELD OF STUDY
The History of Science Society was founded.

1924. PHYSICS / ASTRONOMY
Walter S. Adams (1876-1956), using white dwarf stars, effectively verified the prediction of Einstein's general theory of relativity that there is a reduction in frequency for light that is emitted in an intense gravitational field.

1924. PHYSIOLOGY / INSTRUMENTS AND INSTRUMENTATION
Francis Gano Benedict (1870-1957) invented apparatus for the simultaneous measurement of oxygen consumed, air expired, and heat.

1924. PSYCHOLOGY / INSTRUMENTS AND INSTRUMENTATION
Arnold Lucius Gesell (1880-1961), professor in the Yale Graduate School of Medicine, began the use of cinematography in psychological research. Gesell was noted for the development of innovative means of child study and a pioneer in such uses of motion pictures.

1924 (February 11). GENERAL OR MISCELLANEOUS / PHYSIOLOGY
Physiologist Jacques Loeb (b.1859) died in Hamilton, Bermuda.

1924 (April 24). GENERAL OR MISCELLANEOUS / PSYCHOLOGY
Psychologist Granville Stanley Hall (b.1846) died in Worcester, Massachusetts.

ca.1924. BIOCHEMISTRY AND MOLECULAR BIOLOGY / ZOOLOGY
Edgar Allen (1892-1943), with Edward A. Doisy (1893-1986), discovered estrogen.

1924-1925. GENERAL OR MISCELLANEOUS
In this period, Florence Rena Sabin (1871-1953) was the first woman elected to the presidency of the American Association of Anatomists (1924), first woman elected to the National Academy of Sciences (1925), and first woman to achieve full membership in the Rockefeller Institute (1925).

1925. ASTRONOMY
Edwin Powell Hubble (1889-1953) presented what became the standard classification system for galaxies. It was published the following year in "Extra-Galactic Nebulae," *Astrophysical Journal* 64 (1926): 321-369.

1925. ASTRONOMY
Cecilia Helena Payne (later, Payne-Gaposchkin) (1900-1979) completed a doctoral thesis, *Stellar Atmospheres*, in Radcliffe College (using the facilities of the Harvard College Observatory). Through spectroscopic means, she determined temperatures of stars and the amounts of their chemical elements. Her researches gave the first demonstration that the chemical constitution of the universe was homogeneous.

1925. AWARDS AND PRIZES / MATHEMATICS
The first Chauvenet Prize of the Mathematical Association of America was awarded to Gilbert Ames Bliss (1876-1951) for "Algebraic Functions and Their Divisors" (*Annals of Mathematics* 26:95-124).

1925. BIOCHEMISTRY AND MOLECULAR BIOLOGY
John Jacob Abel (1857-1938) achieved the crystallization of insulin (i.e., an isolation of the substance in pure form).

1925. BIOCHEMISTRY AND MOLECULAR BIOLOGY
Iron as a significant part of red blood cells was discovered by George Hoyt Whipple (1878-1976).

1925. BOTANY
Merritt Lyndon Fernald (1873-1950) published "Persistence of Plants in Unglaciated Areas of Boreal America," *Memoirs of American Academy of Arts and Sciences* 15:239-342, in which he refuted the widespread notion that moving ice had destroyed all plants and animals that it encountered.

1925. BOTANY / GEOLOGY
Edward Charles Jeffrey (1866-1952) published "The Origin and Organization of Coal," *Memoirs of American Academy of Arts and Sciences*, new series 15:1-52. It helped to establish the botanical origin of coal.

1925. GENERAL OR MISCELLANEOUS
Sinclair Lewis (1885-1951) published his novel *Arrowsmith*, in which an individual doing scientific research played the central role.

1925. GENERAL OR MISCELLANEOUS / ASTRONOMY
Astronomer Annie Jump Cannon (1863-1941) received the first honorary degree ever awarded a woman by Oxford University.

1925. GENERAL OR MISCELLANEOUS / FUNDS AND FUNDING
A fund drive for a National Research Endowment (later changed to National Research Fund) was begun by the National Academy of Sciences, instigated by George Ellery Hale (1868-1938). The targeted source was industry and the fund goal of $20,000,000 was conceived as a means of supporting pure science research. Secretary of Commerce Herbert Hoover was the Endowment chair. The Fund did not succeed, in part because of the economic Depression that occurred about the time the Fund effort was operational but also because participation of industry was not widespread.

1925. GENERAL OR MISCELLANEOUS / RELIGION AND THEOLOGY
John Thomas Scopes (1900-1970) was arrested on May 5 and in July put on trial for teaching Darwin's theory of evolution in the public schools of Dayton, Tennessee. He was convicted and fined $100. The conviction was overturned by the state supreme court in 1927 on technical grounds.

1925. GENETICS
Leslie Clarence Dunn (1893-1974) and Edmund Ware Sinnott (1888-1968) published *Principles of Genetics* (New York), an influential textbook. There were new editions in 1932 and 1939; subsequent revisions appeared in 1950 and 1958 with Theodosius Dobzhansky (1900-1975) as an influential coauthor.

1925. GOVERNMENT—FEDERAL
The Patent Office and the Bureau of Mines were transferred from the Department of Interior to the Department of Commerce. Among the science-related agencies already attached to Commerce were the Bureau of Standards, Bureau of Fisheries, Bureau of the Census, and the Coast and Geodetic Survey. At the head of the Department of Commerce was Herbert Hoover (1874-1964), who understood the place of science research and its applications for efficiency.

1925. MATHEMATICS
Julian Lowell Coolidge (1873-1954) published *Introduction to Mathematical Probability* (New York), one of the first English-language works on the topic.

1925. ORGANIZATIONS—ACADEMIC / FUNDS AND FUNDING
The Wisconsin Alumni Research Foundation was incorporated. It grew out of work by Harry Steenbock (1886-1967) on the effects of ultraviolet light on feed and the patents that resulted and which were assigned to the Foundation.

1925. ORGANIZATIONS—OBSERVATORIES
Yale University established a southern station for astronomical observations at Johannesburg, South Africa.

1925. PALEONTOLOGY
Chester Stock (1892-1950) published a notable work on ground sloths and other Pleistocene vertebrates, as *Cenozoic Gravigrade Edentates of Western North America with Special Reference to the Pleistocene Megalonychinae and Mylodontinae of Rancho La Brea* (Carnegie Institution of Washington Publication no. 331).

1925. PHYSICS
Richard L. Doan (1898-?), at a suggestion of Arthur Holly Compton (1892-1962), produced the first photograph of an X-ray grating spectra. Subsequent work by others influenced by Compton (e.g., his students Joyce Alvin Bearden, 1903-1986, and Newell Shiffer Gingrich, 1906-) helped establish X-ray studies as a part of optics.

1925. PHYSICS
Dayton Clarence Miller (1866-1941) reported on his experiments over several years on Mt. Wilson, in which he claimed to have detected evidence of relative motion of the earth to a stationary ether. The experimental results tended toward disproof of the theory of relativity; later examinations of his data suggest a combination of random and systematic variation (the latter due to temperature changes in his experimental location). Miller received a $1000 prize from the American Association for the Advancement of Science for his work.

1925. PHYSICS
Cosmic rays in the upper atmosphere were established experimentally by Robert A. Millikan (1868-1953).

1925. PHYSICS / METEOROLOGY AND CLIMATOLOGY
Merle A. Tuve (1901-1982) and Gregory Breit (1899-1981) measured the height of the ionosphere.

1925. ZOOLOGY / MATHEMATICS
Alfred James Lotka (1880-1949) published *Elements of Physical Biology* (Baltimore), which made a significant use of mathematics in modelling relations between animal groups.

1925 (January 1). ORGANIZATIONS—INDUSTRY / ELECTRICITY AND ELECTRONICS
The research facilities of AT&T became known as Bell Telephone Laboratories. Frank Baldwin Jewett (1879-1949) was made president and Harold DeForest Arnold (1883-1933) director of research.

1925-1959. PSYCHOLOGY
Lewis Madison Terman (1877-1956) published the five-volume *Genetic Studies of Genius* (Stanford, Calif.).

1926. ASTRONOMY
John Charles Duncan (1882-1967) published a textbook, *Astronomy* (New York), which went to a fifth edition in 1955.

1926. BIOCHEMISTRY AND MOLECULAR BIOLOGY
The first enzyme to be crystallized was achieved by James Batcheller Sumner (1887-1955), who crystallized jackbean urease. He also determined that it was a protein.

1926. GENETICS
Thomas Hunt Morgan (1866-1945) published *The Theory of the Gene* (New Haven, Conn.).

1926. GEOLOGY
William Henry Twenhofel (1875-1957) published *Treatise on Sedimentation* (Baltimore). A second edition appeared in 1932.

1926. MICROBIOLOGY AND MICROSCOPY / PATHOLOGY
Paul de Kruif's (1890-1971) *The Microbe Hunters* (New York), a popular book on bacteriology, was published.

1926. ORGANIZATIONS—SOCIETIES AND ASSOCIATIONS
The Association of Academies of Science was founded.

1926. ORGANIZATIONS—SOCIETIES AND ASSOCIATIONS / EUGENICS
The American Eugenics Society was established.

1926. PERIODICALS AND PUBLISHING / BIOLOGY—GENERAL
Raymond Pearl (1879-1940) established the *Quarterly Review of Biology* and for some time served as editor.

1926. PHYSICS
David Mathias Dennison (1900-1976) published the first article applying the matrix mechanics (a component of the new quantum mechanics) to appear in the *Physical Review*. Dennison was at the time studying with Niels Bohr in Copenhagen.

1926. PHYSICS
The familiarity of the American physics community with quantum theory was promoted by publication of the report, by Edwin C. Kemble (1889-1984) and others, "Molecular Spectra in Gases: Report of the Committee on Radiation in Gases," *Bulletin of National Research Council* 11 (December 1926): 69-259.

1926. PHYSICS
Albert Abraham Michelson (1852-1931) carried out measurements of the speed of light over a twenty-two-mile distance in California, between Mt. Wilson and Mt. San Antonio.

1926. PHYSICS
John H. Van Vleck (1899-1980) produced *Quantum Principles and Line Spectra* (Washington, D.C.), a publication of the National Research Council's Committee on Ionization Potentials and Related Subjects.

1926. TECHNOLOGY AND INVENTION / ELECTRICITY AND ELECTRONICS
Sound motion pictures were developed by Western Electric.

1926. MEDICINE / PATHOLOGY
George Richards Minot (1885-1950) and William P. Murphy (1892-1987) discovered that the eating of liver was a means to treat pernicious anemia.

1926 (March 16). SPACE SCIENCE, TECHNOLOGY, AND EXPLORATION
The first liquid-fuel rocket was launched by Robert Hutchings Goddard (1882-1945) in Auburn, Massachusetts. It attained a height of 184 feet and a speed of 60 miles per hour.

1926 (April 11). GENERAL OR MISCELLANEOUS / BOTANY
Plant breeder Luther Burbank (b.1849) died in Santa Rosa, California.

1926 (May 9). EXPLORATION AND SURVEYING
Richard Evelyn Byrd (1888-1957) and Floyd Bennett (1890-1928) made the first flight over the North Pole.

1926-1927. ASTRONOMY
The important textbook, *Astronomy* I: *The Solar System* (Boston, 1926), and II: *Astrophysics and Stellar Astronomy* (Boston, 1927) was published by Henry Norris Russell (1877-1957), Raymond Smith Dugan (1878-1940), and John Quincy Stewart (1894-1972). The first volume was based on Charles A. Young's (1834-1908) earlier manual of astronomy, while the second volume was largely new.

1927. ANTHROPOLOGY AND ETHNOLOGY
Ales Hrdlicka (1869-1943) published "The Neanderthal Phase of Man," *Journal of Royal Anthropological Institute* 57:249-274, in which he argued for the idea of Homo sapiens as descended from *Homo neanderthalensis* and all peoples from a common ancestor.

1927. ASTRONOMY
Ira Sprague Bowen (1898-1973) gave an explanation of the so-called "nebulium" lines in the spectra of gaseous nebulae in terms of transitions in common ions. Previously the origin of the lines were not determined although they resembled an unknown element ("nebulium"). Bowen's work on this topic made it possible for him to characterize the basic physical picture of the gaseous nebulae.

1927. AWARDS AND PRIZES / PHYSICS
Arthur Holly Compton (1892-1962) received the Nobel Prize in physics for his work on wavelength change in X-rays, that is "for his discovery of the effect named after him." (Charles R. Wilson in England also received a physics prize this year.)

1927. CHEMISTRY
Arthur Amos Noyes (1866-1936) and William C. Bray (1879-1946) published the results of many years' work by Noyes, as *A System of Qualitative Analysis for the Rare Elements* (New York).

1927. GENETICS
Hermann Joseph Muller (1890-1967) reported on his induced mutation in *Drosophila*, by use of X-rays, in "Artificial Transmutation of the Gene," *Science* 66:84-87.

1927. GEOLOGY
Geologist William Bowie (1872-1940) published *Isostasy* (New York).

1927. MICROBIOLOGY AND MICROSCOPY / PATHOLOGY
Raymond Alexander Kelser (1892-1952) published *Manual of Veterinary Bacteriology* (Baltimore). Several editions appeared between this year and 1943.

1927. ORGANIZATIONS—INDUSTRY / CHEMISTRY
By this date, Charles Milton Altland Stine (1882-1954), who had been appointed as chemical director at DuPont company in 1924, established at DuPont the program for research which led to the subsequent development of neoprene and nylon.

1927. ORGANIZATIONS—INDUSTRY / ELECTRICITY AND ELECTRONICS
Television transmission was developed by the Bell System. The work was carried out under the general charge of Herbert E. Ives (1882-1953).

1927. ORGANIZATIONS—RESEARCH INSTITUTIONS / SOCIAL SCIENCES--GENERAL
The Brookings Institution (Washington, D.C.) was founded. Its interests were directed especially at the support of research and education in the social sciences.

1927. PHYSICS
Clinton Joseph Davisson (1881-1958) and Lester H. Germer (1896-1971), and English physicist George Thomson, provided experimental proof for the wave character of matter. In their experiments, beams of electrons were diffracted in passing through crystals and thin films.

1927. PHYSICS
A Debate on the Theory of Relativity (New York) was published. Among the contributors was William Duncan Macmillan (1871-1948), Robert D. Carmichael (1879-1967), and Harold T. Davis (1892-1974).

Chronology 221

1927. Physics
Albert Abraham Michelson (1852-1931) published *Studies in Optics* (Chicago).

1927. Physics / Philosophy
Percy W. Bridgman (1882-1961) published *The Logic of Modern Physics* (New York), in which he presented his operational perspective on the understanding of concept and meaning.

1927. Technology and Invention / Electricity and Electronics
Harold Stephen Black (1898-1983) invented the negative-feedback amplifier while with the Bell Telephone Company. The device was a significant contribution to communications in that it made possible distortion-free amplification.

1927. Zoology / Embryology
Thomas Hunt Morgan (1866-1945) published *Experimental Embryology* (New York).

1927. Microbiology and Microscopy
Selman Abraham Waksman (1888-1973) published *Principles of Soil Microbiology* (Baltimore). A second edition appeared in 1932.

1927. Physics
David Mathias Dennison (1900-1976) published a paper in the *Proceedings of Royal Society of London*, Series A vol. 115:483-486, on the specific heat of hydrogen, in which he gave the first quantitative support to the idea of proton spin in the atom.

1927. Psychology
Edward Lee Thorndike (1874-1949), with others, published *The Measurement of Intelligence* (New York).

1927 (March 4). General or Miscellaneous / Chemistry
Chemist Ira Remsen (b.1846) died in Carmel, California.

1927 (August 15). General or Miscellaneous / Chemistry, Physics
Chemist and physicist Bertram Borden Boltwood (b.1870) died in Hancock Point, Maine.

1927 (December 27). GENERAL OR MISCELLANEOUS / GENETICS
Geneticist Theodosius Dobzhansky (1900-1975) arrived in New York from Russia on an International Education Board fellowship to work with Thomas Hunt Morgan (1866-1945).

1927-1928. MEDICINE / INSTRUMENTS AND INSTRUMENTATION
The "iron lung" was developed by Philip Drinker (1893-1972) and Louis A. Shaw (1886-1940).

1927-1930. MATHEMATICS
Jesse Douglas (1897-1965) published a general solution to the mathematical problem of Plateau (concerning minimal surface of a curved space), in *Bulletin of American Mathematical Society*, followed by a fuller account in the Society's *Transactions* in 1931. In 1936, he received the Fields Medal at the International Congress of Mathematicians (Oslo) for this work.

1927-1936. PHYSICS
William Duncan Macmillan (1871-1948) published the three-volume *Theoretical Mechanics* (New York).

1927-1948. HISTORY OF SCIENCE—FIELD OF STUDY
George Alfred Leon Sarton (1884-1956) published his *Introduction to the History of Science* (Baltimore) in three volumes, which covered the subject up to 1400.

1928. ANTHROPOLOGY AND ETHNOLOGY
Franz Boas (1858-1942) published *Anthropology and Modern Life* (New York). Among the features of the book is a dismissal of theories of racial superiority.

1928. ANTHROPOLOGY AND ETHNOLOGY
Margaret Mead (1901-1978) published *Coming of Age in Samoa: A Psychological Study of Primitive Youth for Western Civilisation* (New York).

1928. ASTRONOMY
Charles Edward St. John (1857-1935), Charlotte E. Moore (1898-?), Louise M. Ware, Edward F. Adams, and Harold D. Babcock

(1882-1968) published *Revision of Rowland's Preliminary Table of Solar Spectrum Wavelengths, with an Extension to the Present Limit of the Infra-Red* (Carnegie Institution of Washington Publication no. 396) (Washington, D.C.).

1928. ASTRONOMY
Through solar spectrum studies, Henry Norris Russell (1877-1957) determined the elements of the sun's atmosphere.

1928. EXPLORATION AND SURVEYING
Little America in Antarctica was established by Richard Evelyn Byrd (1888-1957); it was used as a base for exploration of the continent.

1928. GENERAL OR MISCELLANEOUS / MATHEMATICS
Edward V. Huntington (1874-1952) published "The Apportionment of Representatives in Congress," *Transactions of American Mathematical Society* 30:85-110. In 1941, Congress adopted his method of equal proportions for determining representation.

1928. GEOLOGY
Norman Levi Bowen (1887-1956) published *The Evolution of Igneous Rocks* (Princeton), which summarized the highlights of his work on the subject.

1928. MEDICINE / PATHOLOGY
A test for the detection of uterine cancer from the study of cells taken from the vaginal wall was developed by George Nicholas Papanicolaou (1883-1962). In 1943, Papanicolaou and Herbert F. Traut (1894-1963) published *Diagnosis of Uterine Cancer by the Vaginal Smear* (New York). Until the publication of this work, the procedure, which came to be known as the Papanicolaou smear or Pap test, was not widely heeded.

1928. MICROBIOLOGY AND MICROSCOPY / GEOLOGY
Joseph Augustine Cushman (1881-1949) published the textbook, *Foraminifera, Their Classification and Economic Use* (Sharon, Mass.). Containing his classification of the shelled Protozoa, the work eventually had four editions.

1928. ORGANIZATIONS—OBSERVATORIES / INSTRUMENTS AND INSTRUMENTATION
George Ellery Hale (1868-1938) was able to get $6 million from the International Education Board for a 200-inch telescope. The funds went to the California Institute of Technology but the project was to be cooperative with the Mt. Wilson Observatory and its owner, the Carnegie Institution of Washington. The Mt. Wilson and Palomar Observatories resulted. Hale died before the completion of the telescope; it was dedicated in 1948 and was named for Hale. In 1969, the Mt. Wilson and Palomar Observatories were renamed as the Hale Observatories.

1928. ORGANIZATIONS—SOCIETIES AND ASSOCIATIONS / ANTHROPOLOGY AND ETHNOLOGY
The American Association of Physical Anthropologists was established by Ales Hrdlicka (1869-1943), who served as its president. From 1929, the *American Journal of Physical Anthropology* served as its official publication, which Hrdlicka had begun in 1918.

1928. ORGANIZATIONS—SOCIETIES AND ASSOCIATIONS / NUTRITION
The American Institute of Nutrition was established.

1928. PERIODICALS AND PUBLISHING / PHYSIOLOGY
Charles Manning Child (1869-1954) established the journal *Physiological Zoology*.

1928. PHYSICS
Floyd Karker Richtmyer (1881-1939) published *Introduction to Modern Physics* (New York).

1928. PHYSICS
Henry Crew (1859-1953) published *The Rise of Modern Physics* (Baltimore).

1928. PSYCHOLOGY
John Broadus Watson (1878-1858) published *Psychological Care of Infant and Child* (New York).

1928. TECHNOLOGY AND INVENTION / ELECTRICITY AND ELECTRONICS
A patent for color television was granted to Vladimir Zworykin (1889-1982).

1928 (April 2). GENERAL OR MISCELLANEOUS / CHEMISTRY
Chemist Theodore William Richards (b.1868) died in Cambridge, Massachusetts.

1928 (November 15). GENERAL OR MISCELLANEOUS / GEOLOGY
Geologist and cosmologist Thomas Chrowder Chamberlin (b.1843) died in Chicago, Illinois.

1928-1940. ORGANIZATIONS—ACADEMIC / PHYSICS
The University of Michigan conducted Summer Symposia in Theoretical Physics. Physicists from the United States and Europe took part and the series helped to make the University a center for theoretical interests.

1929. ASTRONOMY
Edwin Powell Hubble (1889-1953) established the concept of an expanding universe, by Hubble's Law, that the more distant a galaxy is from the earth the faster it is moving away. He was aided by the availability of measures of velocity of receding spiral nebulae which Vesto Melvin Slipher (1875-1969) had begun to investigate about 1912. Hubble presented his views in "Relation between Distance and Radial Velocity among Extra-galactic Nebulae," *Proceedings of National Academy of Sciences* 15 (March 15, 1929): 168-173.

1929. BIOCHEMISTRY AND MOLECULAR BIOLOGY
The presence of a newly discovered sugar, deoxyribose, was noted by Phoebus A. Levene (1869-1940) in nucleic acids. They later came to be known as deoxyribonucleic acids or DNA.

1929. GENERAL OR MISCELLANEOUS / ORGANIZATIONS—INDUSTRY
General Foods Company was established, the product of a series of mergers.

1929. GENETICS
Theophilus Shickel Painter (1889-1969) in this year and thereafter concluded that humans possess 48 chromosomes. This number was

accepted for many years as true until the correct number of 46 chromosomes was established by others.

1929. GEOLOGY
Adolph Knopf (1882-1966) published "The Mother Lode System of California," *Professional Papers of the U.S. Geological Survey* no. 157.

1929. ORGANIZATIONS / BIOLOGY, PHYSICS
The National Council on Radiation Protection and Measurements was established. Known until 1947 as the Advisory Committee on X-Ray and Radium Protection, in that year it became National Committee on Radiation Protection and between 1957 and 1964 was the National Committee on Radiation Protection and Measurements.

1929. ORGANIZATIONS—ACADEMIC / AERONAUTICS
Theodore von Karman (1881-1963) came from Aachen, Germany, to the California Institute of Technology to take charge of the Guggenheim Aeronautical Laboratory.

1929. ORGANIZATIONS—LIBRARIES
The Library of Congress appointed geologist Alfred Church Lane (1863-1948) as its first consultant in science.

1929. ORGANIZATIONS—RESEARCH INSTITUTIONS
Battelle Memorial Institute was established in Columbus, Ohio.

1929. ORGANIZATIONS—RESEARCH INSTITUTIONS / ZOOLOGY
Jackson Laboratory (Bar Harbor, Maine) was established by Clarence Cook Little (1888-1971), a mammalian geneticist.

1929. ORGANIZATIONS—SOCIETIES AND ASSOCIATIONS / ACOUSTICS
The Acoustical Society of America was founded.

1929. ORGANIZATIONS—SOCIETIES AND ASSOCIATIONS / PHYSICS
The Society of Rheology was founded.

1929. PALEONTOLOGY
George Gaylord Simpson (1902-1984) published *American Mesozoic Mammalia* (New Haven, Conn.).

1929. PERIODICALS AND PUBLISHING
McGraw-Hill publishing began its influential International Series in Physics (an International Chemical Series was underway earlier). Most of the volumes in the physics series dealt with quantum subjects.

1929. PERIODICALS AND PUBLISHING / ZOOLOGY—HUMAN
Raymond Pearl (1879-1940) established the journal *Human Biology* and for some years served as editor.

1929. PHILOSOPHY
Alfred North Whitehead (1861-1947) published *Process and Reality: An Essay in Cosmology* (New York).

1929. PHYSICS
Edward U. Condon (1902-1974) and Philip M. Morse (1903-1985) published *Quantum Mechanics* (New York), the first English-language book on the subject.

1929. PHYSICS / INSTRUMENTS AND INSTRUMENTATION
Robert Jemison Van de Graaff (1901-1967) built the first workable belt-charged electrostatic high-voltage generator, at 80,000 volts, which he had invented. He was at Princeton at the time. In 1933, he devised a generator that produced 7 million volts.

1929. PSYCHOLOGY
Wolfgang Kohler (1887-1967), then in Germany, wrote *Gestalt Psychology* (New York) in order to introduce the subject to psychologists in the United States.

1929. PSYCHOLOGY / NEUROBIOLOGY
The work by Karl Spencer Lashley (1890-1958), *Brain Mechanisms and Intelligence: A Quantitative Study of Injuries to the Brain* (Chicago), was published.

1929. ZOOLOGY / NEUROBIOLOGY
George Ellett Coghill (1872-1941) published *Anatomy and the Problem of Behavior* (Cambridge, Eng.), presenting ideas delivered the previous year in lectures at University College, London. It presented the idea of the dominance of an integrative, central nervous system.

1929. ZOOLOGY / PHYSIOLOGY
Walter B. Cannon (1871-1945) showed that emotions were created in the limbic system of the brain. This conclusion displaced the commonly accepted James-Lange theory that posited emotions as arising from the body's response to stimuli (e.g., that crying was the prelude to a feeling of sadness).

1929. ZOOLOGY / PSYCHOLOGY
Yale University psychologist Robert Mearns Yerkes (1876-1956) and Ada Watterson Yerkes (1873-?) (his wife) published *The Great Apes: A Study of Anthropoid Life* (New Haven, Conn.), long a classic work. By this date, Yerkes had established the facility at Orange Park, Florida that later was called the Yerkes Laboratories of Primate Biology, centerpiece for the Yerkes Regional Primate Research Center. Yerkes retired as director of the Orange Park Laboratories in 1941.

1929 (September 24). INSTRUMENTS AND INSTRUMENTATION / AERONAUTICS
Using instruments that depend on radio signals, Lieutenant (later General) James H. Doolittle (1896-1993) made the first successful "blind" air flight at Mitchell Field, New York.

1929 (November 29). EXPLORATION AND SURVEYING
Lieutenant Commander Richard Evelyn Byrd (1888-1957) and Norwegian-American Bernt Balchen made the first successful air light over the South Pole.

ca.1929. ORGANIZATIONS—ACADEMIC / PHYSIOLOGY
Subsequent to his 1929 appointment in the Yale University medical school, John Farquhar Fulton (1899-1960) established the earliest American primate laboratory for experimental physiology.

1929-1932. PHYSICS / INSTRUMENTS AND INSTRUMENTATION
Ernest O. Lawrence (1901-1958), at the University of California, conceived the idea for the cyclotron in 1929 and the next year his students, Niels Edlefsen (1893-1971) and Milton Stanley Livingston (1905-1986), built the first ones. In 1932, Lawrence and Livingston perfected the device so that it could achieve 1,000,000 volts.

1930. ANTHROPOLOGY AND ETHNOLOGY
Margaret Mead (1901-1978) published *Growing Up in New Guinea: A Comparative Study of Primitive Education* (New York).

1930. ASTRONOMY
Frank Schlesinger (1871-1943) published *Catalogue of Bright Stars, Containing All Important Data Known in June, 1930, Relating to All Stars Brighter Than 6.5 Visual Magnitudes, and to Some Fainter Ones* (New Haven, Conn.). A second edition appeared in 1940 with Louise F. Jenkins (1888-?). In time, the work became one of the most used of all astronomical sources.

1930. ASTRONOMY
Harlow Shapley (1885-1972) published *Star Clusters* (New York).

1930. ASTRONOMY
Clyde William Tombaugh (1906-), at the Lowell Observatory, discovered the long-sought trans-Neptunian planet through an examination of photographs of the sky taken on successive days. It was designated Pluto. Its symbol, a superimposed P and L, also was the initials of Percival Lowell (1855-1916) who had spent the last years of his life in search of the planet, although Pluto most likely was not the Planet-X that Lowell had predicted.

1930. ASTRONOMY
Robert Julius Trumpler (1886-1956) found proof of the existence of diffuse interstellar dust. He argued that this phenomenon gave a Milky Way of about three-fifths the size of previous estimates, explaining dimming as caused by the dust rather than by distance.

1930. AWARDS AND PRIZES / MEDICINE, PHYSIOLOGY
Karl Landsteiner (1868-1943) won the Nobel Prize in medicine or physiology "for his discovery of human blood groups."

1930. BIOCHEMISTRY AND MOLECULAR BIOLOGY
The enzyme pepsin was crystallized by John Howard Northrop (1891-1987).

1930. BOTANY
The Fifth International Botanical Congress adopted the type specimen concept, codified by Albert S. Hitchcock (1865-1935); it was incorporated into the International Rules of Botanical Nomenclature.

1930. CHEMISTRY / PHYSICS
Harold Clayton Urey (1893-1981) and Arthur E. Ruark (1899-1979) published *Atoms, Molecules, and Quanta* (New York), the first all-encompassing English-language work on the structure of the atom.

1930. GOVERNMENT—FEDERAL / PUBLIC HEALTH
The Public Health Service's Hygienic Laboratory was moved to Bethesda, Maryland, and was renamed the National Institutes of Health.

1930. HISTORY OF SCIENCE—FIELD OF STUDY / PHYSIOLOGY
Russell Henry Chittenden's (1856-1943) *The Development of Physiological Chemistry in the United States* (New York) traced the origins of the science in the country to Yale University.

1930. MATHEMATICS
Mathematician Solomon Lefschetz (1884-1972) published *Topology* (New York).

1930. MATHEMATICS / ECONOMICS
Griffith Conrad Evans (1887-1973) published *Mathematical Introduction to Economics* (New York).

1930. ORGANIZATIONS—ACADEMIC
The Institute for Advanced Study was founded at Princeton by Louis Bamberger and Caroline (Bamberger) Fuld, brother and sister.

1930. ORGANIZATIONS—RESEARCH INSTITUTIONS / OCEANOGRAPHY
As part of a general support for oceanography and marine biology, the Rockefeller Foundation gave 2.5 million dollars for support of the Woods Hole Oceanographic Institution, which was incorporated on 6 January 1930. Frank Rattray Lillie (1870-1947) was instrumental in bringing about the establishment of the new facility and served as president from 1930 to 1939. The Institution and its research vessel, *Atlantis*, were in place by the summer of 1931.

1930. ORGANIZATIONS—SOCIETIES AND ASSOCIATIONS / BIOLOGY—GENERAL
The National Association of Biology Teachers was established. Instrumental as founder was Oscar Riddle (1877-1968).

1930. ORGANIZATIONS—SOCIETIES AND ASSOCIATIONS / ECONOMICS, MATHEMATICS
The Econometric Society was established.

1930. ORGANIZATIONS—SOCIETIES AND ASSOCIATIONS / PHYSICS
The American Association of Physics Teachers was founded.

1930. ORGANIZATIONS—SOCIETIES AND ASSOCIATIONS / SPACE SCIENCE, TECHNOLOGY, AND EXPLORATION
The American Rocket Society was established.

1930. TECHNOLOGY AND INVENTION / CHEMISTRY
The use of freon gas in refrigeration was developed by Thomas Midgley, Jr. (1889-1944).

1930. ZOOLOGY / PHYSIOLOGY
Philip Edward Smith (1884-1970) published "Hypophysectomy and a Replacement Therapy in the Rat," *American Journal of Anatomy* 45:205-273. It summarized his pioneering researches on the physiological relations of the pituitary gland.

1930s. ORGANIZATIONS—FOUNDATIONS / FUNDS AND FUNDING
In this time period, the Rockefeller Foundation tended to redirect its aid away from institutional support toward funding for individual research and in specific disciplines or fields. There was a particular concern for work that involved the cooperation of more than one discipline. Warren Weaver (1894-1978), who joined the Foundation in 1932, epitomized this new approach. His organizing activity revolved around the study of "vital processes." In addition to the inclinations of the individuals involved, the redirection of the Foundation in this period was affected by the Great Depression.

early 1930s. PHYSICS / GEOPHYSICS AND GEODESY
During this period, Arthur Holly Compton (1892-1962) redirected his research interests from X-rays to cosmic rays and led expeditions to many parts of the world in order to measure intensity of cosmic rays in different geomagnetic locales. One outcome was to confirm the relations of cosmic ray intensity to geomagnetic latitude and to altitude.

ca.1930. ZOOLOGY / PHYSIOLOGY
Alexis Carrel (1873-1944), with the aid of aviator Charles A. Lindbergh (1902-1974) who devised suitable apparatus, undertook studies on organs kept alive by artificial means. In 1938 Carrel, with Lindbergh, published *The Culture of Organs* (New York).

1930-1933. ORGANIZATIONS—FAIRS AND EXPOSITIONS
Henry Crew (1859-1953) was in charge of the Division of Basic Sciences at the Century of Progress International Exposition in Chicago (1933-1934). His duties included direction of the Hall of Science. Substantial parts of the exhibits were eventually incorporated into the Chicago Museum of Science and Industry.

1931. CHEMISTRY
Lars Onsager (1903-1976) published "Reciprocal Relations in Irreversible Processes," part I in *Physical Review* 37:405-426 and part II in *Physical Review* 38:2265-2279. In 1968, he won the Nobel Prize in chemistry for this work.

1931. CHEMISTRY / ORGANIZATIONS—INDUSTRY
Julius Arthur Nieuwland (1878-1936) published "A New Synthetic Rubber: Chloroprene and Its Polymers," *Journal of American Chemical Society* 53: 4198. Nieuwland had been an academic collaborator with DuPont company chemists whose related research was presented in a companion article. In the 1930s, DuPont sold the synthetic rubber under the name of "DuPrene" or neoprene.

1931. COMPUTERS AND INFORMATION SCIENCE
As of this date, Vannevar Bush (1890-1974) and his students at M.I.T. had devised the differential analyzer, an analog computer. In the period after 1935, Bush developed a more sophisticated machine, the Rockefeller differential analyzer, with support from the Rockefeller

Foundation and also from the Carnegie Institution of Washington. The Rockefeller machine was used for important calculational tasks during World War II.

1931. GEOLOGY
Douglas Wilson Johnson (1878-1944) published *Stream Sculpture on the Atlantic Slope* (New York).

1931. INSTRUMENTS AND INSTRUMENTATION / ELECTRICITY AND ELECTRONICS
Harold Eugene Edgerton (1903-1990) invented the stroboscope. It made possible photography of rapid motion events by stopping the action and had important applications in the development of motion pictures.

1931. ORGANIZATIONS—SOCIETIES AND ASSOCIATIONS / GEOLOGY
Upon the death of Richard F. Penrose, Jr. (1863-1931), the Geological Society of America was designated as recipient of a bequest of about four and a quarter million dollars.

1931. ORGANIZATIONS—SOCIETIES AND ASSOCIATIONS / PHYSICS
The American Institute of Physics was established through the cooperative action of the American Physical Society, Optical Society of America, Acoustical Society of America, and the Society of Rheology.

1931. PERIODICALS AND PUBLISHING / GEOLOGY
The *Journal of Sedimentary Petrology* was begun by Raymond C. Moore (1892-1974).

1931. PHYSICS
Percy W. Bridgman (1882-1961) published his classic *The Physics of High Pressure* (New York). As of this date, Bridgman achieved up to 100,000 atmospheres in his research on materials under pressure.

1931. ZOOLOGY
Gladwyn Kingsley Noble (1894-1940) published *The Biology of the Amphibia* (New York). A pioneering union of natural history and experimental biology, it remained for some years a basic authority for herpetologists.

1931 (May 9). GENERAL OR MISCELLANEOUS / PHYSICS
Physicist Albert A. Michelson (b.1852) died in Pasadena, California.

1931 (October 18). GENERAL OR MISCELLANEOUS / TECHNOLOGY AND INVENTION
Inventor Thomas Alva Edison (b.1847) died in West Orange, New Jersey.

1931-1932. ASTRONOMY
Karl Guthe Jansky (1905-1950), at Bell Telephone Laboratories, studied atmospheric static that interfered with transatlantic telephone calls. It was his conclusion that the detected radio waves came from a source outside the solar system, in the Milky Way galaxy. This was the beginning of radio astronomy, although little immediate notice was given to the discovery, and the field of radio astronomy did not develop until after World War II.

1931-1932. MEDICINE / CHEMISTRY
John P. Peters (1887-1955) and Donald Dexter Van Slyke (1883-1971) published *Quantitative Clinical Chemistry* (Baltimore), in two volumes. A second edition appeared in 1946.

1931-1938. GEOLOGY
Albert Johannsen (1871-1962) published *Descriptive Petrography of the Igneous Rocks* (Chicago), in four volumes, which used a quantitative mineralogical classification.

1932. ASTRONOMY
The International Astronomical Union approved standards for star magnitude based on the work of Frederick Hanley Seares (1873-1964). They were superseded in 1941 by Seares' advanced work.

1932. ASTRONOMY
Robert Grant Aitken (1864-1951), at the Lick Observatory, published *New General Catalogue of Double Stars Within 120° of the North Pole* (Washington, D.C.). The catalogue included 17,180 visible binary stars.

1932. ASTRONOMY
Joseph Haines Moore (1878-1949) published "A General Catalogue of the Radial Velocities of Stars, Nebulae and Clusters," *Publications of the Lick Observatory* 18.

1932. AWARDS AND PRIZES / CHEMISTRY
Irving Langmuir (1881-1957) won the Nobel Prize in chemistry "for his discoveries and investigations in surface chemistry." This field related to the study of chemical forces at the juncture of substances. Langmuir, at General Electric Company, was the first industrial scientist to be honored with the Nobel Prize.

1932. CHEMISTRY
James B. Conant (1893-1978) and his student Paul D. Bartlett (1907-) published "A Quantitative Study of Semicarbazone Formation," *Journal of American Chemical Society* 54:2881-2899, a classic illustration of a type of chemical reaction process.

1932. CHEMISTRY
Harold Clayton Urey (1893-1981) discovered deuterium, a heavy isotope of hydrogen. The achievement earned him a Nobel Prize in 1934.

1932. CHEMISTRY / TECHNOLOGY AND INVENTION
A report on a breakthrough in the production of synthetic fibers was published by Wallace H. Carothers (1896-1937) and Julian W. Hill (1904-?), "Studies of Polymerization and Ring Formation. XV. Artificial Fibers from Synthetic Linear Condensation Superpolymers," *Journal of American Chemical Society* 54 (April 1932): 1579-1587. Further investigation led to the patenting of nylon in 1935 and the achievement of means for its commercial production in 1939.

1932. EVOLUTION
Thomas Hunt Morgan (1866-1945) published *The Scientific Basis of Evolution* (New York).

1932. MATHEMATICS
Mathematician Robert Lee Moore (1882-1974) published *Foundations of Point Set Theory* (New York). A revised edition appeared in 1962.

1932. METEOROLOGY AND CLIMATOLOGY
Meteorologist Carl-Gustaf Arvid Rossby (1898-1957) devised a graphic means of identifying air masses and their dynamic processes. It was known as the Rossby diagram.

1932. ORGANIZATIONS—SOCIETIES AND ASSOCIATIONS / AERONAUTICS
The Institute of Aeronautical Sciences was established.

1932. ORGANIZATIONS—SOCIETIES AND ASSOCIATIONS / MEDICINE
The Harvey Cushing Society was founded by young colleagues of Harvey Williams Cushing (1869-1939). It later became the American Association of Neurological Surgeons.

1932. PHYSICS
John H. Van Vleck (1899-1980) published *The Theory of Electric and Magnetic Susceptibilities* (Oxford).

1932. PHYSICS
Carl David Anderson (1905-1991) discovered the positron as part of his cosmic ray research. In 1933, he produced positrons through gamma-ray bombardment.

1932. PHYSICS / ACOUSTICS
Paul Earls Sabine (1879-1958) published *Acoustics and Architecture* (New York and London).

1932. PHYSICS / ACOUSTICS
Vern Oliver Knudsen (1893-1974) published *Architectural Acoustics* (New York).

1932. PHYSICS / INSTRUMENTS AND INSTRUMENTATION
The cyclotron of Ernest Orlando Lawrence (1901-1958) at Berkeley was used to disintegrate lithium, one of the first important accomplishments of the apparatus invented by Lawrence. (Lithium had, however, been disintegrated elsewhere before this.)

1932. PSYCHOLOGY
Edward C. Tolman (1886-1959) published *Purposive Behavior in Animals and Men* (New York).

1932. ZOOLOGY / ORNITHOLOGY
Waldo Lee McAtee (1883-1962) published "Effectiveness in Nature of the So-called Protective Adaptations in the Animal Kingdom, Chiefly as Illustrated by the Food Habits of Nearctic Birds," *Smithsonian Miscellaneous Collections* 85, no. 7.

1932. ZOOLOGY / PHYSIOLOGY
Allan Winter Rowe (1879-1934) published *The Differential Diagnosis of Endocrine Disorders* (Baltimore).

1932. ZOOLOGY / PHYSIOLOGY
Walter Bradford Cannon's (1871-1945) concept of homeostasis was put into popular form in his *The Wisdom of the Body* (New York). Cannon had introduced the term in a 1926 paper, "Physiological Regulation of Normal States: Some Tentative Postulates Concerning Biological Homeostatics," in *Jubilee Volume for Charles Richet* (Paris), 91-93.

1932. ZOOLOGY / PHYSIOLOGY
Robert W. Bates (1904-), Simon W. Dykshorn, and Oscar Riddle (1877-1968) isolated a substance that could lead to milk secretion (lactation). They published "Prolactin, a New and Third Hormone of the Anterior Pituitary," *Anatomical Record* vol. 54.

1932 (July 1). ORGANIZATIONS—SOCIETIES AND ASSOCIATIONS / PHILOSOPHY
The Philosophy of Science Association was established.

1933. AWARDS AND PRIZES / MEDICINE, PHYSIOLOGY
Thomas Hunt Morgan (1866-1945) won the Nobel Prize in medicine or physiology. He was recognized "for his discoveries concerning the role played by the chromosome in heredity."

1933. CHEMISTRY
Heavy water was produced by Gilbert N. Lewis (1875-1946).

1933. CHEMISTRY / PHYSICS
Richard McLean Badger (1896-1974) arrived at an equation relating force and internuclear distance in a diatomic molecule. It came to be known as "Badger's rule." His "A Relation Between Internuclear

Distances and Bond Force Constants," *Journal of Chemical Physics* 2 (1934): 128-131 relates to this work.

1933. GENERAL OR MISCELLANEOUS
The Emergency Committee in Aid of Displaced German (later, Foreign) Scholars was established. The work of the Committee included the placement of Jewish and other refugee scientists as well as scholars in other fields. Between 1933 and 1939, the Rockefeller Foundation contributed approximately $500,000 to the effort to find and fund academic positions for the refugees.

1933. GENERAL OR MISCELLANEOUS / MATHEMATICS
Amalie Emmy Noether (1882-1935), who is considered to have surpassed in importance all previous women mathematicians, left Germany and came to the United States. During her short life in the United States, she was affiliated with Bryn Mawr College and the Institute for Advanced Study.

1933. GENERAL OR MISCELLANEOUS / PHYSICS
Albert Einstein (1879-1955) left Germany and joined the Institute for Advanced Study at Princeton. He took American citizenship in 1940.

1933. GENETICS
Theophilus Shickel Painter (1889-1969) published "A New Method for the Study of Chromosome Rearrangements and Plotting of Chromosome Maps," *Science* 78:585-586. In this work, Painter drew upon his innovative use of the large salivary gland chromosomes of the fruit fly to confirm that genes were particular physical entities ranked on chromosomes.

1933. GEOLOGY
Walter Herman Bucher (1889-1965) published *The Deformation of the Earth's Crust* (Princeton), in which he dealt with the origin of orogenic belts.

1933. ORGANIZATIONS / ASTRONOMY, COMPUTERS AND INFORMATION SCIENCE
Wallace John Eckert (1902-1971) established the T.J. Watson Astronomical Computing Bureau, a joint venture of Columbia

University, American Astronomical Society, and IBM (whose head, Thomas J. Watson, 1874-1956, supported the venture.)

1933. PERIODICALS AND PUBLISHING / CHEMISTRY, PHYSICS
The *Journal of Chemical Physics* began publication. Its founding and the contraction in size and scope of the *Journal of Physical Chemistry* when it was taken over from Wilder D. Bancroft (1867-1953) by the American Chemical Society at about the same time, signified that the study of relations of chemistry and physics at that time tended to favor the perspective of the physicists. Harold Clayton Urey (1893-1981) was the first editor of the *Journal of Chemical Physics*, serving until 1940.

1933. TECHNOLOGY AND INVENTION / ELECTRICITY AND ELECTRONICS
Edwin H. Armstrong (1890-1954) patented radio frequency modulation (FM). In 1939, he built the first FM station.

1933. ENVIRONMENT AND CONSERVATION
Aldo Leopold (1886-1948) published an innovative textbook, *Game Management* (New York), which drew upon systems ecology and population dynamics.

1933 (May 18). GOVERNMENT—FEDERAL
The Tennessee Valley Authority was established by Congress.

1933 (July 31). GOVERNMENT—FEDERAL
A Science Advisory Board in the federal government was created by an Executive Order of President Franklin D. Roosevelt. Karl T. Compton (1887-1954),president of the Massachusetts Institute of Technology, was made chair. Though deriving its existence and purpose from the National Academy of Sciences-National Research Council, there was little government funding and the chief support in the first year came from the Rockefeller Foundation. The mission of the Board included a study of the entire federal government science structure. It was consulted by some government agencies for advice. At the end of 1935, the Board was replaced by a Government Relations and Science Advisory Committee in the National Academy of Sciences.

1934. ANTHROPOLOGY AND ETHNOLOGY
Ruth Benedict (1887-1948) published a comparative study resulting from her work among the Zuni and Hopi, *Patterns of Culture* (Boston and New York).

1934. AWARDS AND PRIZES / CHEMISTRY
Harold Clayton Urey (1893-1981) won the Nobel Prize in chemistry "for his discovery of heavy hydrogen" (i.e., deuterium).

1934. AWARDS AND PRIZES / MEDICINE, PHYSIOLOGY
George Richards Minot (1885-1950), William Parry Murphy (1892-1987), and George Hoyt Whipple (1878-1976) won the Nobel Prize in medicine or physiology. They were honored "for their discoveries concerning liver therapy in cases of anaemia."

1934. BIOCHEMISTRY AND MOLECULAR BIOLOGY
Robert Russell Bensley (1867-1956), at the University of Chicago, perfected a technique that made it possible to isolate cell mitochondria for purposes of refined chemical analysis.

1934. ORGANIZATIONS—INDUSTRY / ELECTRICITY AND ELECTRONICS
As director of physical research at Bell Telephone Laboratories, Harvey Fletcher (1884-1981) headed a group that devised stereophonic sound.

1934. ORGANIZATIONS—SOCIETIES AND ASSOCIATIONS
The National Association of Science Writers was established.

1934. PARAPSYCHOLOGY
Joseph Banks Rhine (1895-1980) published *Extra-Sensory Perception* (Boston).

1934. PHYSICS
Walter Elsasser (1904-1991) theorized a nuclear shell of protons and neutrons, modeled on electron shells of the atom, as a means of explaining the stability of some nuclei. This was one of several competing models of nuclear structure.

1934. ZOOLOGY / ORNITHOLOGY
Roger Tory Peterson (1908-) published *Field Guide to the Birds: Giving Field Marks of All Species Found in Eastern North America* (Boston and New York).

1934. OCEANOGRAPHY / INSTRUMENTS AND INSTRUMENTATION
A bathysphere was used by (Charles) William Beebe (1877-1962) and Otis Barton (1899-1992) in exploration of the ocean bottom, attaining the depth of 3,028 feet.

1935. ASTRONOMY
Subrahmanyan Chandrasekhar (1910-1995) suggested a limiting stellar mass, above which the temperature of a star starts to increase and resulting in its destruction as a supernova explosion. It came to be known as the Chandrasekhar limit.

1935. BIOCHEMISTRY AND MOLECULAR BIOLOGY
Tobacco mosaic virus was crystallized by Wendell Meredith Stanley (1904-1971), which helped to show that a disease-causing agent might have chemical properties. In 1946, he was awarded a Nobel Prize for the achievement.

1935. BIOCHEMISTRY AND MOLECULAR BIOLOGY
Edward C. Kendall's (1886-1972) work with cortical hormones resulted in the isolation of what later became known as cortisone.

1935. BOTANY
Albert S. Hitchcock (1865-1935) published *Manual of the Grasses of the United States* (Washington, D.C.). Hitchcock was assisted by Agnes Chase (1869-1963); she published a second edition in 1951.

1935. CHEMISTRY
Louis Fieser (1899-1977) published *Experiments in Organic Chemistry* (Boston and New York), an influential laboratory manual. There were later editions, in 1964 becoming *Organic Experiments* (Boston).

1935. CHEMISTRY
Henry Eyring (1901-1981) published "The Activated Complex in Chemical Reactions," *Journal of Chemical Physics* 3:107-115. This was

the first publication on what later became designated as conventional transition-state theory. The work that this paper represented came to have important implications for studies of the rate and the nature of chemical, physical, and biological processes. The transition-state ideas were given extended treatment in Eyring, S. Glasstone, and K.J. Laidler, *The Theory of Rate Processes* (New York, 1941).

1935. CHEMISTRY
Percy Lavon Julian (1899-1975) and Josef Pikl (1904-?) accomplished total synthesis of physostigmine, a natural substance significant in glaucoma treatment. It was reported in "Studies in the Indole Series, V. The Complete Synthesis of Physostigmine (Eserine)," *Journal of American Chemical Society* 57 (1935):755-757.

1935. GENERAL OR MISCELLANEOUS / PHILOSOPHY
Logical positivist Rudolf Carnap (1891-1970) came to the United States from Prague.

1935. GENERAL OR MISCELLANEOUS / PHYSICS
German-born physicist Hans Albrecht Bethe (1906-) came to the United States, by way of England, and to a professorship at Cornell University.

1935. GENERAL OR MISCELLANEOUS / PSYCHOLOGY
Gestalt psychologist Wolfgang Kohler (1887-1967) came to the United States from Germany and became professor at Swarthmore College.

1935. GEOLOGY
Charles Schuchert (1858-1942) published volume 1 of his *Historical Geology of North America*; it bore the title *Historical Geology of the Antillean-Caribbean Region* (New York and London). A second volume appeared posthumously. Volume 3 was not completed.

1935. GEOLOGY / SEISMOLOGY
The work of Charles Francis Richter (1900-1985) and his colleague Beno Gutenberg (1889-1960), on the measurement of the magnitude or strength of earthquakes, produced the Richter scale. In the years 1934-1939, Gutenberg and Richter published "On Seismic Waves," *Beitrage zur Geophysik* 43 (1934), 45 (1935), 47 (1936), 54 (1939).

Chronology

1935. GOVERNMENT—FEDERAL
The National Resources Board, a central planning agency in the U.S. Department of Interior, asked the National Academy of Sciences, the Social Science Research Council, and the American Council on Education to nominate members to a science committee. The committee was chaired by Edwin B. Wilson (1879-1964) and the first meeting was in March 1935. A significant aspect of the committee was the inclusion of the social sciences and education along with the natural sciences. In July 1937, President Franklin D. Roosevelt approved a thorough-going study of the relations of the federal government to science and in 1938-1940 the report, *Research--A National Resource*, was issued.

1935. GOVERNMENT—FEDERAL / AGRICULTURE
The Soil Conservation Service in the Department of Agriculture was established, with Hugh H. Bennett (1881-1960) at its head. A response especially to problems in the West, the Service tended to dwell on application of knowledge to specific erosion situations and related concerns rather than research, which was carried out elsewhere in the Department.

1935. GOVERNMENT—FEDERAL / MEDICINE
The Social Security Act included authorization for expenditure of up to $2,000,000 annually by the National Institutes of Health for research relating to chronic disease and sanitation.

1935. MEDICINE / DISEASE
Hans Zinsser (1878-1940) published *Rats, Lice and History* (Boston), a work on typhus.

1935. ORGANIZATIONS—SOCIETIES AND ASSOCIATIONS / ENVIRONMENT AND CONSERVATION
The Wilderness Society was established. (Among the founders was Aldo Leopold, 1886-1948.)

1935. ORGANIZATIONS—SOCIETIES AND ASSOCIATIONS / GEOLOGY
The Limnological Society of America was founded. (The first president was Chancey Juday, 1871-1944.)

1935. ORGANIZATIONS—SOCIETIES AND ASSOCIATIONS / PSYCHOLOGY
The Psychometric Society was established and in that year became affiliated with the American Psychological Association. The new society began publishing *Psychometrika* in 1936.

1935. ORGANIZATIONS—SOCIETIES AND ASSOCIATIONS / STATISTICS
The Institute of Mathematical Statistics was established.

1935. PHYSICS
Samuel King Allison (1900-1965) and Arthur Holly Compton (1892-1962) published *X-Rays in Theory and Experiment* (New York). Allison was the acknowledged primary author; however, the work was designated as a second edition of Compton's *X-Rays and Electrons: An Outline of Recent X-Ray Theory* (New York, 1926).

1935. PHYSICS
Edward Uhler Condon (1902-1974) and George H. Shortley (1910-1980) published *The Theory of Atomic Spectra* (Cambridge, England), the standard work on the subject for many years thereafter.

1935. PHYSIOLOGY
Rudolf Schoenheimer (1898-1941) began publication of his researches on the use of isotopes in metabolic studies. His studies, conducted with various coworkers, appeared especially in the *Journal of Biological Chemistry*.

1935. MICROBIOLOGY AND MICROSCOPY / PATHOLOGY
Frederick Parker Gay (1874-1939), with others, published *Agents of Disease and Host Resistance* (Springfield, Ill.), the leading work on bacteriology and immunology of the time.

1935 (March 12). GENERAL OR MISCELLANEOUS / PHYSICS
Physicist Michael Idvorsky Pupin (b.1858) died in New York, New York.

1936. AWARDS AND PRIZES / PHYSICS
Carl David Anderson (1905-1991) won the Nobel Prize in physics. He was honored "for his discovery of the positron." The prize was shared with Victor Franz Hess (1883-1964), at that time a professor in Europe;

he came to the United States ca.1958 and became an American citizen. Hess was honored "for his discovery of cosmic radiation."

1936. CHEMISTRY
Vitamin B1 (thiamine) was synthesized by Robert R. Williams (1886-1965).

1936. GENERAL OR MISCELLANEOUS / ENGINEERING AND APPLIED SCIENCE
The Boulder Dam on the Colorado River was completed. (It was renamed Hoover Dam in 1947.)

1936. HISTORY OF SCIENCE—FIELD OF STUDY
Arthur Oncken Lovejoy (1873-1962) published *The Great Chain of Being* (Cambridge, Mass.), a milestone in the study of the history of ideas.

1936. ORGANIZATIONS / SEISMOLOGY
The Carnegie Institution's Seismological Laboratory became part of the California Institute of Technology.

1936. ORGANIZATIONS—FOUNDATIONS
The Ford Foundation was incorporated in Michigan.

1936. ORGANIZATIONS—SOCIETIES AND ASSOCIATIONS / ENVIRONMENT AND CONSERVATION
The National Wildlife Federation was founded.

1936. ORGANIZATIONS—SOCIETIES AND ASSOCIATIONS / PSYCHOLOGY
The Society for the Psychological Study of Social Issues was established and the next year undertook affiliation with the American Psychological Association.

1936. PHYSICS
George Gamow (1904-1968) and Edward Teller (1908-) discovered what became known as the Gamow-Teller selection rule for beta decay.

1936. TECHNOLOGY AND INVENTION / ELECTRICITY AND ELECTRONICS
George Harold Brown (1908-1987), at Radio Corporation of America, invented the turnstile antenna. It was used for broadcasting of television and FM radio.

1936. ZOOLOGY / EMBRYOLOGY
Gregory Goodwin Pincus (1903-1967) published *The Eggs of Mammals* (New York).

1936. ZOOLOGY / PHYSIOLOGY
Wallace Osgood Fenn (1893-1971) presented results of his research on muscle physiology, that explained the processes of intracellular exchanges of potassium and sodium during stimulation and recovery. A related publication, with Doris M. Cobb, was: "Electrolyte Changes in Muscle During Activity," *American Journal of Physiology* 115 (1936): 345-356.

1936. MEDICINE / PHYSIOLOGY
Karl Landsteiner (1868-1943) published *The Specificity of Serological Reactions* (Springfield, Ill., and Baltimore). A German edition had appeared in Berlin in 1933 and a revised edition was published in 1945.

1936. PHYSICS
The first centrifuge separation of isotopes (of chlorine) was achieved by Jesse W. Beams (1898-1977).

1936-1937. PHYSICS
Hans Albrecht Bethe (1906-) published a series of papers that reviewed knowledge up to that point on nuclear physics. These became known as "Bethe's Bible."

1936-1937. ASTRONOMY
Benjamin Boss (1880-1970) published *General Catalogue*, that gave the position of 33,342 stars (Carnegie Institution of Washington, Publication 468).

1937. ASTRONOMY / INSTRUMENTS AND INSTRUMENTATION
The first radiotelescope, 31 feet in diameter, was built by Grote Reber (1911-).

1937. AWARDS AND PRIZES / PHYSICS
Clinton Joseph Davisson (1881-1958) won the Nobel Prize in physics (shared with George P. Thomson of England) "for their experimental discovery of the diffraction of electrons by crystals," work that confirmed electron waves. Davisson's work so honored had been done in 1927, with the collaboration of Lester H. Germer (1896-1971).

1937. EVOLUTION / GENETICS
Theodosius Dobzhansky (1900-1975) published *Genetics and the Origin of Species* (New York), a work of the first importance in the history of evolutionary theory. Subsequent, revised editions, appeared in 1941 and 1951 and in a much changed state as *Genetics of the Evolutionary Process* (New York) in 1970. The book helped substantially to effect the union of genetics and evolution and to make it current among biologists.

1937. GENETICS
Tracy Morton Sonneborn (1905-1981) discovered mating types in Paramecium, in which mating was limited to other clones but not with one's own. This finding made it possible to carry out genetic studies of the organism, which Sonneborn did in the decade that followed.

1937. GOVERNMENT—FEDERAL
The National Resources Committee issued its report, *Technological Trends and National Policy* (Washington, D.C.). Prepared under the guidance of sociologist William F. Ogburn (1886-1959), the report argued for a governmental role in guiding the application of scientific knowledge in order to deter adverse economic effects. The National Resources Committee was a cabinet advisory body.

1937. GOVERNMENT—FEDERAL / MEDICINE
The National Cancer Institute was authorized by Congress. The new Institute addressed its charge through grants-in-aid, advanced specialized training, and fellowships within the Public Health Service. A part of the National Institutes of Health, it signified the research mission of NIH. A National Advisory Cancer Council was established to assist in the formulation of policies for the Institute.

1937. MATHEMATICS
Eric Temple Bell (1883-1960) published *Men of Mathematics* (New York).

1937. MEDICINE / DISEASE
Max Theiler (1899-1972) devised a vaccine for yellow fever. In 1951, he won the Nobel Prize in medicine or physiology for this work.

1937. ORGANIZATIONS—HOSPITALS
The first blood bank in the United States was established at Cook County Hospital (Chicago).

1937. ORGANIZATIONS—INDUSTRY
Edwin Herbert Land (1909-1991) founded the Polaroid Corporation to develop his discovery of a polarizing filter.

1937. PARAPSYCHOLOGY
Joseph Banks Rhine (1895-1980) published *New Frontiers of the Mind* (New York), on his experiments on extrasensory perception.

1937. PERIODICALS AND PUBLISHING / ENVIRONMENT AND CONSERVATION
The *Journal of Wildlife Management* was established. Waldo Lee McAtee (1883-1962) was the first editor.

1937. PHYSICS
The mu meson (muon) was discovered in cosmic radiation by Carl David Anderson (1905-1991).

1937. PHYSICS / CHEMISTRY
Paul Sophus Epstein (1883-1966) published *Textbook of Thermodynamics* (New York).

1937. PSYCHOLOGY
Gordon Willard Allport (1897-1967) published *Personality: A Psychological Interpretation* (New York).

1937. ZOOLOGY / PHYSIOLOGY
Homer William Smith (1895-1962) published *The Physiology of the Kidney* (New York).

1937. ZOOLOGY—HUMAN / GENETICS, PSYCHOLOGY
Horatio Hackett Newman (1875-1957), Frank N. Freeman (1880-1961), and Karl J. Holzinger (1892-1954) published *Twins: A Study of Heredity and Environment* (Chicago).

1937 (April 29). GENERAL OR MISCELLANEOUS / CHEMISTRY
Chemist Wallace Hume Carothers (b.1896) died in Philadelphia, Pennsylvania.

1938. ANTHROPOLOGY AND ETHNOLOGY
Franz Boas (1858-1942), with others, published *General Anthropology* (Boston and New York).

1938. CHEMISTRY
Wendell Mitchell Latimer (1893-1955) published *The Oxidation States of the Elements and Their Potentials in Aqueous Solutions* (New York). A second edition appeared in 1952.

1938. COMPUTERS AND INFORMATION SCIENCE / MATHEMATICS
Claude Elwood Shannon (1916-) published "A Symbolic Analysis of Relay and Switching Circuits," *Transactions of American Institute of Electrical Engineers* 57:713-728. The paper was fundamental to the mathematical theory of information.

1938. GENERAL OR MISCELLANEOUS
A radio adaptation of H.G. Wells' *War of the Worlds*, presented by (George) Orson Welles (1915-1985), caused a national panic.

1938. GENETICS / BOTANY
Karl Sax (1892-1973) began his published studies of the effects of X-rays on chromosomes in plants. In that year, he issued "Chromosome Aberrations Induced by X-rays," *Genetics* 23:494-516. He became noted as one of the founders of such investigations, conducting his experiments especially on *Tradescantia paludosa* (a species of spiderwort).

1938. HISTORY OF SCIENCE—FIELD OF STUDY
Robert K. Merton (1910-) published "Science, Technology & Society in Seventeenth Century England," *Osiris* 4:360-632. It gave rise to what became known as the "Merton thesis," arguing for a strong link between economic, military, and religious (especially Puritan) interests and outlook and the emphases and institutionalization of science in England at the time.

1938. ORGANIZATIONS—ACADEMIC / STATISTICS
Jerzy Neyman (1894-1981) came to the United States from England and assumed a position at the University of California at Berkeley. He established a statistical laboratory at Berkeley and in 1945 held the first of the conferences on mathematical statistics and probability that occurred there every five years until 1970. In 1955, a separate department of statistics was set up at Berkeley.

1938. ORGANIZATIONS—BOTANICAL GARDENS
Fairchild Tropical Gardens, near Miami, were founded by botanist David Grandison Fairchild (1869-1954).

1938. ORGANIZATIONS—FOUNDATIONS / BIOCHEMISTRY AND MOLECULAR BIOLOGY
Warren Weaver (1894-1978) used the term Molecular Biology in reference to a grant program of the Rockefeller Foundation.

1938. ORGANIZATIONS—INDUSTRY / CHEMISTRY
DuPont chemical company researchers discovered Teflon.

1938. ORGANIZATIONS—INDUSTRY / COMPUTERS AND INFORMATION SCIENCE
Bell Telephone Laboratories began development of a computer, completed in 1939 as the Complex Number Computer.

1938. ORGANIZATIONS—INDUSTRY / ELECTRICITY AND ELECTRONICS
The first television receivers were manufactured by DuMont Laboratories. The company was established by Allen Balcom DuMont (1901-1965).

1938. ORGANIZATIONS—SOCIETIES AND ASSOCIATIONS
The American Association of Scientific Workers was founded; it became an affiliate of the American Association for the Advancement of Science. It was intended for the promotion of understanding of the social relations of science and the application of science to the solution of social problems, the first American society with such a focus.

1938. PERIODICALS AND PUBLISHING
The *International Encyclopedia of Unified Science* (Chicago) began publication.

1938. PHYSICS
Richard Chace Tolman (1881-1948) published *Principles of Statistical Mechanics* (Oxford). An innovative work, he based the study on quantum rather than on classical mechanics.

1938. PHYSICS / ASTRONOMY
Hans Albrecht Bethe (1906-) proposed that the energy of stars was caused by nuclear fusion of hydrogen into helium.

1938. PSYCHOLOGY
Burrhus Frederic (B.F.) Skinner (1904-1990) published *The Behavior of Organisms* (New York and London).

1938. ZOOLOGY
Warder Clyde Allee (1885-1955) published *The Social Life of Animals* (New York), in which he argued that cooperation was as fundamental as competition for existence.

1938. ZOOLOGY / PHYSIOLOGY, PSYCHOLOGY
Stanley Smith Stevens (1906-1973) and Hallowell Davis (1896-1992) published *Hearing: Its Psychology and Physiology* (New York).

1938. PHYSICS
Leo Szilard (1898-1964) was in the United States from this date, having gone from Germany to England in 1933. Soon after his arrival, he began research on nuclear fission. In 1939, the self-sustaining character of fission through chain reactions was proven by Leo Szilard and Walter

Zinn (1906-), confirming a 1938 observation in Germany by Otto Hahn and Fritz Strassman.

1938. TECHNOLOGY AND INVENTION / ELECTRICITY AND ELECTRONICS
By this date, Chester Carlson (1906-1968) had developed a workable document reproduction system which was a model for the xerox process. Carlson initially called it electrophotography and he was able to establish a patent for his invention. He was assisted by physicist Otto Kornei (1903-?). The invention was acquired in 1947 by the Haloid Company (Rochester, New York) and it was 1959 before they had a commercially feasible product. Haloid became the Xerox Corporation.

1938 (February 21). GENERAL OR MISCELLANEOUS / ASTRONOMY, PHYSICS
Astrophysicist George Ellery Hale (b.1868) died in Pasadena, California.

1938 (June 24). GOVERNMENT—FEDERAL / CHEMISTRY
The Food, Drug and Cosmetic Act was passed by Congress to replace the 1906 Pure Food Act. The new legislation required detailed labeling of products.

1939. ASTRONOMY
Subrahmanyan Chandrasekhar (1910-1995) published *Introduction to the Study of Stellar Structure* (Chicago).

1939. AWARDS AND PRIZES / PHYSICS
Ernest Orlando Lawrence (1901-1958) won the Nobel Prize in physics. He was recognized "for the invention and development of the cyclotron and for results obtained with it, especially with regard to artificial radioactive elements."

1939. BIOLOGY—GENERAL
Charles Vincent Taylor (1885-1946) organized an international meeting at Stanford University to observe the centennial of the Schleiden and Schwann cell theory. The next year, Forest R. Moulton (1872-1952) edited the resulting work, *The Cell and Protoplasm*, published for the American Association for the Advancement of Science.

Chronology 253

1939. Biology—General / Ecology
Plant ecologist Frederic Edward Clements (1874-1945) and animal ecologist Victor Ernest Shelford (1877-1968) published the textbook, *Bio-Ecology* (New York). This work introduced the idea of the biome as a fundamental unit of the ecological landscape.

1939. Biology—General / Embryology
Ernest Everett Just (1883-1941) published *Biology of the Cell Surface* (Philadelphia), which included his ideas on the important role of ectoplasm in the instigation of animal egg cell development.

1939. Chemistry
Edward Adelbert Doisy (1893-1986) isolated two compounds of vitamin K--i.e., K_1 and K_2 and later delineated the chemical structures of both and synthesized K1. Louis Fieser (1899-1977) achieved a synthesis of Vitamin K_1 in 1939.

1939. Chemistry / Physics
Linus Carl Pauling (1901-1994) published *The Nature of the Chemical Bond, and Structure of Molecules and Crystals* (Ithaca, N.Y.). A revised edition appeared in 1960.

1939. Engineering and Applied Science / Statistics
Walter Andrew Shewhart (1891-1967) published *Statistical Method from the Viewpoint of Quality Control* (Washington, D.C.). This book extended his earlier work on quality control for manufacturing to the realm of measurement for scientific purposes. [For earlier work, see 1924.]

1939. General or Miscellaneous
Vannevar Bush (1890-1974) became president of the Carnegie Institution of Washington.

1939. General or Miscellaneous / Paleontology
Winifred Goldring (1888-1971) was appointed state paleontologist of New York, the first woman to hold such an office.

1939. GENERAL OR MISCELLANEOUS / PHYSICS
Enrico Fermi (1901-1954), after presentation of the Nobel Prize in physics at Stockholm, moved directly to the United States, where he took a position at Columbia University.

1939. GOVERNMENT—FEDERAL / AERONAUTICS
Ames Aeronautical Laboratory was established by the National Advisory Committee for Aeronautics. It later became Ames Research Center (Moffett Field, California), a facility of the National Aeronautics and Space Administration.

1939. GOVERNMENT—FEDERAL / NUCLEAR POWER AND WEAPONS
Albert Einstein (1879-1955), with the urging of Leo Szilard (1898-1964), Edward Teller (1908-), and Eugene Paul Wigner (1902-1995), wrote to President Franklin Roosevelt warning of the possibility of German development of nuclear fission and of atomic weapons. (However, Einstein was not privy to subsequent work that led to development of the bomb.) In this year, Lyman James Briggs (1874-1963), head of the National Bureau of Standards, was appointed by President Franklin Roosevelt to head a top-secret Advisory Committee on Uranium. Other scientist members were Szilard, Teller, and Wigner, along with representatives from the military and the executive office. At least in part due to the fact that the committee did not proceed as quickly in the direction of atomic bomb development as some wished, it was merged into the National Defense Research Committee in June 1940. James B. Conant (1893-1978) was made chair in June 1942 and soon thereafter the committee's functions were superseded by the U.S. Army's Manhattan Project.

1939. MATHEMATICS
Abraham Adrian Albert (1905-1972) published *Structure of Algebras* (Providence, R.I.).

1939. MICROBIOLOGY AND MICROSCOPY / PATHOLOGY
Rene Jules Dubos (1901-1982), at the Rockefeller Institute, discovered tyrothricin, the first antibiotic to be commercially produced.

1939. ORGANIZATIONS—INDUSTRY / CHEMISTRY
Based on earlier work of Wallace Hume Carothers (1896-1937), DuPont Company began the manufacture of nylon at Seaford, Delaware.

1939. ORGANIZATIONS—SOCIETIES AND ASSOCIATIONS
Frank Baldwin Jewett (1879-1949) of Bell Telephone Laboratories became the first industrial scientist elected president of the National Academy of Sciences. He served until 1947.

1939. PALEONTOLOGY
Percy Edward Raymond (1879-1952) published *Prehistoric Life* (Cambridge, Mass.).

1939. PERIODICALS AND PUBLISHING / MATHEMATICS
The *Mathematical Reviews* was established. Begun in reaction to the Nazi control of the German equivalent, and international in focus, newly arrived refugee from Europe, William Feller (1906-1970), became executive editor of the publication.

1939. PHYSICS
Felix Bloch (1905-1983), with Luis W. Alvarez (1911-1988), measured the magnetic moment of the neutron. (Bloch came to the United States from Germany in 1934.)

1939. PHYSICS
John Archibald Wheeler (1911-) and Danish physicist Niels Bohr devised the liquid-drop model of the atomic nucleus.

1939. PHYSICS / INSTRUMENTS AND INSTRUMENTATION
Donald William Kerst (1911-1993) invented the betatron for the acceleration of electrons used to dislodge subatomic particles from the nucleus.

1939. MICROBIOLOGY AND MICROSCOPY / PATHOLOGY
Selman Abraham Waksman (1888-1973), at Rutgers University, and his associates began a systematic search for antibiotic organisms in soil. From this program, Waksman over the years discovered eighteen such substances, including actinomycin, streptomycin, and neomycin. Waksman introduced the term antibiotic in 1941.

1939. PERIODICALS AND PUBLISHING / MATHEMATICS
The influential Princeton Mathematical Series of books began publication. It was established by Albert William Tucker (1905-).

1939 (March). GOVERNMENT—FEDERAL / NUCLEAR POWER AND WEAPONS
George Braxton Pegram (1876-1958) wrote to Admiral S. C. Hooper, head of the Naval Research Committee, promoting the need for research on nuclear energy. This was the first advice to a governmental agency from a scientist suggesting such work.

1939 (March 3). GENERAL OR MISCELLANEOUS / ZOOLOGY
Cytologist and embryologist Edmund Beecher Wilson (b.1856) died in New York, New York.

1939 (April 30). ORGANIZATIONS—FAIRS AND EXPOSITIONS
The New York World's Fair opened with the theme of "The World of Tomorrow." On May 10, 1940, it opened for a second year.

1939 (May 5). ORGANIZATIONS—OBSERVATORIES
The McDonald Observatory, with an 82-inch telescope, one of the largest in the world, was dedicated. Located in Texas, the observatory was the joint project of the University of Chicago and the University of Texas. Otto Struve (1897-1963), of Chicago, was the first director.

1939 (summer). PHYSICS
The capstone to the cosmic ray researches of Arthur Holly Compton (1892-1962) took place at a conference at the University of Chicago.

1939 and 1940. METEOROLOGY AND CLIMATOLOGY
Meteorologist Carl-Gustaf Arvid Rossby (1898-1957), with his associates, devised a simple formula (the Rossby equation) relating to the propagation speed of waves in the atmosphere.

1940. CHEMISTRY
Philip Hauge Abelson (1913-) and Edwin Mattison McMillan (1907-1991) discovered neptunian, the radioactive element 93. It was the first element discovered with an atomic number greater than that of uranium.

1940. CHEMISTRY / NUCLEAR POWER AND WEAPONS
Glenn Theodore Seaborg (1912-) and others at the University of California-Berkeley, using the cyclotron there, discovered element 94, which came to be called plutonium. The discovery suggested the possibility of using the fissionable plutonium as the basis of an atomic bomb. Subsequently, between 1944 and 1958, Seaborg and colleagues at Berkeley participated in the discovery of all of the transuranium elements from 95 to 102.

1940. GENERAL OR MISCELLANEOUS / MATHEMATICS
Mathematician and logician Kurt Friedrich Godel (1906-1978) left Austria and came permanently to the United States, where he became affiliated with the Institute for Advanced Study.

1940. GENETICS
Milislav Demerec (1895-1966) and Berwind P. Kaufmann (1897-1975) published *Drosophila Guide* (Washington, D.C.), an essential source for the study of fruit fly genetics in the schools and colleges. It had eight editions as of 1969.

1940. GOVERNMENT—FEDERAL / BIOLOGY—GENERAL
The Fish and Wildlife Service was established in the U.S. Department of the Interior. The Service incorporated the Bureau of Fisheries and the Bureau of Biological Survey.

1940. HISTORY OF SCIENCE—FIELD OF STUDY / MATHEMATICS
Louis Charles Karpinski (1878-1956) published *Bibliography of Mathematical Works Printed in America Through 1850* (Ann Arbor).

1940. ORGANIZATIONS—SOCIETIES AND ASSOCIATIONS / ENGINEERING AND APPLIED SCIENCE
The Television Engineers Institute of America was established.

1940. ORGANIZATIONS—SOCIETIES AND ASSOCIATIONS / PALEONTOLOGY
The Society of Vertebrate Paleontology was established.

1940. ORGANIZATIONS—SOCIETIES AND ASSOCIATIONS / TECHNOLOGY AND INVENTION
The National Inventors Council was established.

1940. PERIODICALS AND PUBLISHING / MATHEMATICS
The influential book series, Annals of Mathematics Studies, began publication. It was established by Albert William Tucker (1905-) and issued by Princeton University Press.

1940. ZOOLOGY
Libbie Henrietta Hyman (1888-1969) published the first volume of her work on invertebrates, *The Invertebrates, Protozoa Through Ctenophora* (New York). Subsequent issuances were: volume 2, *Platyhelminthes and Rhynchocoela* (1951); 3, *Acanthocephala, Aschelminthes, and Entoprocta* (1951); 4, *Echinodermata* (1955); 5, *Smaller Coelomate Groups* (1959); and 6, *Mollusca I* (1967).

1940. ZOOLOGY—HUMAN / PHYSIOLOGY
Karl Landsteiner (1868-1943), Philip Levine (1900-1987), and Alexander S. Wiener (1907-1976) reported their discovery of the rhesus (or Rh) factor in blood of humans.

1940. GENERAL OR MISCELLANEOUS
George Gamow (1904-1968) published the first of his popular books on science, *Mr. Tompkins in Wonderland; or, Stories of c, G, and h* (New York).

1940 (June 27). GOVERNMENT—FEDERAL
The National Defense Research Committee was established by President Franklin Roosevelt for research related to weapons and was supported by the president's emergency funds. Vannevar Bush (1890-1974), who had promoted the idea, was made Committee chair. The work of the NDRC was done through contracts with universities and industry; the Committee did not have laboratory facilities of its own.

1940 (June 30). GOVERNMENT—FEDERAL / METEOROLOGY AND CLIMATOLOGY
The U.S. Weather Bureau was transferred from the Department of Agriculture to the Department of Commerce.

1940 (November 10). ORGANIZATIONS / MILITARY AND WAR, PHYSICS
The facility that was to become known as the Radiation Laboratory, for research on radar, was established under contract with the National

Defense Research Committee (NDRC) at the Massachusetts Institute of Technology. Lee DuBridge (1901-1994) was made head of the Laboratory. By the end of the War, the so-called Rad Lab employed nearly 4000 individuals.

1940-1941. GENERAL OR MISCELLANEOUS
The work of the National Roster of Scientific and Specialized Personnel, under the direction of Leonard Carmichael (1898-1973), got underway with a national survey.

1940-1945. COMPUTERS AND INFORMATION SCIENCE / ASTRONOMY
As director of the U.S. Nautical Almanac Office, Wallace John Eckert (1902-1971) applied machine computation equipment in the preparation of the *American Ephemeris and Nautical Almanac*.

1940-1945. GENERAL OR MISCELLANEOUS / PHYSICS
During the war period, Wolfgang Pauli (1900-1958), who won the Nobel Prize in physics in 1945, was at the Institute for Advanced Study, Princeton. He became a United States citizen in 1946 but returned to Zurich where he had been a professor since 1928 and remained there for the remainder of his life.

1941. ASTRONOMY
Frederick Hanley Seares (1873-1964), with Frank E. Ross (1874-1966) and Mary C. Joyner (1889-?), published *Magnitudes and Colors of Stars North of +80°* (Papers of Mount Wilson Observatory, vol. 6; Carnegie Institution of Washington Publication, no. 532).

1941. BIOCHEMISTRY AND MOLECULAR BIOLOGY
George Wells Beadle (1903-1989) and Edward Lawrie Tatum (1909-1975) proved that biochemical reactions in cells are controlled by genes. They used the mold *Neurospora* in their research. From this they developed their one-gene, one-enzyme hypothesis.

1941. GEOLOGY / SEISMOLOGY
Charles Francis Richter (1900-1985) and Beno Gutenberg (1889-1960) published the standard reference, *Seismicity of the Earth* (New York). A revised and expanded edition appeared in 1954.

1941. MATHEMATICS / ASTRONOMY
Aurel Wintner (1903-1958) published *Analytical Foundations of Celestial Mechanics* (Princeton, N.J.).

1941. ORGANIZATIONS—RESEARCH INSTITUTIONS / BIOLOGY—GENERAL
Milislav Demerec (1895-1966) became acting director (after 1943, director) of the Carnegie Institution of Washington's Department of Genetics at Cold Spring Harbor, New York, and director of the nearby Biological Laboratory of the Long Island Biological Association. Under his direction, the two facilities functioned effectively as the Cold Spring Harbor Biological Laboratory.

1941. ORGANIZATIONS—SOCIETIES AND ASSOCIATIONS / ENVIRONMENT AND CONSERVATION
The Soil Conservation Society of America was established.

1941. ORGANIZATIONS—SOCIETIES AND ASSOCIATIONS / GOVERNMENT—FEDERAL
The National Academy of Sciences received its first appropriation from the U.S. Congress.

1941. ORGANIZATIONS—SOCIETIES AND ASSOCIATIONS / HISTORY OF SCIENCE—FIELD OF STUDY
The American Institute for the History of Pharmacy was established.

1941. ORGANIZATIONS—SOCIETIES AND ASSOCIATIONS / RELIGION AND THEOLOGY
The American Scientific Affiliation was founded as a creationist organization.

1941. PHYSICS / FUNDS AND FUNDING
The Rockefeller Foundation gave $1,210,000 for support of a 184-inch cyclotron of Ernest O. Lawrence (1901-1958). This was the largest project supported by the Foundation under Warren Weaver's (1894-1978) aegis.

1941. MATHEMATICS
George David Birkhoff (1884-1944) and Ralph Beatley (1892-1989) published *Basic Geometry* (Chicago), a work that proved influential in high school teaching of geometry.

1941 (February 22). GENERAL OR MISCELLANEOUS / PHYSICS
Physicist Dayton Clarence Miller (b.1866) died in Cleveland, Ohio.

1941 (April). ORGANIZATIONS—SOCIETIES AND ASSOCIATIONS / FUNDS AND FUNDING
The National Academy of Sciences established the National Science Fund for the promotion of science.

1941 (April 13). GENERAL OR MISCELLANEOUS / ASTRONOMY
Astronomer Annie Jump Cannon (b.1863) died in Cambridge, Massachusetts.

1941 (June 28). GOVERNMENT—FEDERAL
The U.S. Office of Scientific Research and Development (OSRD) was established by executive order of President Franklin D. Roosevelt for the planning, coordination, and conduct of defense research. It included the already existing National Defense Research Committee (NDRC) and a new Committee on Medical Research. Unlike the earlier NDRC, the new agency was directly funded by Congress rather than from the emergency funds of the President. Vannevar Bush (1890-1974) was the head of OSRD and was responsible directly to the president. Situated in the Office of Emergency Management, OSRD continued in existence until 31 December 1947; during the war, it oversaw the expenditure of more than 500 billion dollars within all sectors of the scientific and development establishment (academic, industrial, governmental).

1941 (October). GOVERNMENT—FEDERAL / NUCLEAR POWER AND WEAPONS
At the urging of members of the scientific community, which were channeled through Vannevar Bush (1890-1974), President Franklin Roosevelt gave approval to the project to test the feasibility of an atomic bomb. An all-out effort was authorized in June 1942. The project was situated in the Manhattan District of the U.S. Army Engineers and in September 1942, Brigadier General Leslie Groves (1896-1970) was put

in administrative command. J. Robert Oppenheimer (1904-1967) was given charge of the development of a bomb.

1941 (November 6). MILITARY AND WAR / NUCLEAR POWER AND WEAPONS
The National Academy of Sciences Committee on Uranium, chaired by Arthur Holly Compton (1892-1962), issued a report on the military potential of atomic energy.

1942. ASTRONOMY
The earliest maps of the universe that located individual radio sources was produced by Grote Reber (1911-).

1942. COMPUTERS AND INFORMATION SCIENCE
The first functioning computer incorporating vacuum tubes (used for mathematical computation) was developed by John V. Atanasoff (1903-) with the assistance of Clifford E. Berry (1918-). It was designated the ABC (Atanasoff Berry Computer). Atanasoff had begun work on the electronic computer in 1937 and by October 1939 had completed an operational model.

1942. GOVERNMENT—FEDERAL / AGRICULTURE
The Agricultural Research Administration, to coordinate the research activities of the U.S. Department of Agriculture, was established.

1942. MATHEMATICS
Solomon Lefschetz (1884-1972) published *Algebraic Topology* (New York).

1942. ORGANIZATIONS—SOCIETIES AND ASSOCIATIONS / MICROBIOLOGY AND MICROSCOPY
The Electron Microscope Society of America was established.

1942. ORGANIZATIONS—SOCIETIES AND ASSOCIATIONS / ZOOLOGY
The American Association of Zoological Parks and Aquariums was established.

1942 (January 8). GENERAL OR MISCELLANEOUS / ASTRONOMY
Astronomer Heber Doust Curtis (b.1872) died in Ann Arbor, Michigan.

1942 (February). ORGANIZATIONS / MILITARY AND WAR, PHYSICS
The Radio Research Laboratory, for radar countermeasures, was established at Harvard University under Office of Scientific Research and Development contract, with Stanford University engineer Frederick E. Terman (1900-1982) in charge.

1942 (September 10). GENERAL OR MISCELLANEOUS / CHEMISTRY
A report of the Baruch-Compton-Conant Commission on the dangers posed by the lack of a supply of rubber led to an agreement to buy all of Mexico's raw rubber for four years. The report also was prelude to the development of production of synthetic rubber. During the years 1942-1956, the United States government, through the Reconstruction Finance Corporation, supported $56 million in research and development efforts relating to synthetic rubber. The United States surpassed the Germans in production of synthetic rubber by 1945. In 1954, industrial synthesis of natural rubber was achieved.

1942 (December 2). PHYSICS / NUCLEAR POWER AND WEAPONS
Enrico Fermi (1901-1954), Herbert L. Anderson (1914-1988), and Walter Zinn (1906-), with others involved in the Manhattan Project, achieved the first controlled nuclear chain reaction at Stagg Field at the University of Chicago. The nuclear pile used the natural radioactivity of uranium isotope 235 in its usual mix with basic uranium.

1943. AWARDS AND PRIZES / MEDICINE, PHYSIOLOGY
Edward Adelbert Doisy (1893-1986) won the Nobel Prize in medicine or physiology "for his discovery of the chemical nature of vitamin K." (Henrik Dam of Denmark also received the prize for related work.)

1943. AWARDS AND PRIZES / PHYSICS
Otto Stern (1888-1969) (German/American) won the Nobel Prize in physics. He was honored "for his contribution to the development of the molecular ray method and his discovery of the magnetic moment of the proton."

1943. ENGINEERING AND APPLIED SCIENCE / ELECTRICITY AND ELECTRONICS
Frederick E. Terman (1900-1982) published *Radio Engineers Handbook* (New York).

1943. MICROBIOLOGY AND MICROSCOPY / GENETICS
Max Delbruck (1906-1981) and Salvador Edward Luria (1912-1991) conducted research that showed that both a virus and the bacteria that they infect can mutate.

1943. PSYCHOLOGY
Clark Leonard Hull (1884-1952) published *Principles of Behavior* (New York), in which he presented his mathematico-deductive theory based on his experiments on learning.

1943. TECHNOLOGY AND INVENTION / MILITARY AND WAR
The radio proximity fuse for the detonation of explosives, developed by Section T of the National Defense Research Committee, was first used in combat with Japanese aircraft as the target. Section T of the NDRC was under the direction of Merle Antony Tuve (1901-1982).

1943 (January 5). GENERAL OR MISCELLANEOUS / AGRICULTURE, CHEMISTRY
Agricultural chemist and botanist George Washington Carver (b.1861?) died in Tuskegee, Alabama.

1943 (March). GOVERNMENT—FEDERAL / NUCLEAR POWER AND WEAPONS
The Manhattan Project's special weapons laboratory at Los Alamos, New Mexico, was opened under the direction of J. Robert Oppenheimer (1904-1967). That year, Hans Albrecht Bethe (1906-) was appointed to head the theoretical physics section at Los Alamos.

1943 (November). GENERAL OR MISCELLANEOUS / NUCLEAR POWER AND WEAPONS
(Emil Julius) Klaus Fuchs (1911-1988), born in Germany and a British citizen, joined the research staff of the Manhattan Project at Los Alamos and immediately undertook to pass secret information to the Soviet Union. It was only several years later, when he was back in Great Britain, that he was suspected and later convicted of spying.

1943-1952. GEOLOGY
William Frederick Foshag (1894-1956), for the first time, followed the life of a volcano, Paricutin, in Mexico, from its birth in 1943 to

extinction in 1952. He published "Birth and Development of Paricutin Volcano, Mexico," *U.S. Geological Society Bulletin* 965-D (1956): 355-489.

1944. AWARDS AND PRIZES / MEDICINE, PHYSIOLOGY
Joseph Erlanger (1874-1965) and Herbert Spencer Gasser (1888-1963) won the Nobel Prize in medicine or physiology. They were chosen "for their discoveries relating to the highly differentiated functions of single nerve fibres."

1944. AWARDS AND PRIZES / PHYSICS
Isidor Isaac Rabi (1898-1988) won the Nobel Prize in physics "for his resonance method for recording the magnetic properties of atomic nuclei."

1944. BIOCHEMISTRY AND MOLECULAR BIOLOGY
Oswald Theodore Avery (1877-1955), in research carried out at Rockefeller Institute Hospital, demonstrated that, in the bacterial cells in his study, deoxyribonucleic acid (DNA) was the active agent in inherited change. (The results of his study, done with Colin M. MacLeod (1909-1972) and Maclyn McCarty (1911-), appeared in the *Journal of Experimental Medicine* 79 (1944):137-158.) This work brought into question the belief that protein was the active agent of inheritance and gave impetus to concentrated studies of the structure and function of DNA.

1944. BIOLOGY—GENERAL
Franz Schrader (1891-1962) published *Mitosis: The Movements of Chromosomes in Cell Division* (New York). A second edition appeared in 1953.

1944. BOTANY / GENETICS
The first species of plant to be artificially synthesized and capable of growing in a natural environment was produced by George Ledyard Stebbins (1906-).

1944. CHEMISTRY
Louis Fieser (1899-1977) and Mary Peters Fieser (1909-) published the textbook, *Organic Chemistry* (Boston).

1944. CHEMISTRY
Robert B. Woodward (1917-1979) and William Von Eggers Doering (1917-) synthesized quinine. This was the first of the important organic syntheses by Woodward.

1944. COMPUTERS AND INFORMATION SCIENCE
Howard Hathaway Aiken (1900-1973) and engineers from the International Business Machines Corporation completed the Automatic Sequence Controlled Calculator, or Harvard Mark I computer. It was the earliest electromechanical computer.

1944. GENERAL OR MISCELLANEOUS / NUCLEAR POWER AND WEAPONS
In 1943, Samuel Abraham Goudsmit (1902-1978) was made head of Project ALSOS with the mission of uncovering German scientific advances (especially in nuclear physics and work on an atomic bomb). The project personnel followed in the wake of the Allied forces into Europe. ALSOS was in the field from April 1944 to May 1945. In November 1944, Goudsmit uncovered what was considered conclusive proof at Strasbourg that German research on an atomic bomb was not greatly advanced.

1944. MATHEMATICS / ECONOMICS
John von Neumann (1903-1957) and Oskar Morgenstern (1902-1977) published *Theory of Games and Economic Behavior* (Princeton).

1944. MEDICINE / SURGERY
Alfred Blalock (1899-1964) and Helen Taussig (1898-1986) performed the first so-called blue-baby operation as a means to overcome congenital circulatory disease.

1944. MICROBIOLOGY AND MICROSCOPY / PATHOLOGY
Selman A. Waksman (1888-1973) developed streptomycin, an antibiotic that was effective against tuberculosis bacteria.

1944. ORGANIZATIONS—INDUSTRY / CHEMISTRY
Dow Chemical Company introduced Styrofoam.

1944. ORGANIZATIONS—RESEARCH INSTITUTIONS / AERONAUTICS
The Jet Propulsion Laboratory at Pasadena, California, was founded.

1944. ORGANIZATIONS—RESEARCH INSTITUTIONS / BIOLOGY—GENERAL
The Worcester Foundation for Experimental Biology was established by Gregory Goodwin Pincus (1903-1967) and Hudson Hoagland (1899-1983). It came to be noted especially for its work in steroid hormones and reproduction in mammals. In this year, Pincus established the annual Laurentian Hormone Conference and, from 1946 to 1967, edited twenty-three volumes of the proceedings as *Recent Progress in Hormone Research*.

1944. ZOOLOGY—HUMAN / EMBRYOLOGY
Miriam F. Menkin and John Rock (1890-1984) achieved the first test-tube fertilization of a human ovum.

1944 (November). GENERAL OR MISCELLANEOUS / NUCLEAR POWER AND WEAPONS
A committee chaired by Zay Jeffries (1888-1965) prepared a report that was the first general consideration of the likely effect on industry and society from nuclear energy. The classified report, titled "Prospectus on Nucleonics," was submitted to the Metallurgical Laboratory, Manhattan Project, University of Chicago.

1945. COMPUTERS AND INFORMATION SCIENCE
John von Neumann (1903-1957) published *First Draft Report on the EDVAC*, the Electronic Discrete Variable Computer.

1945. GENERAL OR MISCELLANEOUS / GOVERNMENT—FEDERAL
In response to a request from President Franklin Roosevelt the previous year, Vannevar Bush (1890-1974) published *Science, The Endless Frontier: A Report to the President* (Washington, D.C.). The document laid out the characteristics of a postwar policy for science, which combined the support of science by government with insulation of scientific research from political influences or controls. Among the important points was his call for an agency later realized in the National Science Foundation.

1945. GENERAL OR MISCELLANEOUS / NUCLEAR POWER AND WEAPONS
James B. Conant (1893-1978), chair of the National Defense Research Committee during the war (1941-1945), went on a mission to the Soviet

Union and there promoted the concept of international control of atomic weapons.

1945. GENERAL OR MISCELLANEOUS / PUBLIC HEALTH
Fluoridation of water supply to prevent tooth decay was introduced into the United States in a study supported by the U.S. Public Health Service. The study community was Grand Rapids, Michigan.

1945. MINERALOGY AND CRYSTALLOGRAPHY
(Fredrik) William Houlder Zachariasen (1906-1979) published *Theory of X-ray Diffraction in Crystals* (New York).

1945. ORGANIZATIONS—RESEARCH INSTITUTIONS / GENETICS
Beginning in this year, summer courses at the Cold Spring Harbor Biological Laboratory (New York) made it an international center for the study of bacteriophage genetics.

1945. ORGANIZATIONS—RESEARCH INSTITUTIONS / MEDICINE
Sloan-Kettering Institute for Cancer Research was founded in New York.

1945. ORGANIZATIONS—SOCIETIES AND ASSOCIATIONS / NUCLEAR POWER AND WEAPONS
The Federation of Atomic Scientists was established. The following year (1946) the name was changed to Federation of American Scientists. Incentive for the organization came from a concern for the consequences of the atomic bomb.

1945. ORGANIZATIONS—SOCIETIES AND ASSOCIATIONS / ZOOLOGY
The Society of Professional Zoologists was established.

1945. PERIODICALS AND PUBLISHING
Bulletin of the Atomic Scientists was established (the first six issues as *Bulletin of the Atomic Scientists of Chicago*).

1945. TECHNOLOGY AND INVENTION / NUCLEAR POWER AND WEAPONS
Enrico Fermi (1901-1954) and Leo Szilard (1898-1964) received the basic patent for the nuclear fission reactor.

1945 (June). MILITARY AND WAR / NUCLEAR POWER AND WEAPONS
A report to the Secretary of War from a group of concerned scientists involved in the Manhattan Project asked that the developing atomic bomb be demonstrated in an uninhabited region as a warning to the Japanese, rather than used in actual combat. The document was known as the Franck Report, for James Franck (1882-1964) who headed the protest group. The same month, an advisory team to the Secretary of War, including J. Robert Oppenheimer (1904-1967), Arthur Holly Compton (1892-1962), Enrico Fermi (1901-1954), and Ernest O. Lawrence (1901-1958), concluded that they could not endorse the demonstration and saw direct military use of the bomb as the viable option. The Franck Report was subsequently published in the *Bulletin of the Atomic Scientist* 1, no. 10 (1946): 1-5.

1945 (July 16). MILITARY AND WAR / NUCLEAR POWER AND WEAPONS
The first atomic bomb was exploded at Alamogordo, near Los Alamos, New Mexico.

1945 (August). MILITARY AND WAR / NUCLEAR POWER AND WEAPONS
The initial, official and authoritative report on the development of the atomic bomb was issued several days after the first bombing of Japan on August 6. The report, a reproduced typescript, was by Henry DeWolf Smyth (1898-1986), *A General Account of the Development of Methods of Using Atomic Energy for Military Purposes under the Auspices of the United States Government, 1940-1945* ([Washington: U.S. War Department], 1945).

1945 (August 6). MILITARY AND WAR / NUCLEAR POWER AND WEAPONS
Carried by the *Enola Gay*, the first atomic bomb, a uranium-235 device, was dropped on Hiroshima. The explosive exceeded the force of 20,000 tons of TNT. The second, plutonium bomb, was dropped on Nagasaki on August 9.

1945 (August 10). GENERAL OR MISCELLANEOUS / SPACE SCIENCE, TECHNOLOGY, AND EXPLORATION
Physicist and rocket engineer Robert Hutchings Goddard (b.1882) died in Baltimore, Maryland.

1945 (December 4). GENERAL OR MISCELLANEOUS / GENETICS
Geneticist Thomas Hunt Morgan (b.1866) died in Pasadena, California.

World War II. GENERAL OR MISCELLANEOUS /
GOVERNMENT—FEDERAL
During the period of the War, the federal portion of public and private expenditure on scientific research and development went from 18 to 83 percent. (In actual dollars, the federal increase rose from $48,000,000 to $500,000,000 per year.)

1946. AWARDS AND PRIZES / ASTRONOMY
The American Astronomical Society established the Henry Norris Russell annual lectureship in honor of Henry Norris Russell (1877-1957).

1946. AWARDS AND PRIZES / CHEMISTRY
John Howard Northrop (1891-1987), Wendell Meredith Stanley (1904-1971), and James Batcheller Sumner (1887-1955) won the Nobel Prize in chemistry. Northrop and Stanley were honored "for their preparation of enzymes and virus proteins in a pure form." Sumner was honored "for his discovery that enzymes can be crystallized."

1946. AWARDS AND PRIZES / MEDICINE, PHYSIOLOGY
Hermann Joseph Muller (1890-1967) won the Nobel Prize in medicine or physiology. He was honored "for the discovery of the production of mutations by means of X-ray irradiation."

1946. AWARDS AND PRIZES / PHYSICS
Percy Williams Bridgman (1882-1961) won the Nobel Prize in physics. He was honored "for the invention of an apparatus to produce extremely high pressures, and for the discoveries he made therewith in the field of high pressure physics."

1946. COMPUTERS AND INFORMATION SCIENCE
Electronic Numerical Integrator and Computer (ENIAC) was completed by John Presper Eckert, Jr. (1919-1995) and John William Mauchly (1907-1980) at the University of Pennsylvania; construction had begun in 1943. An all-electronic computer, it was the prototype for most

computers that followed. Programming was accomplished by changing the wiring and switch configuration.

1946. GENERAL OR MISCELLANEOUS / MEDICINE
Dr. Benjamin Spock (1903-) published *Common Sense Book of Baby and Child Care*, which became his influential *Baby and Child Care* (New York). In the next 45 years, it sold in excess of 30 million copies.

1946. GOVERNMENT—FEDERAL / FUNDS AND FUNDING
The Office of Naval Research was established by Congress; it formalized an agency established the previous year by the navy under conditions of war. Under the direction of Admiral Harold G. Bowen (1883-1965), the ONR became a source of funding for research in academic and other nonmilitary settings. Not limited to military projects, it supported basic research and funded work proposed by civilian scientists.

1946. GOVERNMENT—FEDERAL / NUCLEAR POWER AND WEAPONS
The Atomic Energy Commission was set up under the McMahon Act (the Atomic Energy Act). A significant feature was that atomic energy policy was to be under civilian control. The earlier May-Johnson Bill, which called for military control, had not been enacted. David Lilienthal (1899-1981) was the Commission's first head. J. Robert Oppenheimer (1904-1967) chaired the AEC General Advisory Committee until 1952. The Joint Committee on Atomic Energy was formed along with the Atomic Energy Commission; the Committee was formed to have congressional jurisdiction over the AEC and related matters.

1946. GOVERNMENT—FEDERAL / PHYSICS
Argonne National Laboratory (Illinois) was established. It was an outgrowth of the facility operated at the University of Chicago by the Manhattan Project during the war and was under the aegis of the Atomic Energy Commission.

1946. MATHEMATICS
Gilbert Ames Bliss (1876-1951) published *Lectures on the Calculus of Variations* (Chicago), the culmination of many years of work on the subject, which also drew on the work of others on the subject, making it a definitive production.

1946. ORGANIZATIONS—ACADEMIC / PHYSICAL SCIENCES—GENERAL
Associated Universities, Inc. was formed and operated the Brookhaven National Laboratory (Long Island, New York) under contract with the Atomic Energy Commission. The laboratory at Brookhaven opened in 1947, a project that came about through the cooperative efforts of nine universities.

1946. ORGANIZATIONS—INDUSTRY / INSTRUMENTS AND INSTRUMENTATION
Robert Jemison Van de Graaff (1901-1967), Denis M. Robinson (1907-), and John G. Trump (1907-1985) founded the High Voltage Engineering Corporation, the first commercial organization for the manufacture of particle accelerators. The founders were members of the faculty of the Massachusetts Institute of Technology.

1946. ORGANIZATIONS—RESEARCH INSTITUTIONS
Stanford Research Institute (SRI) was founded. Originally associated with Stanford University, it later became an independent nonprofit research institution.

1946. ORGANIZATIONS—SOCIETIES AND ASSOCIATIONS / ENVIRONMENT AND CONSERVATION
The Ecologists' Union was founded among certain members of the Ecological Society of America. Having the aim of acquiring land for preservation, in 1950 the organization became the Nature Conservancy.

1946. ORGANIZATIONS—SOCIETIES AND ASSOCIATIONS / EVOLUTION
The Society for the Study of Evolution was established. It derived from the 1940 Society for the Study of Speciation and the National Research Council's 1943 Committee on Common Problems of Genetics, Paleontology and Systematics.

1946. ORGANIZATIONS—SOCIETIES AND ASSOCIATIONS / PHYSIOLOGY
The Society of General Physiologists was established.

1946. PHYSICS
Edward Mills Purcell (1912-), and at the same time Felix Bloch (1905-1983), developed means to measure the magnetic fields of atomic nuclei and molecules by nuclear magnetic resonance (NMR) absorption.

Their work provided means to study the molecular structures of many substances. In 1952, Bloch and Purcell shared the Nobel Prize in physics.

1946. PHYSICS / INSTRUMENTS AND INSTRUMENTATION
The first synchro-cyclotron was constructed, under the direction of Edwin M. McMillan (1907-1991), at the University of California, Berkeley.

1946 (March 23). GENERAL OR MISCELLANEOUS / CHEMISTRY
Chemist Gilbert Newton Lewis (b.1875) died in Berkeley, California.

1946 (July 1). MILITARY AND WAR / NUCLEAR POWER AND WEAPONS
The United States began atomic tests at Bikini, Marshall Islands.

1947. AWARDS AND PRIZES / MEDICINE, PHYSIOLOGY
Carl Ferdinand Cori (1896-1984) and Gerty Theresa Radnitz Cori (1896-1957) (both Czechoslovakian/Americans) won the Nobel Prize in medicine or physiology. It was shared with Bernardo Alberto Houssay (Argentinian). Cori and Cori were honored "for their discovery of the course of the catalytic conversion of glycogen."

1947. BOTANY
Botanist and geneticist Ernest Brown Babcock (1877-1954), published *The Genus Crepis* (Los Angeles), in two volumes. The work summarized his far-ranging studies of the genus over a period of years, touching on morphology, genetics, plant distribution, and the relations of species and their origins. (The genus includes the hawk's-beard.)

1947. CHEMISTRY / GEOLOGY
Willard Frank Libby (1908-1980) proposed that knowledge of the half-life of carbon-14 could be used to date biological specimens that are less than 50,000 years old.

1947. EVOLUTION / GENETICS
At a Princeton University Bicentennial Conference on Genetics, Paleontology, and Evolution, a synthetic theory of evolution came to the fore. The conference was organized by Glenn Lowell Jepsen (1903-1974), Ernst Mayr (1904-), and George G. Simpson (1902-1984).

Jepsen, Mayr, and Simpson edited *Genetics, Paleontology, and Evolution* (Princeton, 1949).

1947. GENETICS / MICROBIOLOGY AND MICROSCOPY
Joshua Lederberg (1925-) and Edward Lawrie Tatum (1909-1975) discovered genetic recombination in bacteria *Escherichia coli* as a form of sexual reproduction.

1947. GEOLOGY
Paleontologist Rudolf Ruedemann (1864-1956) published "Graptolites of North America," *Memoir of Geological Society of America*, no. 19. The publication summarized a lifetime of work.

1947. GEOLOGY
Thomas Augustus Jaggar, Jr. (1871-1953) published "The Origin and Development of Craters," *Memoirs of Geological Society of America* 21.

1947. MATHEMATICS
Mathematical statistician Abraham Wald (1902-1950) published *Sequential Analysis* (New York and London).

1947. OCEANOGRAPHY
The first theory of the Gulf Stream was given by Henry Melson Stommel (1920-1992).

1947. ORGANIZATIONS—INDUSTRY / PHOTOGRAPHY
Edwin Herbert Land (1909-1991), founder of the Polaroid Corporation, demonstrated his instant camera.

1947. ORGANIZATIONS—RESEARCH INSTITUTIONS / ZOOLOGY—HUMAN
The Institute for Sex Research was founded by Alfred Charles Kinsey (1894-1956). Affiliated with Indiana University, in 1981 it became known as the Kinsey Institute for Sex Research, Inc.

1947. ORGANIZATIONS—SOCIETIES AND ASSOCIATIONS
The Scientific Research Society of America (Resa) was established under the sponsorship of Sigma Xi and especially to promote the interests of governmental and industrial scientists.

1947. ORGANIZATIONS—SOCIETIES AND ASSOCIATIONS / MINERALOGY AND CRYSTALLOGRAPHY
The American Federation of Mineralogical Societies was established to serve the needs of amateurs in the science.

1947. ORGANIZATIONS—SOCIETIES AND ASSOCIATIONS / ZOOLOGY
The American Society of Protozoologists was established.

1947. PARAPSYCHOLOGY
Joseph Banks Rhine (1895-1980) published *The Reach of the Mind* (New York), on his studies in psychokinesis.

1947. PHYSICS
Willis Eugene Lamb, Jr. (1913-) discovered the so-called Lamb shift, evidence that the predicted existence of hydrogen atoms in two equal energy states was not exactly true, there being a small difference in the energy. His work led others to significant revisions in quantum theory and to the emergence of quantum electrodynamics (QED).

1947. PHYSICS / ELECTRICITY AND ELECTRONICS
The point-contact transistor was invented at Bell Telephone Laboratories, announcement of which was made by John Bardeen (1908-1991) and Walter Houser Brattain (1902-1987) in "The Transistor, a Semi-Conductor Triode," *Physical Review*, 2nd series 74 (July 15, 1948): 230-231. This theoretical breakthrough subsequently was made practical by the junction transistor devised by William Bradford Shockley (1910-1989). In 1956, all three men shared the Nobel Prize in physics. The transistor replaced the older vacuum tube in electronic devices.

1947. TECHNOLOGY AND INVENTION
R. Buckminster Fuller (1895-1983) developed the geodesic dome.

1947 (October). GENERAL OR MISCELLANEOUS
J. Robert Oppenheimer (1904-1967) became director of the Institute for Advanced Study at Princeton.

1947 (October 19). ENGINEERING AND APPLIED SCIENCE / AERONAUTICS
An X-1 research plane, built by Bell Aircraft and piloted by Air Force Captain Charles Yeager (1923-), was the first to achieve supersonic speed.

1948. ASTRONOMY
What was then the smallest of the known satellites of Uranus, Miranda, was discovered by Gerard Peter Kuiper (1905-1973).

1948. ASTRONOMY / CHEMISTRY
Gerard Peter Kuiper (1905-1973) discovered carbon dioxide in the atmosphere of Mars.

1948. ASTRONOMY / INSTRUMENTS AND INSTRUMENTATION
The 200-inch Hale reflecting telescope at Mount Palomar, California, was finished and dedicated. The lens had been cast in 1934.

1948. CHEMISTRY
The structure of strychnine was delineated by Robert Burns Woodward (1917-1979). In 1954, he achieved the synthesis of the substance in a process involving 50 stages.

1948. COMPUTERS AND INFORMATION SCIENCE
Norbert Wiener (1894-1964) published *Cybernetics, or Control and Communication in the Animal and the Machine* (Paris and Cambridge, Mass.).

1948. GENERAL OR MISCELLANEOUS / MINERALOGY AND CRYSTALLOGRAPHY
Isidor Fankuchen (1905-1964) became the first American editor of *Acta Crystallographica* and held the position until his death.

1948. GENERAL OR MISCELLANEOUS / PSYCHOLOGY
Burrhus Frederic (B.F.) Skinner (1904-1990) published the novel *Walden Two* (New York).

Chronology 277

1948. GEOLOGY
Francis Parker Shepard (1897-1985) published *Submarine Geology* (New York); this was the earliest textbook on the topic.

1948. MEDICINE / PHYSIOLOGY
Fuller Albright (1900-1969) and Edward C. Reifenstein, Jr. (1908-1975) published *Parathyroid Glands and Metabolic Bone Disease* (Baltimore).

1948. MICROBIOLOGY AND MICROSCOPY / PATHOLOGY
Plant pathologist Benjamin Minge Duggar's (1872-1956) studies of *Streptomyces aureofaciens* resulted in the antibiotic chlortetracycline (Aureomycin).

1948. ORGANIZATIONS—RESEARCH INSTITUTIONS
The nonprofit research organization, the Rand Corporation, was founded.

1948. ORGANIZATIONS—SOCIETIES AND ASSOCIATIONS / BIOLOGY—GENERAL
The American Institute of Biological Sciences was founded as an organization of biological societies.

1948. ORGANIZATIONS—SOCIETIES AND ASSOCIATIONS / GENETICS
The American Society of Human Genetics was established.

1948. ORGANIZATIONS—SOCIETIES AND ASSOCIATIONS / GEOLOGY
The American Geological Institute, an organization of professional geological societies, was established.

1948. ORGANIZATIONS—SOCIETIES AND ASSOCIATIONS / GEOLOGY
The American Society of Limnology and Oceanography was established.

1948. ORGANIZATIONS—SOCIETIES AND ASSOCIATIONS / ZOOLOGY
The Society of Systematic Zoology was founded.

1948. PALEONTOLOGY
Raymond Cecil Moore (1892-1974) began the *Treatise on Invertebrate Paleontology*, a multi-volume and multi-authored work published by the

Geological Society of America. Publication commenced in 1953, with Moore as editor; he continued until 1974.

1948. PALEONTOLOGY / ZOOLOGY
Johanna Gabrielle Ottilie (Tilly) Edinger (1897-1967) published *Evolution of the Horse Brain* (Memoirs of Geological Society of America, no. 25). The work helped to promote the extensive growth of the field of paleoneurology in the years that followed.

1948. PHYSICS
Maria Goeppert Mayer (1906-1972) put forth the shell theory of the atomic nucleus.

1948. PHYSICS / CHEMISTRY
Ralph Alpher (1921-), Hans A. Bethe (1906-), and George Gamow (1904-1968) published a notable letter on the state of matter prior to the big bang, "The Origin of Chemical Elements," *Physical Review* 73:803-804. (It is said Bethe's name was added by Gamow as a joke, to make the authorship read Alpher-Bethe-Gamow.)

1948. PHYSICS / INSTRUMENTS AND INSTRUMENTATION
Angus Ewan Cameron (1906-1981) and David F. Eggers, Jr. (1922-) published "An Ion Velocitron," *Review of Scientific Instruments* 19:605-607. The instrument was able to display the mass spectrum in its entirety at the same time.

1948. ZOOLOGY—HUMAN / PSYCHOLOGY
Alfred Charles Kinsey (1894-1956) published *Sexual Behavior in the Human Male* (Philadelphia), which became known as the "Kinsey Report." (His research in this area had begun in 1938.)

1948 (January). COMPUTERS AND INFORMATION SCIENCE
IBM dedicated its Selective Sequence Electronic Computer (SSEC), a general-purpose computer that involved design concepts of Wallace John Eckert (1902-1971) and the work of IBM engineers. It was used for astronomical calculations by Eckert and others.

1948-1954. CHEMISTRY / BOTANY
Melvin Calvin (1911-) published a series of articles delineating the process of photosynthesis. The report on the research began with the article (coauthored by Andrew A. Benson, 1917-), "The Path of Carbon in Photosynthesis," *Science* 107 (May 7, 1948): 376-380. Subsequent articles had the same title as here, with differing subtitles.

1949. AWARDS AND PRIZES / CHEMISTRY
William Francis Giauque (1895-1982) won the Nobel Prize in chemistry "for his contributions in the field of chemical thermodynamics, particularly concerning the behaviour of substances at extremely low temperatures."

1949. COMPUTERS AND INFORMATION SCIENCE
The random-access magnetic core memory was devised by Jay Wright Forrester (1918-). It became a basic component of digital computers, for the storage of information.

1949. COMPUTERS AND INFORMATION SCIENCE / MATHEMATICS
Claude Elwood Shannon (1916-) and Warren Weaver (1894-1978), published *The Mathematical Theory of Communication* (Urbana, Ill.).

1949. GENERAL OR MISCELLANEOUS / NUCLEAR POWER AND WEAPONS
J. Robert Oppenheimer (1904-1967) opposed a directed effort to develop a hydrogen bomb. This later played a part in his 1953 loss of security clearance.

1949. GENETICS
Curt Stern (1902-1981) published the textbook, *Principles of Human Genetics* (San Francisco). A third edition appeared in 1973.

1949. GEOLOGY
Carl Owen Dunbar (1891-1979) published a rewritten edition of *Historical Geology* (New York), with subsequent editions in 1960 and 1969. This work of 1949 was Dunbar's rewrite of the previously published part two of *Textbook of Geology*, written with Charles Schuchert (1858-1942).

1949. GEOLOGY
James Gilluly (1896-1980) published "Distribution of Mountain Building in Geologic Time," *Bulletin of Geological Society of America* 60:561-590, his address as GSA president. His orogenic ideas, including the idea that mountain building was a continuous rather than a periodic process, exerted a significant effect at the time.

1949. GOVERNMENT—FEDERAL / NUCLEAR POWER AND WEAPONS
Sandia National Laboratories was established. The laboratories' work related especially to nuclear weapons research.

1949. MEDICINE / MICROBIOLOGY AND MICROSCOPY
John Franklin Enders (1897-1985), Frederick Chapman Robbins (1916-), and Thomas Huckle Weller (1915-) succeeded in the cultivation of poliomyelitis virus in a culture of human nonnerve tissue cells. This was a step in the development of polio vaccine. In 1954, they received the Nobel Prize in medicine or physiology.

1949. ORGANIZATIONS—SOCIETIES AND ASSOCIATIONS
The Society for Social Responsibility in Science was founded.

1949. PHYSICS
Leonard Isaac Schiff (1915-1971) published the textbook, *Quantum Mechanics* (New York). There were subsequent editions in 1955 and 1968, as well as translations into Italian, Japanese, and Russian.

1949. SPACE SCIENCE, TECHNOLOGY, AND EXPLORATION
A facility for testing rockets was set up at Cape Canaveral, Florida. The first launch, of a V2 rocket, occurred on 24 July 1950.

1949. SPACE SCIENCE, TECHNOLOGY, AND EXPLORATION
The first multi-stage rocket was tested by the United States. On 24 February, the WAC-Corporal set a new record for missile altitude of 250 miles; it was launched (at White Sands, New Mexico) as a second stage of a V-2 rocket.

1949. ZOOLOGY / ECOLOGY
Warder Clyde Allee (1885-1955), Alfred E. Emerson (1896-1976), Orlando Park (1901-1969), Thomas Park (1908-1992), and Karl

Patterson Schmidt (1890-1957) published the textbook, *Principles of Animal Ecology* (Philadelphia).

1949. ZOOLOGY / ENTOMOLOGY
Marston Bates (1906-1974) published *The Natural History of Mosquitoes* (New York), a significant contribution to epidemiology.

1949 (April 20). GENERAL OR MISCELLANEOUS / CHEMISTRY, MEDICINE
The ability to produce cortisone in quantities that would meet the needs of many patients was announced. The previous year, Edward Calvin Kendall (1886-1972) and Philip Showalter Hench (1896-1965) succeeded in the use of cortisone for treatment of rheumatoid arthritis.

1949 (August). COMPUTERS AND INFORMATION SCIENCE
The first electronic stored-program computer in the United States, BINAC (Binary Automatic Computer), went into operation. It was developed by John Presper Eckert, Jr. (1919-1995) and John William Mauchly (1907-1980).

1949 (August 1). GEOLOGY
(William) Maurice Ewing (1906-1974) and Frank Press (1924-) presented the first results of their study of the Mohorovicic (Moho) discontinuity between the crust and mantle beneath the Atlantic Ocean. Their study demonstrated a characteristic difference between the crust under the ocean and that of the continents. In this year, the Lamont Geological Observatory (after 1969, Lamont-Doherty) was established by Columbia University. Ewing became director and under his leadership the observatory became one of the world's primary centers for marine geology and geophysics.

1950. AWARDS AND PRIZES / MEDICINE, PHYSIOLOGY
Philip Showalter Hench (1896-1965) and Edward Calvin Kendall (1886-1972) won the Nobel Prize in medicine or physiology. It was shared with Tadeus Reichstein (Polish/Swiss). They were honored "for their discoveries relating to the hormones of the adrenal cortex, their structure and biological effects."

1950. BOTANY
Merritt Lyndon Fernald (1873-1950) produced a completely revised version of Asa Gray's *Manual of the Botany of the Northern United States* (1908 edition) (New York).

1950. BOTANY / EVOLUTION
George Ledyard Stebbins (1906-) published *Variation and Evolution in Plants*. This was a presentation of his pioneering work on the application of the synthetic theory of evolution to plants.

1950. GOVERNMENT—FEDERAL / FUNDS AND FUNDING
An act establishing the National Science Foundation was passed by Congress and approved by the President. This was the outcome of several years of postwar attempts at such legislation, the failures coming on such issues as inclusion or not of the social sciences, whether funds would be distributed geographically or by the merit of projects irrespective of location, and especially over the question of control (whether the agency would be directly responsible to the President or whether the scientific community would have a significant degree of independent judgement). As passed, the bill was somewhat of a compromise--the director was to be a presidential appointee but power was shared with a part-time board whose members turned out to be mainly representatives of influential academic and industrial establishments. There was no provision for geographical distribution and the legislation excluded any concern for military-related research.

1950. ORGANIZATIONS—SOCIETIES AND ASSOCIATIONS
In a controversial election for the presidency of the National Academy of Sciences, Detlev Wulf Bronk (1897-1975) rather than the nominee James B. Conant (1893-1978) was elected. It was the only contested election in the history of the Academy.

1950. ORGANIZATIONS—SOCIETIES AND ASSOCIATIONS / MINERALOGY AND CRYSTALLOGRAPHY
The American Society for X-ray and Electron Diffraction and the Crystallographic Society of America combined to form the American Crystallographic Association. The new organization's first president was Isidor Fankuchen (1905-1964).

1950. PHILOSOPHY
Rudolf Carnap (1891-1970) published his most important work, *The Logical Foundations of Probability* (Chicago).

1950. STATISTICS
Statistician William Gemmell Cochran (1909-1980) and Gertrude M. Cox (1900-1978) published *Experimental Designs* (New York). A second edition appeared in 1957.

1950. ASTRONOMY
The idea that comets are "dirty snowballs" was put forth by Fred Lawrence Whipple (1906-).

1950 (March). GENERAL OR MISCELLANEOUS
Senator Joseph McCarthy named astronomer Harlow Shapley (1885-1972) as one of five apparent Communists associated with the U.S. State Department, but in the same year he was completely cleared by the Senate Foreign Relations Committee. Among Shapley's State Department roles was as a delegate to the 1945 London conference that formulated UNESCO.

1950 (January 31). GOVERNMENT—FEDERAL / NUCLEAR POWER AND WEAPONS
President Harry Truman made known publicly that a hydrogen bomb was being developed under the auspices of the Atomic Energy Commission. This was done in spite of advice to the contrary from the scientific General Advisory Committee of the Atomic Energy Commission, and the Commission itself. By 1951, a major technical obstacle had been overcome by Edward Teller (1908-) and Stanislaw Ulam (1909-1984) that made the bomb practical.

1950 (April). METEOROLOGY AND CLIMATOLOGY / COMPUTERS AND INFORMATION SCIENCE
John von Neumann (1903-1957) and collaborators achieved the first twenty-four-hour weather forecast with the use of a computer (ENIAC).

1950-1954. PHYSICS
Fritz London (1900-1954) published *Superfluids* (New York) in two volumes: volume 1, *Macroscopic Theory of Superconductivity*; and volume 2, *Macroscopic Theory of Superfluid Helium*.

1950-1966. MATHEMATICS
William Feller (1906-1970) published *An Introduction to Probability Theory and Its Applications* (New York) in two volumes.

1951. ASTRONOMY / COMPUTERS AND INFORMATION SCIENCE
Coordinates of the Five Outer Planets, 1653-2060 (Washington, D.C.), by Dirk Brouwer (1902-1966), Wallace John Eckert (1902-1971), and Gerald Maurice Clemence (1908-1974), was the first astronomical problem solved with a high-speed computer.

1951. AWARDS AND PRIZES / CHEMISTRY
Edwin Mattison McMillan (1907-1991) and Glenn Theodore Seaborg (1912-) won the Nobel Prize in chemistry. They were recognized "for their discoveries in the chemistry of the transuranium elements."

1951. BIOCHEMISTRY AND MOLECULAR BIOLOGY
Linus Carl Pauling (1901-1994) and Robert B. Corey (1897-1971) disclosed that in certain instances protein molecules can take on a helical form.

1951. BOTANY / EVOLUTION
Jens Christen Clausen (1891-1969) published *Stages in the Evolution of Plant Species* (Ithaca, N.Y.).

1951. CHEMISTRY
The steroids cortisone and cholesterol were synthesized by Robert B. Woodward (1917-1979).

1951. CHEMISTRY / BIOLOGY—GENERAL
Harold Clayton Urey (1893-1981) and his student Stanley Lloyd Miller (1930-) published the result of experiments that produced amino acids from methane, ammonia, and steam in a closed container by subjecting them to an electrical discharge. The experiment was an attempt to show how life on earth could have originated.

1951. GENETICS / BOTANY
Barbara McClintock (1902-1992), in work on Indian corn, discovered that certain genetic elements changed position from generation to generation, contrary to general belief. This helped in understanding the process of cell differentiation.

1951. GEOLOGY
Marshall Kay (1904-1975) published *North American Geosynclines* (Geological Society of America Memoir 48), a milestone in the understanding of ancient mountain belts.

1951. ORGANIZATIONS—INDUSTRY / COMPUTERS AND INFORMATION SCIENCE
John Presper Eckert, Jr. (1919-1995) and John William Mauchly (1907-1980) had formed the Eckert-Mauchly Computer Corporation in 1947 and as of 1951 had developed UNIVAC I (Universal Automatic Computer), the first commercially viable computer. In 1950, Remington Rand took over the company (which later became Sperry Rand Corporation) and during the year 1951 provided UNIVAC to the Bureau of the Census.

1951. PSYCHOLOGY
Stanley Smith Stevens (1906-1973) edited *Handbook of Experimental Psychology* (New York).

1951. ZOOLOGY / EVOLUTION
William King Gregory (1876-1970) published the important work on vertebrates, *Evolution Emerging: A Survey of Changing Patterns from Primeval Life to Man* (New York), in two volumes.

1951. ZOOLOGY / PHYSIOLOGY
Homer William Smith (1895-1962) published *The Kidney: Structure and Function in Health and Disease* (New York).

1951 (December). ENGINEERING AND APPLIED SCIENCE / NUCLEAR POWER AND WEAPONS
Electrical energy from atomic power was first produced by a nuclear reactor. This took place at the Atomic Energy Commission's reactor testing station at Arco, Idaho.

1952. ASTRONOMY
In a report to the International Astronomical Union on work with the Palomar 200-inch telescope, Walter Baade (1893-1960) suggested a doubling of the accepted size of the universe. The newly projected size was due to a previous error in the Cepheid luminosity scale.

1952. ASTRONOMY
Herbert Rollo Morgan (1875-1957) published "Catalog of 5,268 Standard Stars, 1950.0 based on the Normal System N30," *Astronomical Papers of the American Ephemeris*, 13, part 3.

1952. ASTRONOMY / INSTRUMENTS AND INSTRUMENTATION
Harold D. Babcock (1882-1968) and his son Horace W. Babcock (1912-) developed the solar magnetograph at the Hale Solar Laboratory (Pasadena). With this device, they were able to achieve the first measure of magnetic fields on the sun's surface.

1952. AWARDS AND PRIZES / MEDICINE, PHYSIOLOGY
Selman Abraham Waksman (1888-1973) won the Nobel Prize in medicine or physiology. He was honored "for his discovery of streptomycin, the first antibiotic effective against tuberculosis."

1952. AWARDS AND PRIZES / PHYSICS
Felix Bloch (1905-1983) and Edward Mills Purcell (1912-) won the Nobel Prize in physics. They were honored "for their development of new methods for nuclear magnetic precision measurements and discoveries in connection therewith."

1952. BIOCHEMISTRY AND MOLECULAR BIOLOGY
Alfred Day Hershey (1908-) and Martha Cowles Chase (1927-), with the use of radioisotopes, showed that deoxyribonucleic acid (DNA) was the agent of genetic continuance in the bacteria under study.

1952. BOTANY
Edgar Anderson (1897-1969) published *Plants, Man, and Life* (New York). It was reprinted in 1967.

1952. COMPUTERS AND INFORMATION SCIENCE
EDVAC (electronic discrete variable computer) was completed as a successor to the ENIAC and incorporated ideas of John von Neumann (1903-1957) and colleagues. It was programmable from memory storage.

1952. EVOLUTION
John Thomas Patterson (1873-1960) and Wilson S. Stone (1907-1968) published *Evolution in the Genus Drosophila* (New York), the culmination of wide-ranging field and experimental work on the speciation of genus *Drosophila* that the authors had begun in 1938.

1952. GENETICS / MICROBIOLOGY AND MICROSCOPY
Joshua Lederberg (1925-) published an account of his discovery of transduction in the bacterium *Salmonella typhimurium*. A virus was the agent for transferral of part of the chromosome of one bacterium to another. This finding marked the beginning of understanding that led to genetic engineering.

1952. ORGANIZATIONS—SOCIETIES AND ASSOCIATIONS / MATHEMATICS
The Society for Industrial and Applied Mathematics (SIAM) was established.

1952. ORGANIZATIONS—SOCIETIES AND ASSOCIATIONS / PHYSICS
The Radiation Research Society was established.

1952. ORGANIZATIONS—SOCIETIES AND ASSOCIATIONS / SOCIAL SCIENCES—GENERAL
The Operations Research Society of America was established.

1952. PHYSICS / INSTRUMENTS AND INSTRUMENTATION
Donald A. Glaser (1926-) invented the bubble chamber, used to trace the paths of high-energy atomic particles. The first pictures of interaction at the sub-atomic level, using the bubble chamber, were published the next year.

1952. PHYSIOLOGY
Edmund Newton Harvey (1887-1959) published *Bioluminescence* (New York), a product of a lifetime of studies in this area.

1952 (November 1). MILITARY AND WAR / NUCLEAR POWER AND WEAPONS
The first hydrogen bomb, based on nuclear fusion, was exploded at Eniwetok Atoll in the South Pacific. It was developed by Edward Teller (1908-) and his associates. In 1954, a bomb with the power of 15,000,000 tons of TNT was tested. Einsteinium, an artificial element of atomic number 99, was found by Glenn T. Seaborg (1912-) in the aftermath of the 1952 explosion.

1953. AWARDS AND PRIZES / MEDICINE, PHYSIOLOGY
Fritz A. Lipmann (1899-1986) won the Nobel Prize in medicine or physiology. He was honored "for his discovery of co-enzyme A and its importance for intermediary metabolism." Hans Adolf Krebs in England also won a prize, for discovery of the citric acid cycle.

1953. BIOCHEMISTRY AND MOLECULAR BIOLOGY
Vincent DuVigneaud (1901-1978) achieved the first synthesis of a hormone, oxytocin, a substance secreted by the posterior lobe of the pituitary. For this and work on another pituitary hormone, vasopressin, he won the Nobel Prize in chemistry in 1955.

1953. BIOCHEMISTRY AND MOLECULAR BIOLOGY / GENETICS
The double-helix model of deoxyribonucleic acid (DNA) was developed by James Dewey Watson (1928-) and British scientist Francis H. Crick, which illuminated the way in which DNA functions in determining hereditary characteristics. Watson and Crick published the initial account of their pioneering insight in "A Structure for Deoxyribose Nucleic Acid," *Nature* 171 (April 25, 1953): 737-738. Soon thereafter they also published "Genetical Implications of the Structure of Deoxyribonucleic Acid," *Nature* 171 (May 30, 1953): 964-967.

1953. CHEMISTRY
Paul J. Flory (1910-1985) published *Principles of Polymer Chemistry* (Ithaca, N.Y.).

1953. GENETICS / BOTANY
Ruth Sager (1918-) discovered nonchromosomal genes of an algae and her related experiments were an advance in knowledge of cytoplasmic heredity.

1953. MEDICINE / INSTRUMENTS AND INSTRUMENTATION
The first use of a heart-lung machine in human heart surgery was achieved by John H. Gibbon, Jr. (1903-1973).

1953. MICROBIOLOGY AND MICROSCOPY
Salvador Edward Luria (1912-1991) published *General Virology* (New York).

1953. ORGANIZATIONS—SOCIETIES AND ASSOCIATIONS / ENTOMOLOGY
The Entomological Society of America was established. It came about through the merger of the older Entomological Society of America (founded 1906) and the American Association of Economic Entomologists (founded 1889).

1953. PHYSICS
Murray Gell-Mann (1929-) introduced into high-energy physics the idea of strangeness, a quantum number that explained certain anomalous aspects in the decay of hadrons (particles that responded to the strong force).

1953. PSYCHOLOGY
Eugene Aserinsky (1921-) and Nathaniel Kleitman (1895-?) discovered two phases of sleep, one involving dreams and rapid eye movement (REM) and the other a dreamless phase.

1953. ZOOLOGY—HUMAN / PSYCHOLOGY
Alfred Charles Kinsey (1894-1956) published the pioneering work, *Sexual Behavior in the Human Female* (Philadelphia) which became known as the "Kinsey Report."

1953 (September 28). GENERAL OR MISCELLANEOUS / ASTRONOMY
Astronomer Edwin Powell Hubble (b.1889) died at San Marino, California.

1953 (December). GENERAL OR MISCELLANEOUS / NUCLEAR POWER AND WEAPONS
J. Robert Oppenheimer (1904-1967) lost his security clearance. A hearing was subsequently conducted (at Oppenheimer's request) by the Atomic Energy Commission and the proceedings were published by the

AEC as *In the Matter of J. Robert Oppenheimer: Transcript of Hearings Before the Personnel Security Board* (Washington, D.C., 1954). Although found to be loyal, the Board and the Commission in 1954 reiterated his loss of clearance on grounds of what were referred to as character flaws.

1953 (December 8). GENERAL OR MISCELLANEOUS / NUCLEAR POWER AND WEAPONS
President Dwight Eisenhower delivered a speech to the United Nations General Assembly that stressed the nonmilitary uses of nuclear energy. It laid the basis of the "Atoms for Peace" program for the international availability of nuclear reactors for power and research. A proposed International Atomic Energy Agency came into being in 1957. An effect of these events was to put in the background considerations for banning the atomic bomb while they increased the prospects for nuclear proliferation.

1953 (December 19). GENERAL OR MISCELLANEOUS / PHYSICS
Physicist Robert Andrews Millikan (b.1868) died in San Marino, California.

1954. AWARDS AND PRIZES / CHEMISTRY
Linus Carl Pauling (1901-1994) won the Nobel Prize in chemistry. He was honored "for his research into the nature of the chemical bond and its application to the elucidation of the structure of complex substances."

1954. AWARDS AND PRIZES / MEDICINE, PHYSIOLOGY
John Franklin Enders (1897-1985), Frederick Chapman Robbins (1916-), and Thomas Huckle Weller (1915-) won the Nobel Prize in medicine or physiology. They were honored "for their discovery of the ability of poliomyelitis viruses to grow in cultures of various types of tissues."

1954. BOTANY
Thomas Harper Goodspeed (1887-1966), Helen-Mar Wheeler (1907-), and Paul C. Hutchison (1924-) published a summary work on tobacco, *The Genus Nicotiana: Origins, Relationships and Evolution of Its Species in the Light of Their Distribution, Morphology and Cytogenetics* (Chronica Botanica 16) (Waltham, Mass.).

1954. MEDICINE / DISEASE
The first effective polio vaccine, developed by Jonas Edward Salk (1914-1995) and based on killed virus, was introduced for use in the United States with the inoculation of school children in Pittsburgh, Pennsylvania, on February 23. In April 1955, the Polio Vaccine Evaluation Center at the University of Michigan reported that the Salk vaccine was successful, based on nationwide testing the previous year.

1954. ORGANIZATIONS—INDUSTRY / CHEMISTRY
Percy Lavon Julian (1899-1975), an African-American chemist, established Julian Laboratories in Franklin Park, Illinois, a chemical facility for manufacturing materials related to steroids production. The company was sold in 1961 to Smith, Kline, and French Laboratories.

1954. ORGANIZATIONS—INDUSTRY / ELECTRICITY AND ELECTRONICS
Scientists at Bell Laboratories developed the silicon solar cell (solar battery).

1954. ORGANIZATIONS—SOCIETIES AND ASSOCIATIONS / MEDICINE
The Society for Nuclear Medicine was established.

1954. ORGANIZATIONS—SOCIETIES AND ASSOCIATIONS / NUCLEAR POWER AND WEAPONS
The American Nuclear Society was established with a particular focus on nuclear power.

1954. ORGANIZATIONS--SOCIETIES AND ASSOCIATIONS / SPACE SCIENCE, TECHNOLOGY, AND EXPLORATION
The American Astronautical Society was established.

1954. PHYSICS
Development of the maser (Microwave Amplification by Stimulated Emission of Radiation) was announced by Charles Hard Townes (1915-).

1954. PHYSICS
A foundation for modern quantum field theory was put forth in the Yang-Mills gauge-invariant fields theory developed by Chen Ning Yang (1922-) and Robert L. Mills (1927-).

1954. PHYSICS / ACOUSTICS
Frederick Vinton Hunt (1905-1972) published *Electroacoustics: The Analysis of Transduction, and Its Historical Background* (Cambridge, Mass.).

1954. PHYSICS / INSTRUMENTS AND INSTRUMENTATION
The Bevatron particle accelerator completed at Berkeley was capable of accelerating uranium atoms to 6.5 billion electron volts (GeV). The facility that housed the apparatus, known as the Radiation Laboratory, was renamed the Lawrence Berkeley Laboratory in 1958.

1954. STATISTICS
Leonard Jimmie Savage (1917-1971) published *The Foundations of Statistics* (New York). There was a second edition in 1972.

1954. ZOOLOGY—HUMAN / PHYSIOLOGY
Gregory Goodwin Pincus (1903-1967) began field trials of a female oral contraceptive pill in Haiti and Puerto Rico. The pill was approved for use in the United States in 1959. Pincus, with Min Chueh Chang (1908-1991), had begun studies of the effects of synthesized hormones on laboratory animal reproduction in 1951, while John Rock (1890-1984) and Celso-Ramon Garcia (1921-) also were collaborators with Pincus in development and testing of the contraceptive pill for humans.

1954 (January 21). MILITARY AND WAR / NUCLEAR POWER AND WEAPONS
Nautilus, the first nuclear-powered submarine, was launched. The development of the vessel was directed by Hyman G. Rickover (1900-1986).

1954 (August 30). GENERAL OR MISCELLANEOUS / NUCLEAR POWER AND WEAPONS
President Dwight Eisenhower signed the Atomic Energy Bill that made possible private ownership of nuclear reactors for the production of electrical power. On January 9, 1955, the Atomic Energy Commission announced that private operation of atomic power plants by industry would be possible soon.

Chronology

1954 (November 28). GENERAL OR MISCELLANEOUS / PHYSICS
Physicist Enrico Fermi (b.1901) died in Chicago, Illinois.

1954-1955. GENERAL OR MISCELLANEOUS / PHYSICS
Felix Bloch (1905-1983) was first director general of the European Commission for Nuclear Research (CERN).

1955. ASTRONOMY
Radio emissions from Jupiter were detected by Kenneth L. Franklin (1923-) and Bernard F. Burke (1928-).

1955. AWARDS AND PRIZES / CHEMISTRY
Vincent DuVigneaud (1901-1978) won the Nobel Prize in chemistry. He was honored "for his work on biochemically important sulphur compounds, especially for the first synthesis of a polypeptide hormone."

1955. AWARDS AND PRIZES / GENETICS
The first Kimber Genetics Award of the National Academy of Sciences was given to William Ernest Castle (1867-1962).

1955. AWARDS AND PRIZES / PHYSICS
Polykarp Kusch (1911-1993) and Willis Eugene Lamb, Jr. (1913-) won the Nobel Prize in physics. Kusch was honored "for his precision determination of the magnetic moment of the electron," Lamb "for his discoveries concerning the fine structure of the hydrogen spectrum."

1955. COMPUTERS AND INFORMATION SCIENCE
Arthur W. Horton, Jr. (1898-1969) and Henry Earle Vaughan (1912-) published "Transmission of Digital Information over Telephone Circuits," *Bell System Technical Journal* 34 (May 1955): 511-528.

1955. ENGINEERING AND APPLIED SCIENCE / ELECTRICITY AND ELECTRONS
John Robinson Pierce (1910-), an engineer at Bell Telephone Laboratories, was the first to publish on the possibility of satellite radio communication.

1955. GENERAL OR MISCELLANEOUS
Karl Sax (1892-1973) published *Standing Room Only: The Challenge to Overpopulation* (Boston).

1955. GENERAL OR MISCELLANEOUS / ECOLOGY
Marston Bates (1906-1974) published *The Prevalence of People* (New York), drawing attention to likely ecological results of uncontrolled population growth.

1955. ORGANIZATIONS—INDUSTRY / ELECTRICITY AND ELECTRONICS
William Bradford Shockley (1910-1989) established the first semiconductor company in Santa Clara County, California, the region known by about 1970 as Silicon Valley. Previously, Shockley had been employed by Bell Laboratories.

1955. ORGANIZATIONS—RESEARCH INSTITUTIONS / GENETICS
Beginning in this year, summer courses at the Cold Spring Harbor Biological Laboratory (New York) made it an international center for the study of bacterial genetics.

1955. ORGANIZATIONS—SOCIETIES AND ASSOCIATIONS / CHEMISTRY, GEOLOGY
The Geochemical Society was established.

1955. PERIODICALS AND PUBLISHING / BIOLOGY—GENERAL
The *Journal of Biophysical and Biochemical Cytology* (later *Journal of Cell Biology*) was founded.

1955. PHYSICS
At the Lawrence Radiation Laboratory (Berkeley), Owen Chamberlain (1920-) and Emilio Gino Segre (1905-1989) used the Bevatron particle accelerator to create an antiproton. This particle had been predicted by Paul Dirac in 1930. In 1959, they received the Nobel Prize in physics for their discovery.

1955. PHYSICS
Arthur Leonard Schawlow (1921-) and Charles H. Townes (1915-) published *Microwave Spectroscopy* (New York).

1955. PHYSICS
Maria Goeppert Mayer (1906-1972) and German physicist J. Hans D. Jensen published the textbook, *Elementary Theory of Nuclear Shell Structure* (New York).

1955. PHILOSOPHY
Nelson Goodman (1906-) published *Fact, Fiction, and Forecast* (Cambridge, Mass.). In this work, he presented what became known as "Goodman's paradox" or "new riddle of induction."

1955 (February 20). GENERAL OR MISCELLANEOUS / MICROBIOLOGY AND MICROSCOPY
Bacteriologist Oswald Theodore Avery (b.1877) died in Nashville, Tennessee.

1955 (April 18). GENERAL OR MISCELLANEOUS / PHYSICS
Physicist Albert Einstein (b.1879) died in Princeton, New Jersey.

1955 (July 29). SPACE SCIENCE, TECHNOLOGY, AND EXPLORATION
The United States announced plans to launch artificial satellites as part of the activities of the International Geophysical Year. A target date of 1957 was set.

1955 (August 8-20). GENERAL OR MISCELLANEOUS / NUCLEAR POWER AND WEAPONS
Through United States initiative, an International Conference on the Peaceful Uses of Atomic Energy was held in Geneva. There were nearly 1500 delegates and about 450 scientific papers.

1956. AWARDS AND PRIZES
George Gamow (1904-1968) won the UNESCO Kalinga Prize for his popular books on science.

1956. AWARDS AND PRIZES / MEDICINE, PHYSIOLOGY
Andre F. Cournand (1895-1988) and Dickinson Woodruff Richards (1895-1973) won the Nobel Prize in medicine or physiology. It was shared with the German Werner Forssmann. They were honored "for their discoveries concerning heart catherization and pathological changes in the circulatory system."

1956. AWARDS AND PRIZES / PHYSICS
John Bardeen (1908-1991), Walter H. Brattain (1902-1987), and William Bradford Shockley (1910-1989) won the Nobel Prize in physics. They were honored "for their researches on semiconductors and their discovery of the transistor effect."

1956. BIOCHEMISTRY AND MOLECULAR BIOLOGY
Test tube replication of biologically inactive DNA was achieved by Arthur Kornberg (1918-). He used enzyme from the bacterium *Escherichia coli* acting on nucleic-acid bases, or nucleotides. Synthesis of biologically active DNA was announced by Kornberg and associates in 1967.

1956. BIOCHEMISTRY AND MOLECULAR BIOLOGY
Transfer RNA was discovered by Paul Berg (1926-).

1956. CHEMISTRY
Robert B. Woodward (1917-1979) and associates synthesized reserpine.

1956. COMPUTERS AND INFORMATION SCIENCE
FORTRAN (FORmula TRANslation), the first computer programming language, was devised by John Backus (1924-) and associates at IBM. It was extensively used for scientific programming.

1956. COMPUTERS AND INFORMATION SCIENCE
John McCarthy (1927-) devised a computer language for artificial intelligence called LISP (List Processor).

1956. GEOLOGY
The 1952 discovery by cartographer Marie Tharp (1920-) of the rift valley, a seismically active aspect of the Mid-Atlantic Ridge, was confirmed by Bruce C. Heezen (1924-1977) and announced by Heezen and (William) Maurice Ewing (1906-1974) at a meeting of the American Geophysical Union.

1956. PERIODICALS AND PUBLISHING / PHYSICS
The American Institute of Physics began a program of regular translation of eight Soviet journals in physics into English.

1956. PERIODICALS AND PUBLISHING / PSYCHOLOGY
The journal *Contemporary Psychology* was established.

1956. PHILOSOPHY / PHYSICS
The posthumous work of Hans Reichenbach (1891-1953), *The Direction of Time* (Berkeley-Los Angeles) was published under the editorship of Maria Reichenbach.

1956. PHYSICS
Tsung-Dao Lee (1926-) and Chen Ning Yang (1922-) proposed that the law of conservation of parity, in which nature recognizes no difference between left and right, does not apply to the weak interaction. They put forth ideas on how the proposal could be tested and it subsequently was confirmed by experiments at Columbia University by Chien-Shiung Wu (1912-). In 1957, Lee and Yang received the Nobel Prize in physics for their discovery; Wu, a woman, was not included.

1956. PHYSICS
Predicted in 1930 (by Wolfgang Pauli), neutrinos were seen for the first time by Clyde Lorrain Cowan (1919-1974) and Frederick Reines (1918-).

1956. PHYSICS
The anti-neutron was discovered by scientists using the Bevatron at the University of California, Berkeley.

1956. ZOOLOGY
Alfred Sherwood Romer (1894-1973) published *The Osteology of the Reptiles* (Chicago).

1956. ZOOLOGY—HUMAN / EVOLUTION, PSYCHOLOGY
The most significant work of Charles Judson Herrick (1868-1960), *The Evolution of Human Nature* (Austin, Texas), was published.

1956 (February 3). GENERAL OR MISCELLANEOUS / PSYCHOLOGY
Comparative psychologist Robert Mearns Yerkes (b.1876) died in New Haven, Connecticut.

1956 (February 25). GENERAL OR MISCELLANEOUS / BOTANY
Botanist Elmer Drew Merrill (b.1876) died in Forest Hills, Massachusetts.

1956 (March 22). GENERAL OR MISCELLANEOUS / HISTORY OF SCIENCE—FIELD OF STUDY
Historian of science George Alfred Leon Sarton (b.1884) died in Cambridge, Mass.

1957. ASTRONOMY
William Alfred Fowler (1911-), E. Margaret Burbidge (1919-), Geoffrey R. Burbidge (1925-), and English astronomer Fred Hoyle published a paper that outlined the production of all natural elements in stellar reactions.

1957. AWARDS AND PRIZES / PHYSICS
The Acoustical Society of America first awarded the Wallace Clement Sabine Medal, to Vern Oliver Knudsen (1893-1974).

1957. AWARDS AND PRIZES / PHYSICS
Tsung-Dao Lee (1926-) and Chen Ning Yang (1922-) won the Nobel Prize in physics. They were honored "for their penetrating investigation of the so-called parity laws which has led to important discoveries regarding the elementary particles."

1957. BIOCHEMISTRY AND MOLECULAR BIOLOGY
Matthew Meselson (1930-), Franklin W. Stahl (1929-), and Jerome Ruben Vinograd (1913-1976) published "Equilibrium Sedimentation of Macromolecules in Density Gradients," *Proceedings of National Academy of Sciences* 43:581-588. The technique was significant to the development of molecular biology, demonstrating that, through use of ultracentrifuge, it was possible to distinguish components of minutely differing molecular weight in a mixture.

1957. ENGINEERING AND APPLIED SCIENCE / NUCLEAR POWER AND WEAPONS
The first large-scale nuclear power plant in the United States, at Shippingport, Pennsylvania, went into operation.

1957. MEDICINE / DISEASE
A polio vaccine that used live but weakened virus was introduced by Albert Bruce Sabin (1906-1993). The vaccine could be administered orally and it came to supplant that of Jonas Edward Salk (1914-1995).

1957. ORGANIZATIONS—OBSERVATORIES
The National Radio Astronomy Observatory was established as a facility administered by Associated Universities, Inc. and located at Green Bank, West Virginia.

1957. ORGANIZATIONS—RESEARCH INSTITUTIONS / COMPUTERS AND INFORMATION SCIENCE
System Development Corporation (SDC) was founded as a nonprofit entity by the Rand Corporation. The company was founded to serve the information systems needs of the Air Defense Command. In 1969, it became a for-profit organization and in 1981 merged with the Burroughs Corporation.

1957. ORGANIZATIONS—SOCIETIES AND ASSOCIATIONS / BIOLOGY—GENERAL, PHYSICS
The Biophysical Society was established.

1957. ORGANIZATIONS—SOCIETIES AND ASSOCIATIONS / ENVIRONMENT AND CONSERVATION
The Institute of Environmental Sciences was established.

1957. ORGANIZATIONS—SOCIETIES AND ASSOCIATIONS / NUCLEAR POWER AND WEAPONS
Sane, National Committee for a Sane Nuclear Policy was founded.

1957. PHYSICS
E. Richard Cohen (1922-), Kenneth M. Crowe (1926-), and Jesse W.M. DuMond (1892-1976) published *Fundamental Constants of Physics* (New York).

1957. PHYSICS
A theoretical explanation of superconductivity (in which a number of metals lose all resistance to electrical current at very low temperature) was proposed by John Bardeen (1908-1991), Leon N. Cooper (1930-),

and John R. Schrieffer (1931-). Known as the BCS theory, and based on quantum theory, it won the three scientists the Nobel Prize in physics in 1972.

1957. PSYCHOLOGY / PHYSICS
Psychophysicist Stanley Smith Stevens (1906-1973) put forth the power function that correlated increase in sensation intensity with an increase in stimulus intensity. His "On the Psychophysical Law," *Psychological Review* 64:153-181 is related to this work.

1957 (February 8). GENERAL OR MISCELLANEOUS / MATHEMATICS
Mathematician John von Neumann (b.1903) died in Washington, D.C.

1957 (February 18). GENERAL OR MISCELLANEOUS / ASTRONOMY
Astronomer Henry Norris Russell (b.1877) died in Princeton, New Jersey.

1957 (March 11). GENERAL OR MISCELLANEOUS / EXPLORATION AND SURVEYING
Explorer Richard Evelyn Byrd (b.1888) died in Boston, Massachusetts.

1957 (July 29). GENERAL OR MISCELLANEOUS / NUCLEAR POWER AND WEAPONS
Legislation for United States membership in the International Atomic Energy Agency was signed by President Dwight Eisenhower.

1957 (August 16). GENERAL OR MISCELLANEOUS / CHEMISTRY
Physical chemist Irving Langmuir (b.1881) died in Falmouth, Massachusetts.

1957 (September 19). MILITARY AND WAR / NUCLEAR POWER AND WEAPONS
The country's first underground atomic testing was carried out in Nevada.

1957 (October 4). GENERAL OR MISCELLANEOUS / SPACE SCIENCE, TECHNOLOGY, AND EXPLORATION
The Soviet Union launched *Sputnik*, the first satellite.

1957-1959. ORGANIZATIONS / GEOPHYSICS AND GEODESY

Begun with the suggestion by Lloyd Viel Berkner (1905-1967) for an international polar year, the International Geophysical Year was originally planned for 1957-58 and later expanded to the end of 1959. The IGY committee, created by the International Council of Scientific Unions, was headed by Sydney Chapman (Great Britain) as president with Berkner as vice-president. Internationally, the endeavor came to involve more than 30,000 individuals, the largest such undertaking ever attempted. The most costly component of the United States effort, rocket and satellite concerns, was directed by Berkner.

1957-1973. GOVERNMENT—FEDERAL

The President's Science Advisory Committee (PSAC) was in existence during this period. A predecessor science advisory committee had been created on April 20, 1951 by the president in the Office of Defense Mobilization, but when it was refurbished in 1957 as PSAC it was relocated in the White House. PSAC and the position of Special Assistant to the President for Science and Technology (Science Advisor) were created in November 1957 by President Dwight Eisenhower following the Soviet launching of *Sputnik* in October. James R. Killian, Jr. (1904-1988) was appointed as the first full-time science advisor to the president.

1958. AWARDS AND PRIZES / MEDICINE, PHYSIOLOGY

George Wells Beadle (1903-1989), Joshua Lederberg (1925-), and Edward Lawrie Tatum (1909-1975) received the Nobel Prize in medicine or physiology. Beadle and Tatum were honored "for their discovery that genes act by regulating definite chemical events," Lederberg "for his discoveries concerning genetic recombination and the organization of the genetic material of bacteria."

1958. GEOLOGY

Bruce C. Heezen (1924-1977) reported his concept of an expanding earth that accompanied new crust from the rift valley of the Mid-Oceanic Ridge. He further suggested that such earth expansion was a way in which continental drift could take place. (His idea was presented in a paper, "Geologie sous-marine et deplacements des continents" at Nice, France.) The expanding earth idea ultimately proved to be wrong, but the phenomenon of new seafloor material at the ridge was later used to support Harry Hess's (1906-1969) concept of seafloor spreading.

1958. GOVERNMENT—FEDERAL / MILITARY AND WAR
The Jason Division of the Institute for Defense Analyses was formed to focus scientific talent (especially that of physicists) on issues of defense. The group was a particular target of the New Left in protest against the war in Vietnam.

1958. GOVERNMENT—FEDERAL / SPACE SCIENCE, TECHNOLOGY, AND EXPLORATION
The National Aeronautics and Space Administration was established when President Dwight Eisenhower signed legislation on 29 July. The employees of the National Advisory Committee for Aeronautics were transferred to the new agency.

1958. MEDICINE / DISEASE
An effective vaccine for measles was produced by John Franklin Enders (1897-1985) and colleagues. The development of the vaccine was reported by Samuel L. Katz (1927-) and Enders in "Immunization of Children with Live Attenuated Measles Virus," *A.M.A. Journal of Diseases of Children* 98 (November 1959): 605-607.

1958. ORGANIZATIONS—OBSERVATORIES
The Kitt Peak National Observatory was founded.

1958. ORGANIZATIONS—SOCIETIES AND ASSOCIATIONS / SPACE SCIENCE, TECHNOLOGY, AND EXPLORATION
The National Academy of Sciences' Space Sciences Board was established.

1958. PHYSICS
Arthur Leonard Schawlow (1921-) and Charles Hard Townes (1915-) published "Infrared and Optical Masers," *Physical Review* 2nd series 112 (December 15, 1958): 1940-1949. Their work was a theoretical step in the development of the laser (Light Amplification by Stimulated Emission of Radiation).

1958 (January 16). GENERAL OR MISCELLANEOUS / SPACE SCIENCE, TECHNOLOGY, AND EXPLORATION
Secretary of State John F. Dulles asked for an international agreement for the peaceful uses of outer space.

1958 (January 31). SPACE SCIENCE, TECHNOLOGY, AND EXPLORATION / PHYSICS
The first United States satellite, Explorer I, was orbited by the Army under the direction of Werner von Braun (1912-1977). What came to be known as the Van Allen radiation belts that encircle the earth were first discovered by a Geiger counter in Explorer I, placed there by James Alfred Van Allen (1914-). Further knowledge of the radiation belts was derived from subsequent satellite exploration.

1958 (February 7). GOVERNMENT—FEDERAL / MILITARY AND WAR
The U.S. Defense Department established the Advanced Research Projects Agency. The Agency promoted basic and applied research, which was carried out both within the government and in other sectors.

1958 (March 17). SPACE SCIENCE, TECHNOLOGY, AND EXPLORATION
The U.S. Navy launched the satellite *Vanguard I*.

1958 (July 1). PERIODICALS AND PUBLISHING / PHYSICS
The first number of *Physical Review Letters* was published, under the auspices of the American Physical Society. Promoted by Samuel A. Goudsmit (1902-1978), it was an attempt to promote prompt publication of new research in the rapidly growing field. Goudsmit was editor of the *Letters* as well as of *The Physical Review*.

1958 (August 5). MILITARY AND WAR / NUCLEAR POWER AND WEAPONS
The nuclear submarine *Nautilus* completed the first undersea crossing of the North Pole.

1958 (August 27). GENERAL OR MISCELLANEOUS / PHYSICS
Physicist Ernest Orlando Lawrence (b.1901) died in Palo Alto, California.

1958 (September 2). GOVERNMENT—FEDERAL
The National Defense Education Act, providing federal support for science education, was signed by President Dwight Eisenhower.

1958 (September 25). GENERAL OR MISCELLANEOUS / PSYCHOLOGY
Behavioral psychologist John Broadus Watson (b.1878) died in New York, New York.

1958 (October 12). SPACE SCIENCE, TECHNOLOGY, AND EXPLORATION
Although the *Pioneer* rocket attained a record altitude of 79,193 miles, it failed in the attempt to circle the moon.

1958-1966. GEOLOGY
Project Mohole, to drill into the earth's mantle under the ocean, was funded by the National Science Foundation. The first core samples were produced in 1958. (The agency charged with determination of the drilling location was chaired by Harry Hammond Hess, 1906-1969.)

1959. ASTRONOMY
Frank Donald Drake (1930-) produced a high-resolution map of the galaxy, using the techniques of radio astronomy.

1959. AWARDS AND PRIZES / MEDICINE, PHYSIOLOGY
Arthur Kornberg (1918-) and Severo Ochoa (1905-1993) won the Nobel Prize in medicine or physiology. They were honored "for their discovery of the mechanisms in the biological synthesis of ribonucleic acid and deoxiribonucleic acid."

1959. AWARDS AND PRIZES / PHYSICS
Owen Chamberlain (1920-) and Emilio Gino Segre (1905-1989) won the Nobel Prize in physics. They were honored "for their discovery of the antiproton."

1959. BIOCHEMISTRY AND MOLECULAR BIOLOGY
Robert Bruce Merrifield (1921-) conceptualized solid-phase peptide synthesis, a means of automating the preparation of peptides, the chainlike molecules that are the building units of proteins. The procedure was first used in 1962.

1959. BIOCHEMISTRY AND MOLECULAR BIOLOGY / EVOLUTION
Christian Boehmer Anfinsen (1916-) published *The Molecular Basis of Evolution* (New York).

1959. BIOLOGY—GENERAL
Leo Szilard (1898-1964) published the influential paper "On the Nature of the Aging Process," *Proceedings of National Academy of Sciences* 45:30-45.

1959. MATHEMATICS
Mathematician and statistician Henry Scheffe (1907-1977) published *The Analysis of Variance* (New York).

1959. ORGANIZATIONS—OBSERVATORIES
Otto Struve (1897-1963) became director of the nearly completed National Radio Astronomy Observatory at Green Bank, West Virginia.

1959 (January 3). GOVERNMENT—FEDERAL
The U.S. House Committee on Science and Astronautics was established. The name later was changed (1975) to the Committee on Science and Technology. In addition to legislative purview over particular agencies, the committee came to have a general oversight for research and development activities in the federal establishment.

1959 (August 7). SPACE SCIENCE, TECHNOLOGY, AND EXPLORATION
The National Aeronautics and Space Administration launched its first satellite, *Explorer IV*.

1959-1960. COMPUTERS AND INFORMATION SCIENCE
The business computer language, COBOL (COmmon Business Oriented Language), was developed. Grace Murray Hopper (1906-1992) was a major contributor to the undertaking, which was carried out under the auspices of the Department of Defense.

1960. ASTRONOMY
Allan Rex Sandage (1926-), utilizing the Hale telescope on Mount Palomar, discovered an astronomical body emitting radio waves that later was identified as a quasar (quasi stellar radio source).

1960. AWARDS AND PRIZES / CHEMISTRY
Willard Frank Libby (1908-1980) won the Nobel Prize in chemistry. He was honored "for his method to use carbon 14 for age determination in archaeology, geology, geophysics, and other branches of science."

1960. AWARDS AND PRIZES / PHYSICS
Donald Arthur Glaser (1926-) won the Nobel Prize in physics. He was honored "for the invention of the bubble chamber."

1960. BIOCHEMISTRY AND MOLECULAR BIOLOGY
The sequence of all 124 amino acids in ribonuclease were determined by Stanford Moore (1913-1982) and William Howard Stein (1911-1980).

1960. CHEMISTRY
Robert B. Woodward (1917-1979) synthesized chlorophyll.

1960. MEDICINE / DISEASE
Albert Bruce Sabin (1906-1993) published "Oral, Live Poliovirus Vaccine for Elimination of Poliomyelitis," *A.M.A. Archives of Internal Medicine* 106 (July 1960): 5-9.

1960. ORGANIZATIONS—RESEARCH INSTITUTIONS / AERONAUTICS
Aerospace Corporation (a nonprofit research corporation) was founded at the request of and with the support of the U.S. Air Force.

1960. PHYSICS
The theory of relativity was confirmed in experiments by Robert Vivian Pound (1919-) and Glen A. Rebka (1931-) involving the measurement of changes in frequency of gamma rays (i.e., gravitational red shift) between the bottom and top of a water tower.

1960. RELIGION AND THEOLOGY
Harlow Shapley (1885-1972) edited the book, *Science Ponders Religion* (New York).

1960 (April 1). METEOROLOGY AND CLIMATOLOGY / INSTRUMENTS AND INSTRUMENTATION
The first weather satellite, TIROS 1 (Television and Infra-Red Observation Satellite), was launched.

1960 (May). Physics / Instruments and Instrumentation
The first operational laser, a device under development by scientists in both the United States and the Soviet Union, was demonstrated by Theodore Harold Maiman (1927-). Maiman's report appeared as "Stimulated Optical Radiation in Ruby," *Nature* 187 (August 6, 1960): 493-494.

1960 (June 25). General or Miscellaneous / Astronomy
Astronomer Walter Baade (b.1893) died in Göttingen, Germany.

1960 (August 10). General or Miscellaneous
A treaty that recognized Antarctica as a "peaceful scientific preserve" was consented to by the U.S. Senate. The other parties to the treaty included Belgium, Japan, Norway, South Africa, and Great Britain.

1960 (August 12). Space Science, Technology, and Exploration
The first communication satellite, *Echo 1*, was launched by the National Aeronautics and Space Administration.

1960 (December). Geology
Harry Hammond Hess (1906-1969), in a preprinted paper, proposed the idea of seafloor spreading, in relation to the beginnings and development of midocean ridges. The idea was of crucial importance in the development of plate tectonics later in the 1960s and 1970s. Hess did not publish his ideas until 1962, but Robert S. Dietz (1914-), with access to Hess's preprint, in 1961 published the first article on the subject and introduced the term seafloor spreading.

1960s and 1970s. Religion and Theology
During this period, religiously based evolution changed tactics from earlier times. Rather than attempts to outlaw evolution, the emphasis was on the elevation of a creationist perspective to a par with that of the Darwinian.

1960s. General or Miscellaneous
The U.S. Department of Defense's Project Hindsight showed that the accepted model of pure research leading directly to technology was more complex and less direct or timely than assumed.

1961. AWARDS AND PRIZES / CHEMISTRY
Melvin Calvin (1911-) won the Nobel Prize in chemistry. He was honored "for his research on the carbon dioxide assimilation in plants."

1961. AWARDS AND PRIZES / MEDICINE, PHYSIOLOGY
Georg von Bekesy (1899-1972) won the Nobel Prize in medicine or physiology. He was honored "for his discoveries of the physical mechanism of stimulation within the cochlea."

1961. AWARDS AND PRIZES / PHYSICS
Robert Hofstadter (1915-1990) won the Nobel Prize in physics. It was shared with Rudolf Ludwig Moessbauer (1929-) (German, who in 1961 became professor at California Institute of Technology). Hofstadter was honored "for his pioneering studies of electron scattering in atomic nuclei and for his thereby achieved discoveries concerning the structure of the nucleons." Moessbauer was honored "for his researches concerning the resonance absorption of gamma-radiation and his discovery in this connection of the effect which bears his name."

1961. BIOCHEMISTRY AND MOLECULAR BIOLOGY
Arthur Kornberg (1918-) published *The Enzymatic Synthesis of DNA* (New York).

1961. BIOCHEMISTRY AND MOLECULAR BIOLOGY
The first genetic code was produced by Marshall Warren Nirenberg (1927-) who demonstrated an RNA code for a specific amino acid.

1961. PHYSICS
Murray Gell-Mann (1929-) introduced what he called the "Eightfold Way" as a means to classify elementary particles. The scheme had some gaps and was confirmed in 1964 when the Omega-minus particle was discovered.

1961. GEOLOGY / RELIGION AND THEOLOGY
Henry M. Morris (1918-) and John C. Whitcomb, Jr. (1924-) published *Genesis Flood: The Biblical Record and Its Scientific Implications* (Philadelphia), which gave the creationist movement a major promotion.

1961 (May 5). SPACE SCIENCE, TECHNOLOGY, AND EXPLORATION
Alan Bartlett Shepard, Jr. (1923-) became the first American in space. He rode the Mercury *Freedom 7* spacecraft in a 116-mile high suborbital flight that lasted for 15 minutes.

1961 (May 25). GOVERNMENT—FEDERAL / SPACE SCIENCE, TECHNOLOGY, AND EXPLORATION
In an address to Congress, President John F. Kennedy proposed that the United States set its sights on the landing of a manned vehicle on the moon and its safe return, by the end of the decade. This was the beginning of Project Apollo.

1961 (August 20). GENERAL OR MISCELLANEOUS / PHYSICS
Physicist Percy Williams Bridgman (b.1882) died in Randolph, New Hampshire.

1962. ASTRONOMY
Earl C. Slipher (1883-1964) published *The Photographic Story of Mars* (Cambridge, Mass.). The work included 512 photographs of the planet.

1962. ASTRONOMY
The first discovery of an X-ray source outside the solar system, in Scorpio, was made by Riccardo Giacconi (1931-), Herbert Gursky (1930-), Francis R. Paolini (1930-), and Bruno B. Rossi (1905-1993). It was reported in "Evidence for X-Rays from Sources Outside the Solar System," *Physical Review Letters* 9 (1962):439-443.

1962. AWARDS AND PRIZES
Linus Carl Pauling (1901-1994) won the Nobel Prize for peace. He was honored for work in pointing out the dangers of radioactive fallout in weapons testing and in war. Pauling, who had won the Nobel Prize for chemistry in 1954, became the first person to win an unshared Nobel prize in two different fields.

1962. AWARDS AND PRIZES / MEDICINE, PHYSIOLOGY
James D. Watson (1928-) won the Nobel Prize in medicine or physiology. It was shared with Francis H.C. Crick and Maurice Wilkins of Great Britain. They were honored "for their discoveries concerning

the molecular structure of nucleic acids and its significance for information transfer in living material."

1962. BIOCHEMISTRY AND MOLECULAR BIOLOGY
The first result on the relations between specified nucleotides and amino acids forming proteins was reported by Robert G. Martin (1935-) and others in "Ribonucleotide Composition of the Genetic Code," *Biochemical and Biophysical Research Communications* 6 (January 14, 1962): 410-414. This was a milestone in understanding the genetic code.

1962. ENVIRONMENT AND CONSERVATION
Rachel Louise Carson (1907-1964) published *Silent Spring* (Boston), demonstrating the widespread and interconnected environmental damages done by the use of chemical pesticides. The best-selling book was attacked by the chemical industry that provided agricultural sprays. In 1963, a report from the President's Science Advisory Committee supported Carson's contentions.

1962. HISTORY OF SCIENCE—FIELD OF STUDY
Thomas S. Kuhn (1922-) published *The Structure of Scientific Revolutions* (Chicago).

1962. ORGANIZATIONS—SOCIETIES AND ASSOCIATIONS / NUCLEAR POWER AND WEAPONS
The lobbying organization, Council for a Livable World, was founded by Leo Szilard (1898-1964). Its focus was on issues of nuclear arms and foreign policy.

1962. PHYSICS / INSTRUMENTS AND INSTRUMENTATION
Stanford Linear Accelerator (SLAC) was established. It became operational in 1966.

1962. SPACE SCIENCE, TECHNOLOGY, AND EXPLORATION
The first interplanetary exploration took place. On 27 August, *Mariner 2* was launched by the Jet Propulsion Laboratory and on 14 December passed within 22,000 miles of Venus.

1962. ZOOLOGY—HUMAN / EVOLUTION, GENETICS
Theodosius Dobzhansky (1900-1975) published *Mankind Evolving: The Evolution of the Human Species* (New Haven, Conn.), drawing upon genetics, evolutionary theory, anthropology, and sociology; it approached the question in both the biological and the cultural domain.

1962. GOVERNMENT—FEDERAL
President John F. Kennedy created the Office of Science and Technology (OST) in the Executive Office of the President. The director of OST also was chair of the President's Science Advisory Committee (PSAC). Both PSAC and OST were terminated in 1973 by President Richard Nixon.

1962 (February 20). SPACE SCIENCE, TECHNOLOGY, AND EXPLORATION
John H. Glenn, Jr. (1921-) made the first American orbit of earth, aboard *Mercury 6*.

1962 (March 15). GENERAL OR MISCELLANEOUS / PHYSICS
Physicist Arthur Holly Compton (b.1892) died in Berkeley, California.

1962 (July 10). SPACE SCIENCE, TECHNOLOGY, AND EXPLORATION
Telstar, the earliest amplifying communications satellite, was launched. The project involved the National Aeronautics and Space Administration and AT&T.

1963. ASTRONOMY
The first quasar (quasi-stellar radio source) was recognized as a new type of astronomical object by Maarten Schmidt (1929-).

1963. AWARDS AND PRIZES / PHYSICS
Maria Goeppert Mayer (1906-1972) and Eugene Paul Wigner (1902-1995) won the Nobel Prize in physics. J. Hans D. Jensen in Germany also won the prize. Goeppert Mayer and Jensen were honored "for their discoveries concerning nuclear shell structure," Wigner "for his contributions to the theory of the atomic nucleus and the elementary particles, particularly through the discovery and application of fundamental symmetry principles."

1963. BIOCHEMISTRY AND MOLECULAR BIOLOGY
Jerome Ruben Vinograd (1913-1976) and Swiss scientist Roger Weil (1928-) published "The Cyclic Helix and Cyclic Coil Forms of Polyoma Viral DNA," *Proceedings of National Academy of Sciences* 50:730-738. During this year, Renato Dulbecco (1914-) and Marguerite Vogt (1913-) together also published first accounts of closed circular DNA.

1963. BIOLOGY—GENERAL / ECOLOGY
Victor Ernest Shelford (1877-1968) published *The Ecology of North America* (Urbana, Illinois).

1963. ORGANIZATIONS—RESEARCH INSTITUTIONS / BIOLOGY—GENERAL
Jonas Edward Salk (1914-1995) established the Salk Institute for Biological Studies at La Jolla, California.

1963. ORGANIZATIONS—SOCIETIES AND ASSOCIATIONS /
ENGINEERING AND APPLIED SCIENCE
The Institute of Electrical and Electronics Engineers was established by the merger of the American Institute of Electrical Engineers and the Institute of Radio Engineers. The roots of the new organization went back to 1884.

1963. ORGANIZATIONS—SOCIETIES AND ASSOCIATIONS / RELIGION
AND THEOLOGY
The Creation Research Society was founded. The organization required that members have a graduate degree in science.

1963. ORGANIZATIONS—SOCIETIES AND ASSOCIATIONS / SPACE
SCIENCE, TECHNOLOGY, AND EXPLORATION
The American Institute of Aeronautics and Astronautics was formed when the Institute of Aerospace Science and the American Rocket Society combined.

1963. STATISTICS
William Gemmell Cochran (1909-1980) published *Sampling Techniques* (New York).

Chronology 313

1963 (April 5). AWARDS AND PRIZES / PHYSICS
J. Robert Oppenheimer (1904-1967) received the U.S. Atomic Energy Commission's Enrico Fermi Prize of $50,000 for his contributions to nuclear energy.

1963 (August 5). GENERAL OR MISCELLANEOUS / NUCLEAR POWER AND WEAPONS
A nuclear test ban treaty was signed by the United States, Great Britain, and the Soviet Union. The terms banned testing in the atmosphere, underwater, and in outer space. The treaty was affirmed by the U.S. Senate on September 24 and took effect on October 10.

1964. ASTRONOMY / PHYSICS
Arno A. Penzias (1933-) and Robert Woodrow Wilson (1936-) accidentally discovered cosmic microwave background radiation, which emanated from the universe in general. As a remnant of the Big Bang, it gave support to that theory of the history of the universe. In 1978, Penzias and Wilson won the Nobel Prize in physics for the discovery.

1964. AWARDS AND PRIZES / MEDICINE, PHYSIOLOGY
Konrad E. Bloch (1912-) won the Nobel Prize in medicine or physiology. Feodor F.K. Lynen, in Germany, also won the prize. They were honored "for their discoveries concerning the mechanism and regulation of the cholesterol and fatty acid metabolism."

1964. AWARDS AND PRIZES / PHYSICS
Charles Hard Townes (1915-) won the Nobel Prize in physics. It was shared with the Russians Nikolai G. Basov and Alexander M. Prokhorov. They were honored "for fundamental work in the field of quantum electronics, which ... led to the construction of oscillators and amplifiers based on the maser-laser principle."

1964. BIOCHEMISTRY AND MOLECULAR BIOLOGY
The sequence of nucleotides in DNA was shown by Charles Yanofsky (1925-) and English scientist Sydney Brenner to be in linear relation to the order of amino acids in proteins, a significant event in understanding protein synthesis.

1964. COMPUTERS AND INFORMATION SCIENCE
IBM introduced the first word processor, the Magnetic Tape/Selectric Typewriter.

1964. GENERAL OR MISCELLANEOUS
In the presidential campaign of this year, a new precedent for political involvement was evident in Scientists and Engineers for Johnson.

1964. ORGANIZATIONS—SOCIETIES AND ASSOCIATIONS / ENGINEERING AND APPLIED SCIENCE
The National Academy of Engineering was established. It was founded under the charter of the National Academy of Sciences as a participant in the NAS role as advisor to the United States government.

1964. ORGANIZATIONS—SOCIETIES AND ASSOCIATIONS / RELIGION AND THEOLOGY
The Bible-Science Association was founded.

1964. PHYSICS
Murray Gell-Mann (1929-) and George Zweig (1937-) proposed the idea of quarks, particles of fractional charge that make up protons, neutrons, and other so-called elementary particles.

1964. PHYSICS
Sheldon Lee Glashow (1932-) proposed the existence of a fourth (what was referred to as a charmed) quark. Though initially doubted, its existence was confirmed when the J (Psi) particle was discovered in 1974.

1964. PHYSICS
Experiments conducted by James Watson Cronin (1931-) and Val Logsdon Fitch (1923-) showed asymmetry in the course of decay of neutral K mesons. This was disproof of the law of symmetry that was thought to be in effect for the behavior of subatomic particles. For their work, Cronin and Fitch shared the Nobel Prize in physics in 1980.

1964. ZOOLOGY
D. Dwight Davis (1908-1965) published *The Giant Panda* (Chicago), a work notable in the field of comparative anatomy.

1964 (January 11). MEDICINE / PATHOLOGY
The U.S. Surgeon General issued a committee report, *Smoking and Health*, that gave conclusive proof that tobacco smoking causes lung cancer. On January 13, the Federal Trade Commission announced that cigarette packages would be required to carry a warning regarding the health hazards of smoking.

1964 (March 18). GENERAL OR MISCELLANEOUS / MATHEMATICS
Mathematician Norbert Wiener (b.1894) died in Stockholm, Sweden.

1964 (May 30). GENERAL OR MISCELLANEOUS / BIOLOGY, PHYSICS
Physicist and biologist Leo Szilard (b.1898) died in La Jolla, California.

1965. AWARDS AND PRIZES / CHEMISTRY
Robert Burns Woodward (1917-1979) won the Nobel Prize in chemistry. He was honored "for his outstanding achievements in the art of organic synthesis."

1965. AWARDS AND PRIZES / PHYSICS
Richard Phillips Feynman (1918-1988) and Julian Seymour Schwinger (1918-1994) won the Nobel Prize in physics. It was shared by Sin-Itiro Tomonaga in Japan. They were honored "for their fundamental work in quantum electrodynamics, with deep-ploughing consequences for the physics of elementary particles."

1965. CHEMISTRY / PHYSICS
Roald Hoffmann (1937-) and Robert B. Woodward (1917-1979), applying quantum theory to chemistry, developed rules of orbital symmetry.

1965. COMPUTERS AND INFORMATION SCIENCE
BASIC computer language (Beginners All-purpose Symbolic Instruction Code) was developed by John G. Kemeny (1926-1992) and Thomas E. Kurtz (1928-).

1965. PALEONTOLOGY
Marshall Kay (1904-1975) and Edwin H. Colbert (1905-) published the textbook, *Stratigraphy and Life History* (New York).

1965. STATISTICS
Lester E. Dubins (1920-) and Leonard Jimmie Savage (1917-1971) published *How to Gamble If You Must: Inequalities for Stochastic Processes* (New York).

1965. ZOOLOGY—HUMAN / MEDICINE
Gregory Goodwin Pincus (1903-1967) published *Control of Fertility* (New York).

1965. BIOCHEMISTRY AND MOLECULAR BIOLOGY
The entire structure of a molecule of transfer-RNA was delineated by Robert W. Holley (1922-1993).

1965 (April 6). SPACE SCIENCE, TECHNOLOGY, AND EXPLORATION
The first commercial satellite, *Early Bird*, was launched by NASA. It was intended for the transmittal of telephone and television signals.

1965 (June 3). SPACE SCIENCE, TECHNOLOGY, AND EXPLORATION
The space flight *Gemini 4* was launched. During its mission, the initial American spacewalk took place; the Soviet Union had first performed the feat in the spring.

1965 (July 14). SPACE SCIENCE, TECHNOLOGY, AND EXPLORATION
Mariner IV approached within 6200 miles of Mars, the first such exploration of the planet.

1965 (December 5). GENERAL OR MISCELLANEOUS / PHYSIOLOGY
Physiologist Joseph Erlanger (b.1874) died in St. Louis, Missouri.

1965-1966. GEOLOGY
The large collection of oceanographic data at the Lamont Geological Observatory of Columbia University was used by its staff to confirm a central hypothesis supporting the concept of plate tectonics. (William) Maurice Ewing (1906-1974) was director of the Observatory and overcame his initial opposition to continental drift and seafloor spreading to accept it by the end of 1966.

1966. AWARDS AND PRIZES / CHEMISTRY
Robert Sanderson Mulliken (1896-1986) won the Nobel Prize in chemistry. He was honored "for his fundamental work concerning chemical bonds and the electronic structure of molecules by the molecular orbital method."

1966. AWARDS AND PRIZES / MEDICINE, PHYSIOLOGY
Charles Brenton Huggins (1901-) and (Francis) Peyton Rous (1879-1970) won the Nobel Prize in medicine or physiology. Huggins was honored "for his discoveries concerning hormonal treatment of prostatic cancer," Rous "for his discovery of tumor inducing viruses." Rous's prize came fifty-five years after the research was done.

1966. METEOROLOGY AND CLIMATOLOGY / INSTRUMENTS AND INSTRUMENTATION
ESSA I, the first weather satellite that could observe the entire earth, was launched by the Environmental Science Services Administration.

1966. ZOOLOGY—HUMAN / PHYSIOLOGY, PSYCHOLOGY
William Howell Masters (1915-) and Virginia Eshelman Johnson (1925-) published *Human Sexual Response*. It was the first general study of the anatomy and physiology of human sexual activity.

1966 (February 23). GOVERNMENT—FEDERAL / ENVIRONMENT AND CONSERVATION
A special message from President Lyndon Johnson to Congress called for efforts to restore and protect the environment, including pollution-control research. In November 1966, he signed the Clean Waters Restoration Act.

1966 (May 30). SPACE SCIENCE, TECHNOLOGY, AND EXPLORATION
Surveyor I, a spacecraft without personnel aboard, made the first United States soft landing on the moon. The Soviet Union had accomplished the feat three months earlier.

1966 (September 6). GENERAL OR MISCELLANEOUS
Founder of what became the Planned Parenthood Federation of America, Margaret Sanger (b.1879), died at Tucson, Arizona.

1967. AWARDS AND PRIZES / MEDICINE, PHYSIOLOGY
Haldan Keffer Hartline (1903-1983) and George Wald (1906-) won the Nobel Prize in medicine or physiology. It was shared with Ragnar A. Granit in Sweden. They were honored "for their discoveries concerning the primary physiological and chemical visual processes in the eye."

1967. AWARDS AND PRIZES / PHYSICS
Hans Albrecht Bethe (1906-) won the Nobel Prize in physics. He was honored "for his contributions to the theory of nuclear reactions, especially his discoveries concerning the energy production in stars."

1967. BIOLOGY—GENERAL
Robert Helmer MacArthur (1930-1972) and Edward O. Wilson (1929-) published *The Theory of Island Biogeography* (Princeton). This work established species equilibrium theory.

1967. GEOLOGY
Marshall Kay (1904-1975) organized the International Gander Conference on Stratigraphy and Structure Bearing on the Origin of the North Atlantic. Held in Newfoundland, the conference coincided with the year in which the plate tectonics theory was made public. The conference volume was edited by Kay, *North Atlantic: Geology and Continental Drift* (Tulsa, Oklahoma, 1969).

1967. GEOLOGY
A theory of plate tectonics was proposed by W. Jason Morgan (1935-) and in Britain by Daniel P. McKenzie and Robert L. Parker.

1967 (January 16). GENERAL OR MISCELLANEOUS / PHYSICS
Physicist Robert Jemison Van de Graaff (b.1901) died in Boston, Massachusetts.

1967 (January 27). SPACE SCIENCE, TECHNOLOGY, AND EXPLORATION
Ground testing of *Apollo I* resulted in a fire in which three astronauts died.

1967 (February 18). GENERAL OR MISCELLANEOUS / PHYSICS
Physicist J. Robert Oppenheimer (b.1904) died in Princeton, New Jersey.

1967 (April 5). GENERAL OR MISCELLANEOUS / GENETICS
Geneticist Hermann Joseph Muller (b.1890) died in Indianapolis, Indiana.

1967 (May 18). GENERAL OR MISCELLANEOUS / RELIGION AND THEOLOGY
The Tennessee state law that made illegal the teaching of evolution in the public schools, and under which John Thomas Scopes (1900-1970) was tried in 1925, was repealed.

1967 (August 22). GENERAL OR MISCELLANEOUS / ZOOLOGY
Endocrinologist Gregory Goodwin Pincus (b.1903) died in Boston, Massachusetts.

1967 (November 14). GOVERNMENT—FEDERAL / ENVIRONMENT AND CONSERVATION
Congress passed and President Lyndon Johnson signed the Air Quality Act.

1967-1968. PHYSICS
Steven Weinberg (1933-) and Pakistani physicist Abdus Salam independently proposed a theory to unify the weak and the electromagnetic forces, called "electroweak theory." Their ideas drew upon related work of Sheldon Lee Glashow (1932-). The predicted particles were discovered at the Centre Europeen de Recherche Nucleaire (CERN) in 1983. The three shared the 1979 Nobel Prize for their efforts.

1967 and 1970. HISTORY OF SCIENCE—FIELD OF STUDY / TECHNOLOGY AND INVENTION
Lewis Mumford (1895-1990) published his two-volume work, *The Myth of the Machine* (New York), a critical historical study of the relations of technology and civilization.

1967-1977. CHEMISTRY
Louis Fieser (1899-1977), with his wife Mary Peters Fieser (1909-), issued the series, *Reagents for Organic Synthesis* (New York), in six volumes. After his death in 1977, Mary Peters Fieser prepared additional volumes in the series.

1968. AWARDS AND PRIZES / CHEMISTRY
Lars Onsager (1903-1976) won the Nobel Prize in chemistry. He was honored "for the discovery of the reciprocal relations bearing his name, which are fundamental for the thermodynamics of irreversible processes."

1968. AWARDS AND PRIZES / MEDICINE, PHYSIOLOGY
Robert William Holley (1922-1993), Har Gobind Khorana (1922-), and Marshall Warren Nirenberg (1927-) won the Nobel Prize in medicine or physiology. They were honored "for their interpretation of the genetic code and its function in protein synthesis."

1968. AWARDS AND PRIZES / PHYSICS
Luis Walter Alvarez (1911-1988) won the Nobel Prize in physics. He was honored "for his decisive contributions to elementary particle physics, in particular the discovery of a large number of resonance states, made possible through his development of the technique of using hydrogen bubble chamber and data analysis."

1968. BIOLOGY—GENERAL / GEOLOGY
The work of Elso S. Barghoorn (1915-1984) and associates uncovered traces of amino acids in rocks some three billion years old, evidence that life forms were present at least that early.

1968. GENERAL OR MISCELLANEOUS / BIOCHEMISTRY AND MOLECULAR BIOLOGY
James D. Watson (1928-) published *Double Helix: Being a Personal Account of the Discovery of the Structure of DNA* (New York).

1968. GENERAL OR MISCELLANEOUS / RELIGION AND THEOLOGY
An Arkansas anti-evolution law was declared unconstitutional by the U.S. Supreme Court.

1968. GOVERNMENT—FEDERAL / PHYSICS
The Atomic Energy Commission's Fermi National Accelerator Laboratory (Fermilab) was founded at Batavia, Illinois.

1968. OCEANOGRAPHY / GEOLOGY
The oceanographic vessel *Glomar Challenger* became operational. The Deep Sea Drilling Project, over 15 years, resulted in a great many core samples from the ocean bottom, and supplied support for the theory of plate tectonics.

1968 (August 19). GENERAL OR MISCELLANEOUS / PHYSICS
Physicist George Gamow (b.1904) died in Boulder, Colorado.

1968 (December). SPACE SCIENCE, TECHNOLOGY, AND EXPLORATION
Apollo 8 was launched on 21 December and became the first spacecraft with personnel aboard to circumnavigate the moon. The crew on the ten orbit flight consisted of Frank Borman (1928-), James A. Lovell, Jr. (1928-), and William A. Anders (1933-).

1968-1978. GENETICS / EVOLUTION
Sewall Wright (1889-1988) published *Evolution and the Genetics of Populations* (Chicago) in four volumes.

1969. AWARDS AND PRIZES / MEDICINE, PHYSIOLOGY
Max Delbruck (1906-1981), Alfred Day Hershey (1908-), and Salvador Edward Luria (1912-1991) won the Nobel Prize in medicine or physiology. They were honored "for their discoveries concerning the replication mechanism and the genetic structure of viruses."

1969. AWARDS AND PRIZES / PHYSICS
Murray Gell-Mann (1929-) won the Nobel Prize in physics. He was honored "for his contributions and discoveries concerning the classification of elementary particles and their interactions."

1969. BIOCHEMISTRY AND MOLECULAR BIOLOGY
Robert Bruce Merrifield (1921-) and Bernd Gutte synthesized ribonuclease, complex enzyme that has a part in cleavage of nucleotides in RNA. They utilized Merrifield's solid-phase peptide synthesis method.

1969. BIOCHEMISTRY AND MOLECULAR BIOLOGY / GENETICS
A single gene was successfully isolated by Jonathan Beckwith (1935-) and colleagues at the Harvard Medical School. They worked with a bacterial gene involved in the metabolism of sugar.

1969. GENERAL OR MISCELLANEOUS
A study by Edward U. Condon (1902-1974) for the U.S. Air Force resulted in a report that found no evidence that unidentified flying objects were space vehicles from outer space.

1969. GOVERNMENT—FEDERAL / MILITARY AND WAR
The U.S. Congress passed a provision in the military authorization bill (referred to as the Mansfield amendment, for Senate Majority Leader Mike Mansfield, who had introduced the provision) that prohibited the Defense Department from funding research not directly related to a particular military concern. The amendment was omitted from the authorization the next year, but the sentiment in the defense establishment on the limits of acceptably funded academic research lingered.

1969. MATHEMATICS
Stefan Bergman (1895-1977) published a second, revised edition of *Integral Operators in the Theory of Linear Partial Differential Equations* (New York). The first edition was published in 1961.

1969. MEDICINE / SURGERY
The first artificial heart implant in a human was carried out by surgeon Denton Arthur Cooley (1920-). The plastic heart was designed by Argentinian Domingo Liotta (1924-). The artificial heart was replaced soon thereafter by a human heart but the patient died.

1969. ORGANIZATIONS—SOCIETIES AND ASSOCIATIONS
The Union of Concerned Scientists was founded.

1969. ORGANIZATIONS—SOCIETIES AND ASSOCIATIONS
SESPA (Scientists and Engineers for Social and Political Action) was founded; it became known as Science for the People.

1969. PSYCHOLOGY
Arthur R. Jensen (1923-) contended that the average lower scores of American blacks than whites on certain mental tests were explainable in genetic terms. The presentation resulted in widespread controversy.

Chronology

1969 (July 20). SPACE SCIENCE, TECHNOLOGY, AND EXPLORATION
As part of the mission of *Apollo 11*, launched on 16 July, Neil Armstrong (1930-) and Edwin (Buzz) Aldrin (1930-) landed on the moon. Armstrong was the first human to stand on the lunar surface. A second lunar landing took place, with a different crew, on 19 November.

1969 (August 17). GENERAL OR MISCELLANEOUS / PHYSICS
Physicist Otto Stern (b.1888) died in Berkeley, California.

1969 (August 25). GENERAL OR MISCELLANEOUS / GEOLOGY
Geologist and oceanographer Harry Hammond Hess (b.1906) died in Woods Hole, Massachusetts.

1969 (November 8). GENERAL OR MISCELLANEOUS / ASTRONOMY
Astronomer Vesto Melvin Slipher (b.1875) died in Flagstaff, Arizona.

1969 (November 25). GENERAL OR MISCELLANEOUS / MILITARY AND WAR
President Richard Nixon ordered that the entire United States supply of germ warfare material be destroyed.

1969-1970. BIOCHEMISTRY AND MOLECULAR BIOLOGY / BIOTECHNOLOGY
Hamilton Othanel Smith (1931-) and Daniel Nathans (1928-) discovered a restriction enzyme (endonuclease) and demonstrated that it could be used to cut a DNA molecule at particular sites. This was a significant step in research on recombinant DNA.

1970. AWARDS AND PRIZES / AGRICULTURE
Norman Ernest Borlaug (1914-) won the Nobel Prize for peace. He was honored for his leadership role in the Green Revolution. His accomplishments involved development of a high-yield, dwarf, disease-resistant wheat that could be grown in many climates.

1970. AWARDS AND PRIZES / MEDICINE, PHYSIOLOGY
Julius Axelrod (1912-) won the Nobel Prize in medicine or physiology. It was shared with Sir Bernard Katz of Great Britain and Ulf S. von Euler of Sweden. They were honored "for their discoveries concerning

the humoral transmittors in the nerve terminals and the mechanism for their storage, release and inactivation."

1970. BIOCHEMISTRY AND MOLECULAR BIOLOGY
The first complete synthesis of a (yeast) gene (analine-transfer RNA) was accomplished by Har Gobind Khorana (1922-) and associates at the University of Wisconsin. It was constructed directly from the component chemicals, unlike earlier workers who used a natural gene as template.

1970. BIOCHEMISTRY AND MOLECULAR BIOLOGY
Howard Martin Temin (1934-1994) and David Baltimore (1938-) independently discovered reverse transcriptase in virus. The enzyme is capable of directing genetic information from RNA to DNA, a conclusion that contradicted the accepted idea that the transcription relation always was from DNA to RNA.

1970. COMPUTERS AND INFORMATION SCIENCE
IBM developed the floppy disk memory.

1970. GOVERNMENT—FEDERAL / ENVIRONMENT AND CONSERVATION
The U.S. Environmental Protection Agency was established as an independent agency, bringing together programs previously scattered in the government structure.

1970. GOVERNMENT—FEDERAL / METEOROLOGY AND CLIMATOLOGY, OCEANOGRAPHY
The National Oceanic and Atmospheric Administration was established in the U.S. Department of Commerce.

1970. MEDICINE / CHEMISTRY
Linus Carl Pauling (1901-1994) published *Vitamin C and the Common Cold* (San Francisco).

1970. GENERAL OR MISCELLANEOUS / RELIGION AND THEOLOGY
Equal treatment of Biblical accounts of creation with Darwinian theory in textbooks was promulgated by the California State Board of Education.

Chronology

1971. AWARDS AND PRIZES / MEDICINE, PHYSIOLOGY
Earl Wilbur Sutherland, Jr. (1915-1974) won the Nobel Prize in medicine or physiology. He was honored "for his discoveries concerning the mechanisms of the action of hormones."

1971. CHEMISTRY
Robert B. Woodward (1917-1979) synthesized vitamin B12.

1971. GEOLOGY / OCEANOGRAPHY
Bruce C. Heezen (1924-1977) and Charles D. Hollister (1936-) published *The Face of the Deep* (New York), a textual and photographic presentation of knowledge of the seafloor.

1971. ORGANIZATIONS—ACADEMIC / PHYSICS
Theoretical physicist and 1993 Nobel Prize recipient Paul A.M. Dirac (1902-1984), after his retirement from Cambridge University in 1969, became associated with Florida State University.

1971. ORGANIZATIONS—SOCIETIES AND ASSOCIATIONS
The Association for Women in Science was founded.

1971. ZOOLOGY / ENTOMOLOGY
Edward O. Wilson (1929-) published *The Insect Societies* (Cambridge, Mass.).

1971 (March 23). GENERAL OR MISCELLANEOUS / AERONAUTICS
The U.S. Senate voted to deny funding for a supersonic plane (SST).

1971 (June 15). GENERAL OR MISCELLANEOUS / CHEMISTRY
Chemist and virologist Wendell Meredith Stanley (b.1904) died in Salamanca, Spain.

1971 (November 13). SPACE SCIENCE, TECHNOLOGY, AND EXPLORATION
The first orbiting of another planet was accomplished when *Mariner 9* spacecraft orbited Mars.

1972. AWARDS AND PRIZES / CHEMISTRY
Christian Boehmer Anfinsen (1916-), Stanford Moore (1913-1982), and William Howard Stein (1911-1980) won the Nobel Prize in chemistry. Anfinsen was honored "for his work on ribonuclease, especially concerning the connection between the amino acid sequence and the biologically active conformation." Moore and Stein were honored "for their contribution to the understanding of the connection between chemical structure and catalytic activity of the active centre of the ribonuclease molecule."

1972. AWARDS AND PRIZES / MEDICINE, PHYSIOLOGY
Gerald Maurice Edelman (1929-) won the Nobel Prize in medicine or physiology. The prize also was won by Rodney R. Porter of England. They were honored "for their discoveries concerning the chemical structure of antibodies."

1972. AWARDS AND PRIZES / PHYSICS
John Bardeen (1908-1991), Leon Neil Cooper (1930-), and John Robert Schrieffer (1931-) won the Nobel Prize in physics. They were honored "for their jointly developed theory of superconductivity, usually called the BCS-theory." Bardeen was the first individual to win a second Nobel Prize in the same field, having previously won in 1956.

1972. BIOLOGY—GENERAL / ECOLOGY
Robert Helmer MacArthur (1930-1972) published *Geographical Ecology: Patterns in the Distribution of Species* (New York).

1972. BOTANY
A work completed just before his death by Ralph Erskine Cleland (1892-1971), *Oenothera: Cytogenetics and Evolution* (New York), was published. It summarized a number of years of his work on the cytogenetic mechanism and the evolutionary relations of Oenothera (evening primrose) species. This botanical group had been the basis of Hugo de Vries's idea of mutation as the means of evolution (which Cleland began to study in the 1920s).

1972. GOVERNMENT—FEDERAL
Congress established the Office of Technology Assessment for purposes of helping the legislative branch with policymaking in the area of science and technology. The office became operational in January 1974.

1972. PHYSICS
Steven Weinberg (1933-) published *Gravitation and Cosmology: Principles and Applications of the General Theory of Relativity* (New York).

1972. PHYSICS
Murray Gell-Mann (1929-) introduced quantum chromodynamics (QCD), a theory that attempted (by analogy to light) to account for the combination of quarks through a classification that designated quarks in terms of colors that combined to create white.

1972. SPACE SCIENCE, TECHNOLOGY, AND EXPLORATION
Landsat I was launched by NASA. The satellite was designed to provide photographs of earth as a means of studying the planet's features and resources.

1972. SPACE SCIENCE, TECHNOLOGY, AND EXPLORATION
Pioneer 10 was launched and was destined, on June 13, 1983, to be the first human object to leave the solar system.

1972 (January 5). GOVERNMENT—FEDERAL / SPACE SCIENCE, TECHNOLOGY, AND EXPLORATION
Approval was given by President Richard Nixon for development by NASA of a space shuttle. The reusable space vehicle was budgeted at 5.5 billion dollars over six years.

1972 (February 20). GENERAL OR MISCELLANEOUS / PHYSICS
Physicist Maria Goeppert Mayer (b.1906) died in San Diego, California.

1972 (June 14). GENERAL OR MISCELLANEOUS / ENVIRONMENT AND CONSERVATION
Due to the adverse environmental effects of DDT, the U.S. Environmental Protection Agency announced a ban on the use of the pesticide, to become effective on December 31.

1972 (October 20). GENERAL OR MISCELLANEOUS / ASTRONOMY
Astronomer Harlow Shapley (b.1885) died in Boulder, Colorado.

1973. AWARDS AND PRIZES / PHYSICS
Ivar Giaever (1929-) won the Nobel Prize in physics. It was shared with Leo Esaki of Japan and Brian D. Josephson of Great Britain. Esaki and Giaever were honored "for their experimental discoveries regarding tunneling phenomena in semiconductors and superconductors, respectively." Josephson was honored for related work.

1973. BIOCHEMISTRY AND MOLECULAR BIOLOGY
Michael Stuart Brown (1941-) and Joseph Leonard Goldstein (1940-) discovered low-density lipoprotein (LDL) receptors. Knowledge of the functions of these molecules, which are involved in the movement of cholesterol, had an important impact on the study of the cholesterol metabolism.

1973. BIOCHEMISTRY AND MOLECULAR BIOLOGY / BIOTECHNOLOGY
Stanley Norman Cohen (1935-) and Herbert W. Boyer (1936-) demonstrated the feasibility of genetic engineering. They inserted a piece of genetic material (fragment of DNA) from one bacterium into another and showed that it was reproduced by the recipient. Cohen and Boyer, with others, published "Construction of Biologically Functional Bacterial Plasmids In Vitro," *Proceedings of National Academy of Sciences* 70 (November 1973): 3240-3244, thus initiating recombinant DNA technologies.

1973. GENERAL OR MISCELLANEOUS / SOCIOLOGY
Robert King Merton (1910-) published *The Sociology of Science: Theoretical and Empirical Investigations* (Chicago, Ill.).

1973. GOVERNMENT—FEDERAL
President Richard Nixon abolished the President's Science Advisory Committee and the Office of Science and Technology. The mechanism for science advice in the White House was reestablished under President Gerald Ford in 1976.

1973. GOVERNMENT—FEDERAL / MEDICINE
The Centers for Disease Control was made a separate agency within the Public Health Service, at the same level as the National Institutes of Health, as a result of a reorganization of the Public Health Service. CDC's beginnings, however, went back to 1946, when it was established as the Communicable Disease Center.

1973 (May 14). SPACE SCIENCE, TECHNOLOGY, AND EXPLORATION
The first *Skylab*, in which crew members performed medical and other experiments, was launched by a Saturn rocket. The first crew joined the spacecraft on May 25.

1973 (August 16). GENERAL OR MISCELLANEOUS / MICROBIOLOGY AND MICROSCOPY
Biochemist and microbiologist Selman Abraham Waksman (b.1888) died in Hyannis, Massachusetts.

1974. ANTHROPOLOGY AND ETHNOLOGY / ZOOLOGY
The most complete skeleton of a human ancestor, which came to be known as "Lucy," was discovered in Ethiopia by Donald Carl Johanson (1943-) and associates. An early and bipedal hominid more than 3 million years old, the remains represented Australopithecus afarensis, a new species.

1974. AWARDS AND PRIZES / CHEMISTRY
Paul J. Flory (1910-1985) won the Nobel Prize in chemistry. He was honored "for his fundamental achievements, both theoretical and experimental, in the physical chemistry of the macromolecules."

1974. AWARDS AND PRIZES / MEDICINE, PHYSIOLOGY
Belgian/American Albert Claude (1898-1983) and George Emil Palade (1912-), together with Christian de Duve in Belgium, won the Nobel Prize in medicine or physiology. They were honored "for their discoveries concerning the structural and functional organization of the cell."

1974. BIOCHEMISTRY AND MOLECULAR BIOLOGY / BIOTECHNOLOGY
A committee of the National Academy of Sciences, headed by Paul Berg (1926-), asked for a moratorium on genetics engineering research of

certain types until the implications could be better understood. The Academy convened an international conference at Asilomar, California, in February 1975, which was chaired by Berg, to address the question of guidelines for recombinant DNA research.

1974. ENVIRONMENT AND CONSERVATION / CHEMISTRY
The warning was issued by Frank Sherwood Rowland (1927-) and Mario J. Molina (1943-) that chlorofluorocarbons may contribute to the destruction of the ozone layer.

1974. GOVERNMENT—FEDERAL / NUCLEAR POWER AND WEAPONS
The Nuclear Regulatory Commission was established by Congress and through executive order, taking over licensing and other regulatory functions of the Atomic Energy Commission.

1974. PHYSICS
Howard M. Georgi (1947-) and Sheldon Lee Glashow (1932-) proposed what was the first of the grand unified theories (GUTs). Their work appeared as "Unity of All Elementary-Particle Forces," *Physical Review Letters* 32:438-441.

1974. PHYSICS
Burton Richter (1931-), using the Stanford Positron-Electron Accelerating Ring (SPEAR), discovered a subatomic particle believed to include a charmed quark and a charmed antiquark. Richter called it "psi." At about the same time, the same particle was found by Samuel C.C. Ting (1936-) and associates at the Brookhaven National Laboratory; they called it "J." In 1976, Richter and Ting received the Nobel Prize in physics.

1974. ZOOLOGY / ICHTHYOLOGY AND PISCICULTURE
Carl Leavitt Hubbs (1894-1979), Robert Rush Miller (1916-), and Laura C. Hubbs (1893-?) published "Hydrographic History and Relict Fishes of the North-Central Great Basin," *Memoirs of California Academy of Sciences* 7:1-259.

Chronology

1974 (May 4). GENERAL OR MISCELLANEOUS / GEOPHYSICS AND GEODESY, OCEANOGRAPHY
Geophysicist and oceanographer (William) Maurice Ewing (b.1906) died in Galveston, Texas.

1974 (June 28). GENERAL OR MISCELLANEOUS
Electrical engineer and government science administrator Vannevar Bush (b.1890) died in Belmont, Massachusetts.

1974 (September 10). ASTRONOMY
Leda, thirteenth moon of Jupiter, was found by Charles T. Kowal (1940-).

1975. AWARDS AND PRIZES / MEDICINE, PHYSIOLOGY
David Baltimore (1938-), Renato Dulbecco (1914-), and Howard Martin Temin (1934-1994) won the Nobel Prize in medicine or physiology. They were honored "for their discoveries concerning the interaction between tumour viruses and the genetic material of the cell."

1975. AWARDS AND PRIZES / PHYSICS
Ben Roy Mottelson (1926-) and (Leo) James Rainwater (1917-1986) won the Nobel Prize in physics. It was shared with Aage Bohr in Denmark. They were honored "for the discovery of the connection between collective motion and particle motion in atomic nuclei and the development of the theory of the structure of the atomic nucleus based on this connection." (Mottelson was born in the United States and went to Europe in the early 1950s; he became a Danish citizen.)

1975. BIOLOGY—GENERAL / SOCIAL SCIENCES—GENERAL
Edward Osborne Wilson (1929-) published *Sociobiology: The New Synthesis* (Cambridge, Mass.).

1975. COMPUTERS AND INFORMATION SCIENCE
The Altair 8800, the first personal computer, was introduced as a kit into the United States.

1975 (July 15). SPACE SCIENCE, TECHNOLOGY, AND EXPLORATION
The joint Apollo/Soyuz space mission of the United States and the Soviet Union was begun. On July 17, the American *Apollo* spaceship linked

with the Soviet *Soyuz* vehicle and remained together in orbit for two days.

1975 (December 18). GENERAL OR MISCELLANEOUS / GENETICS
Geneticist Theodosius Dobzhansky (b.1900) died in Davis, California.

1976. AWARDS AND PRIZES / CHEMISTRY
William Nunn Lipscomb, Jr. (1919-) won the Nobel Prize in chemistry. He was honored "for his studies on the structure of boranes illuminating problems of chemical bonding."

1976. AWARDS AND PRIZES / MEDICINE, PHYSIOLOGY
Baruch Samuel Blumberg (1925-) and Daniel Carleton Gajdusek (1923-) won the Nobel Prize in medicine or physiology. They were honored "for their discoveries concerning new mechanisms for the origin and dissemination of infectious diseases." Blumberg's work was chiefly on hepatitis, Gajdusek's on kuru.

1976. AWARDS AND PRIZES / PHYSICS
Burton Richter (1931-) and Samuel Chao Chung Ting (1936-) won the Nobel Prize in physics. They were honored "for their pioneering work in the discovery of a heavy elementary particle of a new kind." Their discovery required significant changes in the quark theories. Richter named the particle "psi," while Ting labeled it "J."

1976. CHEMISTRY
Equilibrium in Solutions; Surface and Colloid Chemistry (Cambridge, Mass.), by George Scatchard (1892-1973), was published posthumously.

1976. GOVERNMENT—FEDERAL
Congress passed the National Science and Technology Policy, Organization, and Priorities Act, which established the Office of Science and Technology Policy (OSTP) in the Executive Office of the President.

1976. ORGANIZATIONS—INDUSTRY / BIOTECHNOLOGY
The first company for the output of products by genetic engineering, Genentech, was established in San Francisco, California.

1976 (July 20). SPACE SCIENCE, TECHNOLOGY, AND EXPLORATION
Viking 1 soft-landed on Mars. In September, *Viking 2* also landed, in another locale. The two space vehicles provided surface images and other data.

1977. ASTRONOMY
The rings of Uranus were observed from NASA's airborne observatory, Kuiper, by James L. Elliot (1943-), Edward Dunham, and Douglas J. Mink (1951-), as the planet moved in front of a star.

1977. AWARDS AND PRIZES / MEDICINE, PHYSIOLOGY
Roger C.L. Guillemin (1924-), Andrew Victor Schally (1926-), and Rosalyn Sussman Yalow (1921-) won the Nobel Prize in medicine or physiology. Guillemin and Schally were honored "for their discoveries concerning the peptide hormone production of the brain," Yalow "for the development of radioimmunoassays of peptide hormones."

1977. AWARDS AND PRIZES / PHYSICS
Philip Warren Anderson (1923-), John Hasbrouck Van Vleck (1899-1980), together with Sir Nevill Mott of Great Britain, won the Nobel Prize in physics. They were honored "for their fundamental theoretical investigations of the electronic structure of magnetic and disordered systems."

1977. COMPUTERS AND INFORMATION SCIENCE
The first assembled and successful personal computer, Apple II, was introduced.

1977. GOVERNMENT—FEDERAL / NUCLEAR POWER AND WEAPONS, PHYSICS
The U.S. Department of Energy was established. Among the agencies consolidated into the new department was the Energy Research and Development Administration, which in 1975 had absorbed some of the functions of the discontinued Atomic Energy Commission.

1977. MEDICINE / DISEASE
Although AIDS was not officially recognized until 1981, two homosexual men diagnosed in this year with Kaposi's sarcoma in New York probably were the first victims of the disease.

1977. ORGANIZATIONS—SOCIETIES AND ASSOCIATIONS / MICROBIOLOGY AND MICROSCOPY
The American organization of workers on streptococci took the name Lancefield Society, to honor the major contributor to the field, Rebecca Craighill Lancefield (1895-1981). In 1981, the corresponding international society likewise changed its name.

1977. SPACE SCIENCE, TECHNOLOGY, AND EXPLORATION / ASTRONOMY
Voyagers 1 and *2* were launched to explore Jupiter and the outer planets. In March 1979, *Voyager 1* discovered a ring around Jupiter. *Voyager 2* passed the next to most distant planet, Neptune, in 1989.

1977 (June 16). GENERAL OR MISCELLANEOUS / SPACE SCIENCE, TECHNOLOGY, AND EXPLORATION
Engineer and leader in rocket development Werner von Braun (b.1912) died in Alexandria, Virginia.

1977 (August 12). SPACE SCIENCE, TECHNOLOGY, AND EXPLORATION
Space shuttle *Enterprise* completed its first test flight, taken aloft by a Boeing 747.

1978. AWARDS AND PRIZES / MEDICINE, PHYSIOLOGY
Daniel Nathans (1928-) and Hamilton Othanel Smith (1931-) won the Nobel Prize in medicine or physiology. It was shared with Werner Arber of Switzerland. They were honored "for the discovery of restriction enzymes and their application to problems of molecular genetics."

1978. AWARDS AND PRIZES / PHYSICS
Arno Allan Penzias (1933-) and Robert Woodrow Wilson (1936-) won the Nobel Prize in physics. It was shared with Pyotr L. Kapitsa of the Soviet Union. Penzias and Wilson were honored "for their discovery of cosmic microwave background radiation," which gave support to the Big Bang theory of the creation of the universe. Kapitsa was honored for work in low-temperature physics.

1978. ENVIRONMENT AND CONSERVATION / CHEMISTRY
Chlorofluorocarbons were banned as spray propellants because of their effects on the earth's protective ozone layer.

Chronology 335

1978. SPACE SCIENCE, TECHNOLOGY, AND EXPLORATION / OCEANOGRAPHY
The oceanographic satellite, *Seasat*, was launched by NASA.

1978 (January 14). GENERAL OR MISCELLANEOUS / MATHEMATICS
Mathematician Kurt Friedrich Godel (b.1906) died in Princeton, New Jersey.

1978 (April 7). GENERAL OR MISCELLANEOUS / NUCLEAR POWER AND WEAPONS
The decision to postpone production of a neutron bomb was announced by President Jimmy Carter. (The device depends on radiation rather than explosion for its effect.) In August 1981, President Ronald Reagan announced that production of the bomb would begin.

1978 (June 22). ASTRONOMY
Charon, the satellite of Pluto, was discovered by James W. Christy (1938-) from photographic evidence.

1978 (November 15). GENERAL OR MISCELLANEOUS / ANTHROPOLOGY AND ETHNOLOGY
Anthropologist Margaret Mead (b.1901) died in New York, New York.

1978 (December 11). GENERAL OR MISCELLANEOUS / BIOCHEMISTRY AND MOLECULAR BIOLOGY
Biochemist Vincent DuVigneaud (b.1901) died in Scarsdale, New York.

1979. AWARDS AND PRIZES / CHEMISTRY
Herbert Charles Brown (1912-) won the Nobel Prize in chemistry. It was shared with Georg Wittig in Germany. They were honored "for their development of the use of boron- and phosphorus-containing compounds, respectively, into important reagents in organic synthesis."

1979. AWARDS AND PRIZES / MEDICINE, PHYSIOLOGY
Allan MacLeod Cormack (1924-) won the Nobel Prize in medicine or physiology. It was shared with Sir Godfrey Hounsfield in Great Britain. They were honored "for the development of computer assisted tomography," or computerized axial tomography, known as CAT.

1979. AWARDS AND PRIZES / PHYSICS
Sheldon Lee Glashow (1932-) and Steven Weinberg (1933-) won the Nobel Prize in physics. It was shared with Abdus Salam of Pakistan and Great Britain. They were honored "for their contributions to the theory of the unified weak and electromagnetic interaction between elementary particles, including, inter alia, the prediction of the weak neutral current."

1979. BIOCHEMISTRY AND MOLECULAR BIOLOGY / BIOTECHNOLOGY
David V. Goeddel (1951-), and others, published "Expression in Escherichia coli of Chemically Synthesized Genes for Human Insulin," *Proceedings of National Academy of Sciences* 76 (January 1979): 106-110, a report on the genetically engineered production of insulin.

1979. COMPUTERS AND INFORMATION SCIENCE
The computer language, ADA (for Lady Ada Lovelace), was developed by Jean D. Ichbiah (ca.1940-) and associates, for use by the U.S. military.

1979. ORGANIZATIONS—SOCIETIES AND ASSOCIATIONS / ENGINEERING AND APPLIED SCIENCE
The American Association of Engineering Societies was founded, superseding the Engineers Joint Council.

1979 (March 28). GENERAL OR MISCELLANEOUS / NUCLEAR POWER AND WEAPONS
A serious accident occurred at the nuclear power plant at Three Mile Island, near Harrisburg, Pennsylvania. The result of a problem with the cooling system and operator error, there was partial melting of the core and release of radioactive gases into the atmosphere.

1979 (July 8). GENERAL OR MISCELLANEOUS / CHEMISTRY
Organic chemist Robert Burns Woodward (b.1917) died in Cambridge, Massachusetts.

1979 (December 7). GENERAL OR MISCELLANEOUS / ASTRONOMY, PHYSICS
Astrophysicist Cecilia Helena Payne-Gaposchkin (b.1900) died in Cambridge, Massachusetts.

1980. ASTRONOMY / INSTRUMENTS AND INSTRUMENTATION
The Very Large Array (VLA) radio telescope at Socorro, New Mexico, became operational. It was under the administration of the National Radio Astronomy Observatory.

1980. AWARDS AND PRIZES / CHEMISTRY
Paul Berg (1926-) and Walter Gilbert (1932-) won the Nobel Prize in chemistry. Frederick Sanger of Great Britain also won a prize. Berg was honored "for his fundamental studies of the biochemistry of nucleic acids, with particular regard to recombinant-DNA." Gilbert and Sanger were honored "for their contributions concerning the determination of base sequences in nucleic acids."

1980. AWARDS AND PRIZES / MEDICINE, PHYSIOLOGY
Baruj Benacerraf (1920-) (Venezuelan/American) and George Davis Snell (1903-) won the Nobel Prize in medicine or physiology. It was shared with Jean Dausset of France. They were honored "for their discoveries concerning genetically determined structures on the cell surface that regulate immunological reactions."

1980. AWARDS AND PRIZES / PHYSICS
James Watson Cronin (1931-) and Val Logsdon Fitch (1923-) won the Nobel Prize in physics. They were honored "for the discovery of violations of fundamental symmetry principles in the decay of neutral K-mesons."

1980. GENERAL OR MISCELLANEOUS / ASTRONOMY
Carl Edward Sagan (1934-) coproduced and narrated the television series "Cosmos" and published a book based on the series.

1980. GEOLOGY / ASTRONOMY
Walter Alvarez (1940-), with Luis W. Alvarez (1911-1988), Frank Asaro (1927-), and Helen V. Michel, speculated (based on the discovery of the heavy metal iridium at the boundary of Cretaceous and Tertiary) that the extinction of the dinosaurs may have been caused by the collision of a large body with the earth that set up a large amount of dust and soot in the atmosphere. The extraterrestrial object was assumed to have had an unusual amount of iridium. In time, this idea became accepted by many experts.

1980. GOVERNMENT—FEDERAL
Synthetic Fuels Corporation was established by Congress to promote development of liquid and gas fuels from coal and shale. The corporation was discontinued in 1986.

1980. SPACE SCIENCE, TECHNOLOGY, AND EXPLORATION / ASTRONOMY
The Solar Maximum Mission (SMM) satellite was launched by the National Aeronautics and Space Administration for investigation of solar activity, particularly solar flares. SMM experienced technical difficulties but was repaired by space shuttle astronauts in 1984.

1980. TECHNOLOGY AND INVENTION / BIOTECHNOLOGY
The Supreme Court permitted patenting of a microbe developed by General Electric for oil cleanup. (The bacterium was not a product of recombinant DNA technology.)

1980 (January 8). GENERAL OR MISCELLANEOUS / COMPUTERS AND INFORMATION SCIENCE
Physicist, engineer, and computer pioneer John William Mauchly (b.1907) died at Ambler, Pennsylvania.

1980 (February 20). General or Miscellaneous / Parapsychology
Joseph Banks Rhine (b.1895) (investigator of parapsychology) died in Hillsborough, North Carolina.

1980 (September 8). GENERAL OR MISCELLANEOUS / CHEMISTRY
Chemist Willard Frank Libby (b.1908) died in Los Angeles, California.

1980 (October 27). GENERAL OR MISCELLANEOUS / PHYSICS
Physicist John Hasbrouck Van Vleck (b.1899) died in Cambridge, Massachusetts.

1980s. GENERAL OR MISCELLANEOUS / ORGANIZATIONS—INDUSTRY
In the 1980s, half of the scientists in the United States worked in business and industry.

1981. AWARDS AND PRIZES / CHEMISTRY
Roald Hoffmann (1937-) won the Nobel Prize in chemistry. It was shared with Kenichi Fukui in Japan. They were honored "for their

theories, developed independently, concerning the course of chemical reactions."

1981. AWARDS AND PRIZES / MEDICINE, PHYSIOLOGY
David Hunter Hubel (1926-), Roger Wolcott Sperry (1913-1994), and Torsten N. Wiesel (1924-) won the Nobel Prize in medicine or physiology. Hubel and Wiesel were honored "for their discoveries concerning information processing in the visual system." Sperry was honored "for his discoveries concerning the functional specialization of the cerebral hemispheres."

1981. AWARDS AND PRIZES / PHYSICS
Nicolaas Bloembergen (1920-) and Arthur Leonard Schawlow (1921-) won the Nobel Prize in physics. It was shared with Kai M. Siegbahn of Sweden. Bloembergen and Schawlow were honored "for their contribution to the development of laser spectroscopy." Siegbahn was honored for work relating to high-resolution electron spectroscopy.

1981. BIOCHEMISTRY AND MOLECULAR BIOLOGY / BIOTECHNOLOGY
The first successful transfer of genes from one mammal to another (from a rabbit into a mouse) was accomplished at Ohio University (Athens, Ohio) by Thomas E. Wagner (1942-) and associates.

1981. MEDICINE / DISEASE
Acquired immune deficiency syndrome (AIDS) was first recognized by the Centers for Disease Control.

1981 (January 5). GENERAL OR MISCELLANEOUS / CHEMISTRY
Chemist Harold Clayton Urey (b.1893) died in La Jolla, California.

1981 (March 9). GENERAL OR MISCELLANEOUS /BIOLOGY—GENERAL
Biologist Max Delbruck (b.1906) died in Pasadena, California.

1981 (April 12). SPACE SCIENCE, TECHNOLOGY, AND EXPLORATION
The Space Transportation System, space shuttle *Columbia*, was first successfully tested in flight. It landed on April 14.

1981 (August 12). COMPUTERS AND INFORMATION SCIENCE
IBM introduced its first personal computer, called IBM Personal Computer.

1982. AWARDS AND PRIZES / PHYSICS
Kenneth Geddes Wilson (1936-) won the Nobel Prize in physics. He was honored "for his theory for critical phenomena in connection with phase transitions."

1982 (January 5). GENERAL OR MISCELLANEOUS / RELIGION AND THEOLOGY
The requirement that schools in Arkansas teach both creationism and the theory of evolution was overturned by a federal judge.

1982 (March 28). GENERAL OR MISCELLANEOUS / CHEMISTRY
Chemist William Francis Giauque (b.1895) died in Berkeley, California.

1982 (October 29). BIOCHEMISTRY AND MOLECULAR BIOLOGY / BIOTECHNOLOGY
The Food and Drug Administration approved sale of the first commercial genetic engineering product. The permit was granted to Eli Lily and Company to market human insulin produced by bacteria.

1982 (December 2). MEDICINE / SURGERY
An artificial heart (of aluminum and plastic), Jarvik 7, was substituted for a malfunctioning one by William C. DeVries (1943-), Willem Johan Kolff (1911-), and associates. The pioneering surgery was performed on Barney Clark who survived for 112 days.

1983. ASTRONOMY
A satellite for the detection of infrared radiation from objects in space (the Dutch-American Infrared Astronomy Satellite, IRAS) uncovered evidence of the formation of planets around other stars.

1983. AWARDS AND PRIZES / CHEMISTRY
Henry Taube (1915-) won the Nobel Prize in chemistry. He was honored "for his work on the mechanisms of electron transfer reactions, especially in metal complexes."

Chronology   341

1983. AWARDS AND PRIZES / MEDICINE, PHYSIOLOGY
Barbara McClintock (1902-1992) won the Nobel Prize in medicine or physiology. She was honored "for her discovery of mobile genetic elements."

1983. AWARDS AND PRIZES / PHYSICS
Subrahmanyan Chandrasekhar (1910-1995) (Indian/American) and William Alfred Fowler (1911-) won the Nobel Prize in physics. Chandrasekhar was honored "for his theoretical studies of the physical processes of importance to the structure and evolution of the stars." Fowler was honored "for his theoretical and experimental studies of the nuclear reactions of importance in the formation of the chemical elements in the universe."

1983. PHYSICS / INSTRUMENTS AND INSTRUMENTATION
Fermilab's Tevatron particle accelerator became operational.

1983 (March 17). GENERAL OR MISCELLANEOUS / PHYSIOLOGY
Physiologist Haldan Keffer Hartline (b.1903) died in Fallston, Maryland.

1983 (March 23). MILITARY AND WAR
The Strategic Defense Initiative program (also referred to as Star Wars) was proposed in a televised speech by President Ronald Reagan.

1983 (April 4). SPACE SCIENCE, TECHNOLOGY, AND EXPLORATION
The space shuttle *Challenger* was first launched. (This was the space program's second shuttle.)

1983 (July 1). GENERAL OR MISCELLANEOUS / TECHNOLOGY AND INVENTION
Engineer, architect, inventor R(ichard) Buckminster Fuller (b.1895) died in Los Angeles, California.

1983 (September 10). GENERAL OR MISCELLANEOUS / PHYSICS
Physicist Felix Bloch (b.1905) died in Zurich, Switzerland.

1984. AWARDS AND PRIZES / CHEMISTRY
Robert Bruce Merrifield (1921-) won the Nobel Prize in chemistry. He was honored "for his development of methodology for chemical

synthesis on a solid matrix," work that related to production of peptide chains.

1984. ZOOLOGY / EVOLUTION
DNA studies by Charles G. Sibley (1917-) and Jon E. Ahlquist concluded that humans are most closely related evolutionarily to chimpanzees of all the great apes. The separation of humans from apes was dated at about 5 or 6 million years ago. Their research appeared in "The Phylogeny of the Hominoid Primates, as Indicated by DNA-RNA Hybridization," *Journal of Molecular Evolution* 20, no 1:2-16.

1985. AWARDS AND PRIZES / CHEMISTRY
Herbert Aaron Hauptman (1917-) and Jerome Karle (1918-) won the Nobel Prize in chemistry. They were honored "for their outstanding achievements in the development of direct methods for the determination of crystal structures."

1985. AWARDS AND PRIZES / MEDICINE, PHYSIOLOGY
Michael Stuart Brown (1941-) and Joseph Leonard Goldstein (1940-) won the Nobel Prize in medicine or physiology. They were honored "for their discoveries concerning the regulation of cholesterol metabolism."

1986. AWARDS AND PRIZES / CHEMISTRY
Dudley Robert Herschbach (1932-) and Yuan Tseh Lee (1936-) won the Nobel Prize in chemistry. It was shared with John C. Polanyi of Canada. They were honored "for their contributions concerning the dynamics of chemical elementary processes."

1986. AWARDS AND PRIZES / MEDICINE, PHYSIOLOGY
Stanley Cohen (1922-), and Rita Levi-Montalcini (1909-) of Italy and the United States, won the Nobel Prize in medicine or physiology. They were honored "for their discoveries of growth factors."

1986. BIOCHEMISTRY AND MOLECULAR BIOLOGY / BIOTECHNOLOGY
The world's first license to sell a living organism created by genetic engineering was granted by the U.S. Department of Agriculture to Biologics Corporation of Omaha, Nebraska. The item was a virus that was a vaccine for prevention of a disease in swine.

1986. BIOCHEMISTRY AND MOLECULAR BIOLOGY / BIOTECHNOLOGY
The Food and Drug Administration approved the first human vaccine produced through genetic engineering. The vaccine was for hepatitis B, made by yeast.

1986 (January 28). SPACE SCIENCE, TECHNOLOGY, AND EXPLORATION
The space shuttle *Challenger* exploded during launching.

1987. AWARDS AND PRIZES / CHEMISTRY
Donald James Cram (1919-) and Charles John Pedersen (1904-1989) won the Nobel Prize in chemistry. The prize also was awarded to Jean-Marie Lehn in France. They were honored "for their development and use of molecules with structure-specific interactions of high selectivity."

1987. AWARDS AND PRIZES / MEDICINE, PHYSIOLOGY
Susumu Tonegawa (1939-), Japanese-American, won the Nobel Prize in medicine or physiology. He was honored "for his discovery of the 'genetic principle for generation of antibody diversity.'"

1987. GENERAL OR MISCELLANEOUS / RELIGION AND THEOLOGY
The argument that creation science should have equal time was rejected by the U.S. Supreme Court.

1987. TECHNOLOGY AND INVENTION / BIOTECHNOLOGY
The U.S. Patent and Trademark Office ruled that all animals could be patented, having extended patent protection for plants in 1985.

1988. AWARDS AND PRIZES / MEDICINE, PHYSIOLOGY
Gertrude Belle Elion (1918-) and George Herbert Hitchings (1905-) won the Nobel Prize in medicine or physiology. It was shared with Sir James W. Black in Britain. They were honored "for their discoveries of important principles for drug treatment."

1988. AWARDS AND PRIZES / PHYSICS
Leon Max Lederman (1922-), Melvin Schwartz (1932-), and Jack Steinberger (1921-) (German/American) won the Nobel Prize in physics.

They were honored "for the neutrino beam method and the demonstration of the doublet structure of the leptons through the discovery of the muon neutrino."

1988. BIOCHEMISTRY AND MOLECULAR BIOLOGY / BIOTECHNOLOGY
The U.S. Patent and Trademark Office issued the first patent for a vertebrate. It was patent no. 4,736,866, issued to the Harvard Medical School for a genetically engineered mouse, developed by Philip Leder (1934-) and Timothy A. Stewart, for use in cancer research.

1988. ZOOLOGY—HUMAN / GENETICS
Scientists in the United States began work on what was labeled the Human Genome Project. A committee of the National Academy of Sciences endorsed a national effort to map the human genome. Later in the year, the National Institutes of Health created an office of human genome research and James D. Watson (1928-) was appointed its head.

1988 (January 4). GENERAL OR MISCELLANEOUS / CHEMISTRY
Organic chemist Carl Shipp ("Speed") Marvel (b. 1894) died in Tucson, Arizona. During his career, he published some 500 scientific papers.

1988 (February 15). GENERAL OR MISCELLANEOUS / PHYSICS
Theoretical physicist Richard Phillips Feynman (b.1918) died in Los Angeles, California.

1988 (March 3). GENERAL OR MISCELLANEOUS / GENETICS
Geneticist Sewall Wright (b.1889) died in Madison, Wisconsin.

1988 (September 1). GENERAL OR MISCELLANEOUS / PHYSICS
Physicist Luis Walter Alvarez (b.1911) died in Berkeley, California.

1989. AWARDS AND PRIZES / CHEMISTRY
Sidney Altman (1939-) and Thomas Robert Cech (1947-) won the Nobel Prize in chemistry. They were honored "for their discovery of catalytic properties of RNA." Their work had been done independently.

1989. Awards and Prizes / Medicine, Physiology
John Michael Bishop (1936-) and Harold Elliot Varmus (1939-) won the Nobel Prize in medicine or physiology. They were honored "for their discovery of the cellular origin of retroviral oncogenes."

1989. Awards and Prizes / Physics
Hans Georg Dehmelt (1922-) (German/American) and Norman Foster Ramsey, Jr. (1915-) won the Nobel Prize in physics. It was shared with Wolfgang Paul in Germany. Dehmelt and Paul were honored "for the development of the ion trap technique." Ramsey was honored "for the invention of the separated oscillatory fields method and its use in the hydrogen maser and other atomic clocks."

1989. Chemistry / Physics
A claim was made for the achievement of nuclear fusion at room temperature (cold fusion) by B. Stanley Pons (1943-) and Martin Fleischmann (1927-), at the University of Utah (Fleischmann was associated with the University of Southampton, England). The discovery was not confirmed by others.

1990. Awards and Prizes / Chemistry
Elias James Corey (1928-) won the Nobel Prize in chemistry. He was honored "for his development of the theory and methodology of organic synthesis."

1990. Awards and Prizes / Medicine, Physiology
Joseph E. Murray (1919-) and Edward Donnall Thomas (1920-) won the Nobel Prize in medicine or physiology. They were honored "for their discoveries concerning 'Organ and Cell Transplantation in the Treatment of Human Disease.'" (Murray was the first to transplant kidneys; Thomas worked in the area of bone marrow transplant.)

1990. Awards and Prizes / Physics
Jerome Isaac Friedman (1930-), Henry Way Kendall (1926-), and Richard E. Taylor (1929-) (Canadian/American) won the Nobel Prize in physics. They were honored "for their pioneering investigations concerning deep inelastic scattering of electrons on protons and bound neutrons, which have been of essential importance for the development of the quark model in particle physics."

1990 (April). ASTRONOMY / INSTRUMENTS AND INSTRUMENTATION
The Hubble Space Telescope was launched as the culmination of years of planning. Though the telescope was functional, it soon was discovered that the primary mirror was flawed and other problems arose. (These difficulties were corrected as part of a later space mission.)

Chronology: Note on Sources

The data given in the Chronology were selectively compiled from standard reference and secondary historical works. Certain publications were consulted initially to develop the overall structure of the historical outline, while others provided confirming and supplementary data. In a work such as this, the primary goal is not originality, it is to reflect the collective wisdom and judgement of an ongoing community of scholars. I am both obligated and privileged to acknowledge and to praise the authors of the works on which the Chronology has been built. The mission of the compiler has been, at once, to use the results of historians' efforts, over many years, to understand the history of science in America, and to aid future scholars and students in the quest to extend and deepen that knowledge.

Some of the secondary historical works consulted constitute part of the central bibliography for the history of science in the United States. Though by no means an exhaustive or systematic array of titles, their citation here may aid newer students of the subject who are seeking initial guidance in the literature; the list also will acknowledge the sources for the Chronology itself.

Among the chief historical works used for the Chronology are the following: Ralph S. Bates, *Scientific Societies in the United States*, 3rd edition (Cambridge, Mass.: MIT Press, 1965); Robert V. Bruce, *The Launching of Modern American Science, 1846-1876* (New York: Knopf, 1987); George H. Daniels, *American Science in the Age of Jackson* (New York and London: Columbia University Press, 1968); A. Hunter Dupree, *Science in the Federal Government: A History of Policies and Activities to 1940* (Cambridge, Mass.: Harvard University Press, 1957); John C. Greene, *American Science in the Age of Jefferson* (Ames: Iowa State University Press, 1984); Brooke Hindle, *The Pursuit of Science in Revolutionary America, 1735-1789* (Chapel Hill, N.C.: University of North Carolina Press, 1956); Elizabeth B. Keeney, *The Botanizers: Amateur Scientists in Nineteenth-Century America* (Chapel Hill and London: University of North Carolina Press, 1992); Daniel J. Kevles, *The Physicists: The History of a Scientific Community in Modern America* (New York: Alfred A. Knopf, 1978); Robert E. Kohler, *Partners in Science: Foundations and Natural Scientists, 1900-1945* (Chicago and London: University of Chicago Press, 1991); John W. Servos, *Physical Chemistry from Ostwald to Pauling: The Making of a*

Science in America (Princeton, N.J.: Princeton University Press, 1990); Raymond P. Stearns, *Science in the British Colonies of America* (Urbana: University of Illinois Press, 1970); David D. Van Tassel and Michael G. Hall, editors, *Science and Society in the United States* (Homewood, Ill.: The Dorsey Press, 1966); Trevor I. Williams, editor, *Science: A History of Discovery in the Twentieth Century* (Oxford and New York: Oxford University Press, 1990).

The *Dictionary of Scientific Biography*, edited by Charles C. Gillispie and prepared under the auspices of the American Council of Learned Societies (New York: Scribner's, 1970-) was systematically consulted for all American entries and constitutes a significant source for the Chronology. It also is the core source for the "Scientist Cohorts" section that follows. Other essential sources among reference works consulted are: W.F. Bynum, E.J. Browne and Roy Porter, editors, *Dictionary of the History of Science* (Princeton, N.J.: Princeton University Press, 1981); *A Dictionary of Twentieth Century World Biography*, Asa Briggs, consultant editor, (Oxford and New York: Oxford University Press, 1992); John H. Gribbin, "Selected Heralds of 20th Century Science," *Science & Technology Libraries* 13, no.2 (1992):89-109; Alexander Hellemans and Bryan Bunch, *The Timetables of Science: A Chronology of the Most Important People and Events in the History of Science* (New York: Simon and Schuster, 1988); Arthur M. Schlesinger, Jr., general editor, *The Almanac of American History* (New York: Putnam, 1983); Bernard S. Schlessinger and June H. Schlessinger, editors, *The Who's Who of Nobel Prize Winners 1901-1990*, 2nd edition (Phoenix, Arizona: Oryx Press, 1991); Edward Vernoff and Rima Shore, *The International Dictionary of 20th Century Biography* (New York: New American Library, 1987). The Nobel prize citations quoted in the Chronology are taken from the 1991/92 edition of Nobelstiftelsen, *Nobel Foundation Directory* ([Stockholm]: Nobel Foundation, 1979/80-).

Other particularly helpful reference sources included general encyclopedias, particularly the *Encyclopedia Americana*, International edition (New York: Americana Corporation, 1964) and *New Encyclopaedia Britannica*, 15th edition (Chicago: Encyclopaedia Britannica, 1994). Also quite useful were various editions of the *Encyclopedia of Associations* (Detroit: Gale Research Co., 1961-); the *Research Centers Directory* (Detroit: Gale Research Co., [1965]-); and *United States Government Organization Manual* (Washington, D.C.: [Office / Division of] Federal Register, 1935-). Biographical reference works included *Current Biography* (New York: H.W. Wilson Co., 1940-, vol. 1-) and *Current Biography Yearbook* (New York: H.W. Wilson Co., 1955-); *Dictionary of American Biography*, prepared under the auspices of the American Council of Learned Societies (New York: Scribner's, 1928-) and *Concise Dictionary of American Biography* (New York: Scribner's, 1964); Clark A. Elliott, *Biographical Dictionary of American Science: The Seventeenth through the Nineteenth Centuries* (Westport, Conn.: Greenwood Press, 1979); and *Who Was Who in American History—Science and Technology* (Chicago: Marquis Who's Who, 1976). Further useful reference sources were: Isaac Asimov, *Asimov's Chronology of Science and Discovery* (New York: Harper and Row, 1989); *History of Science and Technology: A Narrative Chronology*, Edgardo Marcorini, editorial director (New York and Oxford: Facts on File, 1988); Max Meisel, *A Bibliography of American Natural History: The Pioneer Century, 1769-1865* (Brooklyn, N.Y.: Premier Publishing Co., 1924-29; reprinted: New York and London: Hafner Publishing Co., 1967); and Milton Moskowitz, Michael Katz and Robert Levering, editors, *Everybody's Business: An Almanac: The Irreverent Guide to Corporate America* (San Francisco: Harper & Row, Publishers, 1980).

The foregoing were supplemented by a number of important but less frequently accessed sources. Categorically, these additional sources included other reference works (of various types), library catalogues including online sources, science and technology dictionaries, and journal articles and monographic works on particular topics.

2

Chronology: Scientist Cohorts by Decade Reaching Age 25

Following are the names of selected leading scientists in the United States, grouped by the decade in which each reached twenty-five years. This age was chosen as the time when most would have entered upon his or her career. Within each decade, the names are listed alphabetically. Each scientist is briefly identified by year of birth and death and by area of study or contribution. The United States has, of course, always been an immigrant society and no such list could be compiled that limited itself to the native-born. Consequently, though scientists are grouped by age, in some cases the individual was not on the American scene at age twenty-five and may not have arrived until rather later. Scientists emigrating from Europe in the 1930s and 1940s, for example, often were mature researchers by the time they arrived in the United States. There also were individuals significant in American science who spent only a portion of their career in the United States. Such persons are included if they were in the country for at least five years. Although foreign birth or residence is noted below when the fact was a salient feature of the individual's life and career, absence of such a notation does not mean the person necessarily was born in the United States.

Lives are central to the history of science and one of the best and most rewarding ways by which to study the subject. For this reason, and to facilitate access to additional information, reference is given with each name to one or more standard works in which biographical information on the individual can be found. The reference codes (based on titles) and their bibliographical equivalents are as follows:

BES =
Daintith, John and others. *Biographical Encyclopedia of Scientists*. 2nd ed. Bristol and Philadelphia: Institute of Physics Publishing, 1994.

CBDS =
Williams, Trevor I., ed. *Collins Biographical Dictionary of Scientists*. 4th ed. Glasgow: Harper Collins, 1994.

DAB =
Dictionary of American Biography. New York: Scribner's, 1928-. (The basic set and supplements; the latter are indicated by DABs, numbers 1-10, now covering deaths through 1980.)

DSB =
Dictionary of Scientific Biography. New York: Scribner's, 1970-1980, 1990.

DTCWB =
Dictionary of Twentieth-Century World Biography. Rev. ed. New York and Oxford: Oxford University Press, 1992.

ID20thCB =
Vernoff, Edward and Rima Shore. *International Dictionary of 20th-Century Biography*. New York: New American Library, 1987.

M-H:MS&E =
McGraw-Hill Modern Scientists and Engineers. New York: McGraw-Hill, 1980.

NTCS =
McMurray, Emily J., editor. *Notable Twentieth-Century Scientists*. New York: Gale Research, 1995.

1630-1639 Cohort (reached age 25)

Winthrop, John, Jr. (1606-1676. Medicine, Natural philosophy)
[DAB; DSB v.14]

1650-1659 Cohort (reached age 25)

Starkey [or Stirk], George (1628-1665. Alchemy, Medicine. Went to England in 1650)
[DSB v.12]

1670-1679 Cohort (reached age 25)

Banister, John (1650-1692. Anthropology, Botany, Entomology, Malacology. Came to U.S. in 1678)
[DAB; DSB v.1]

1690-1699 Cohort (reached age 25)

Logan, James (1674-1751. Dissemination of knowledge)
[DAB; DSB v.8]

1700-1709 Cohort (reached age 25)

Catesby, Mark (1683-1749. Natural history. In U.S. 1712-1719, 1722-1726)
[DAB; DSB v.3]

1710-1719 Cohort (reached age 25)

Colden, Cadwallader (1688-1776. Botany, Medicine, Physics)
[DAB; DSB v.3]

1720-1729 Cohort (reached age 25)

Bartram, John (1699-1777. Botany)
[DAB; DSB v.1]

Evans, Lewis (1700-1756. Cartography, Geography, Geology, Geomorphology)
[DAB; DSB v.4]

Godfrey, Thomas (1704-1749. Technology)
[DAB; DSB v.5]

Greenwood, Isaac (1702-1745. Natural philosophy, Education)
[DAB; DSB v.5]

1730-1739 Cohort (reached age 25)

Franklin, Benjamin (1706-1790. Electricity, General physics, Meteorology, Oceanography)
[BES; CBDS; DAB; DSB v.5]

Kinnersley, Ebenezer (1711-1778. Electricity)
[DAB; DSB v.7]

Winthrop, John (1714-1779. Astronomy, Mathematics)
[DAB; DSB v.14]

1750-1759 Cohort (reached age 25)

Dixon, Jeremiah (1733-1779. Astronomy. In U.S. 1763-1768)
[DSB v.4]

Mason, Charles (1728-1786. Astronomy, Geodesy. In U.S. 1763-1768)
[DSB v.9]

Priestley, Joseph (1733-1804. Chemistry, Electricity, Natural philosophy, Theology. Came to U.S. in 1794)
[BES; CBDS; DAB; DSB v.11]

Rittenhouse, David (1732-1796. Astronomy, Natural philosophy, Technology)
[DAB; DSB v.11]

1760-1769 Cohort (reached age 25)

Bartram, William (1739-1823. Botany, Ornithology)
[DAB; DSB v.1]

Jefferson, Thomas (1743-1826. Agriculture, Botany, Cartography, Ethnology, Meteorology, Paleontology, Surveying, Technology, Diplomacy)
[DAB; DSB v.7]

Peale, Charles Willson (1741-1827. Museum direction)
[DAB; DSB v.10]

1770-1779 Cohort (reached age 25)

Rush, Benjamin (1746-1813. Chemistry, Medicine, Psychiatry)
[BES; CBDS; DAB; DSB v.11]

Thompson, Benjamin (Count Rumford) (1753-1814. Physics. Moved to Europe in 1776)
[BES; CBDS; DAB; DSB v.13]

1780-1789 Cohort (reached age 25)

Cooper, Thomas (1759-1839. Chemistry. Came to U.S. in 1793)
[DAB; DSB v.3]

Maclure, William (1763-1840. Geology)
[DAB; DSB v.8]

Morse, Jedidiah (1761-1826. Geography)
[DAB; DSB v.9]

Wells, William Charles (1757-1817. Medicine, Meteorology, Natural philosophy, Physiology. Moved permanently to Great Britain in ca.1784)
[DAB; DSB v.14]

Wistar, Caspar (1761-1818. Anatomy)
[DAB; DSB v.14]

1790-1799 Cohort (reached age 25)

Barton, Benjamin Smith (1766-1815. Botany, Ethnography, Medicine, Zoology)
[DAB; DSB v.1]

Bowditch, Nathaniel (1773-1838. Astronomy)
[DAB; DSB v.2]

Hassler, Ferdinand Rudolph (1770-1843. Geodesy. Came to U.S. in 1805)
[DAB; DSB v.6]

Hosack, David (1769-1835. Botany, Medicine)
[DAB; DSB v.6]

Maclean, John (1771-1814. Chemistry)
[DAB; DSB v.8]

Pursh, Frederick [Friedrich Traugott Pursch] (1774-1820. Botany)
[DAB; DSB v.11]

Smithson, James Louis Macie (1765-1829. Chemistry, Benefactor. Never in U.S.)
[BES; CBDS; DSB v.12]

Wilson, Alexander (1766-1813. Ornithology. Came to U.S. in 1794)
[BES; DAB; DSB v.14]

1800-1809 Cohort (reached age 25)

Adrain, Robert (1775-1843. Mathematics)
[DAB; DSB v.1]

Cleaveland, Parker (1780-1858. Mineralogy)
[DAB; DSB v.3]

Darlington, William (1782-1863. Botany)
[DAB; DSB v.3]

Eaton, Amos (1776-1842. Botany, Geology, Scientific and applied education)
[DAB; DSB v.4]

Farrar, John (1779-1853. Mathematics, Physics, Education)
[DAB; DSB v.4]

Featherstonhaugh, George William (1780-1866. Geology)
[DSB v.15]

Hare, Robert (1781-1858. Chemistry)
[BES; CBDS; DAB; DSB v.6]

Lesueur, Charles Alexandre (1778-1846. Natural history. In U.S. 1816 to 1837)
[DAB; DSB v.8]

Peale, Rembrandt (1778-1860. Natural history, Technology)
[DAB; DSB v.15]

Rafinesque, Constantine Samuel (1783-1840. Archaeology, Natural history. In U.S. 1802-1804 and came permanently in 1815)
[DAB; DSB v.11]

Silliman, Benjamin (1779-1864. Chemistry, Geology, Mineralogy)
[BES; CBDS; DAB; DSB v.12]

Troost, Gerard (1776-1850. Geology, Mineralogy, Natural history, Paleontology. Came to U.S. in ca.1810)
[DAB; DSB v.13]

Young, John Richardson (1782-1804. Physiology)
[DAB; DSB v.14]

1810-1819 Cohort (reached age 25)

Audubon, John James (1785-1851. Ornithology)
[BES; CBDS; DAB; DSB v.1]

Beaumont, William (1785-1853. Physiology)
[BES; CBDS; DAB; DSB v.1]

Bond, William Cranch (1789-1859. Astronomy)
[BES; DAB; DSB v.2]

Espy, James Pollard (1785-1860. Meteorology)
[BES; DAB; DSB v.4]

Green, Jacob (1790-1841. Biology, Botany, Chemistry, Dissemination of knowledge)
[DAB; DSB v.5]

Hitchcock, Edward (1793-1864. Geology)
[DAB; DSB v.6]

Holbrook, John Edwards (1794-1871. Herpetology, Ichthyology)
[DAB; DSB v.15]

Lea, Isaac (1792-1886. Malacology)
[DAB; DSB v.8]

Mitchell, Elisha (1793-1857. Natural history)
[DAB; DSB v.9]

Nuttall, Thomas (1786-1859. Botany, Natural history, Ornithology)
[DAB; DSB v.10]

Redfield, William C. (1789-1857. Meteorology, Paleontology)
[BES; DAB; DSB v.11]

Say, Thomas (1787-1834. Conchology, Entomology)
[DAB; DSB v.12]

Schoolcraft, Henry Rowe (1793-1864. Ethnology)
[DAB; DSB v.12]

Vanuxem, Lardner (1792-1848. Geology)
[DAB; DSB v.13]

1820-1829 Cohort (reached age 25)

Bonaparte, Lucien Jules Laurent (Charles Lucien) (1803-1857.
Zoology. In U.S. ca.1822-1828)
[DSB v.2]

Clark, Alvan (1804-1887. Astronomical instrumentation)
[DAB; DSB v.3]

Conrad, Timothy Abbott (1803-1877. Malacology, Paleontology)
[DSB v.3]

Dunglison, Robley (1798-1869. Physiology, Lexicography, Medical education)
[DAB; DSB v.4]

Emmons, Ebenezer (1799-1863. Geology)
[DAB; DSB v.4]

Harlan, Richard (1796-1843. Comparative anatomy)
[DAB; DSB v.6]

Henry, Joseph (1797-1878. Physics)
[BES; CBDS; DAB; DSB v.6]

Mather, William Williams (1804-1859. Geology)
[DAB; DSB v.9]

Morton, Samuel George (1799-1851. Anthropology)
[DAB; DSB v.9]

Peale, Titian Ramsay (1799-1885. Natural history)
[DAB; DSB v.10]

Rogers, William Barton (1804-1882. Geology)
[DAB; DSB v.11]

Saxton, Joseph (1799-1873. Scientific instrumentation)
[DAB; DSB v.12]

Torrey, John (1796-1873. Botany)
[BES; DAB; DSB v.13]

1830-1839 Cohort (reached age 25)

Agassiz, Jean Louis Rodolphe (1807-1873. Geology, Ichthyology, Paleontology. Came to U.S. in 1846)
[BES; CBDS; DAB; DSB v.1]

Bache, Alexander Dallas (1806-1867. Physics)
[BES; DAB; DSB v.1]

Chapman, Alvan Wentworth (1809-1899. Botany)
[DAB; DSB v.3]

Dana, James Dwight (1813-1895. Geology)
[BES; DAB; DSB v.3]

Desor, Pierre Jean Edouard (1811-1882. Glacial geology, Paleontology, Stratigraphy. In U.S. 1846-1852)
[DSB v.4]

Draper, John William (1811-1882. Chemistry; History)
[BES; CBDS; DAB; DSB v.4]

Engelmann, George (1809-1884. Botany)
[BES; DAB; DSB v.15]

Fitch, Asa (1809-1879. Economic entomology, Medicine)
[DAB; DSB v.5]

Gould, Augustus Addison (1805-1866. Conchology, Medicine)
[DAB; DSB v.5]

Gray, Asa (1810-1888. Botany)
[BES; CBDS; DAB; DSB v.5]

Guyot, Arnold Henri (1807-1884. Geography, Glacial geology. Came to U.S. in ca.1848)
[BES; DAB; DSB v.5]

Hall, James, Jr. (1811-1898. Geology, Paleontology)
[BES; CBDS; DAB; DSB v.6]

Houghton, Douglass (1809-1845. Geology, Medicine)
[DAB; DSB v.6]

Isaacs, Charles Edward (1811-1860. Medicine)
[DSB v.7]

Jackson, Charles Thomas (1805-1880. Chemistry, Geology, Medicine, Mineralogy)
[BES; DAB; DSB v.7]

Kellogg, Albert (1813-1887. Botany)
[DAB; DSB v.7]

Kirkwood, Daniel (1814-1895. Astronomy)
[BES; DAB; DSB v.7]

Lesquereux, Leo (1806-1889. Botany, Paleontology. Came to U.S. in 1848)
[DAB; DSB v.8]

Loomis, Elias (1811-1889. Astronomy, Mathematics, Meteorology)
[BES; DAB; DSB v.8]

Lyman, Chester Smith (1814-1890. Astronomy, Geology)
[DAB; DSB v.8]

Maury, Matthew Fontaine (1806-1873. Meteorology, Oceanography, Physical geography)
[BES; CBDS; DAB; DSB v.9]

Owen, David Dale (1807-1860. Geology)
[DAB; DSB v.10]

Peirce, Benjamin (1809-1880. Astronomy, Mathematics)
[DAB; DSB v.10]

Peters, Christian Heinrich Friedrich (1813-1890. Astronomy. Came to U.S. in 1854)
[DAB; DSB v.10]

Rogers, Henry Darwin (1808-1866. Geology)
[DAB; DSB v.11]

Sylvester, James Joseph (1814-1897. Mathematics. In U.S. 1841-1843, 1876-1883)
[BES; CBDS; DAB; DSB v.13]

Wyman, Jeffries (1814-1874. Anatomy, Physiology)
[DAB; DSB v.14]

1840-1849 Cohort (reached age 25)

Baird, Spencer Fullerton (1823-1887. Zoology, Scientific administration)
[BES; CBDS; DAB; DSB v.1]

Bradford, Joshua Taylor (1818-1871. Medicine, Surgery)
[DSB v.2]

Brown-Sequard, Charles-Edouard (1817-1894. Physiology. In U.S. ca.1852 to ca.1877, for periods of time)
[BES; CBDS; DSB v.2]

Ferrel, William (1817-1891. Mathematical geophysics)
[BES; DAB; DSB v.4]

Genth, Frederick Augustus (1820-1893. Chemistry, Mineralogy. Came to U.S. in 1848)
[DAB; DSB v.5]

Gibbs, Oliver Wolcott (1822-1908. Chemistry)
[DAB; DSB v.5]

Gould, Benjamin Apthorp (1824-1896. Astronomy)
[BES; DAB; DSB v.5]

Horsford, Eben Norton (1818-1893. Chemistry)
[DAB; DSB v.6]

Lane, Jonathan Homer (1819-1880. Physics)
[DSB v.8]

LeConte, John (1818-1891. Natural history, Physics)
[DAB; DSB v.8]

LeConte, Joseph (1823-1901. Geology, Natural history, Physiology)
[DAB; DSB v.8]

Leidy, Joseph (1823-1891. Biology)
[DAB; DSB v.8]

Lesley, J. Peter (1819-1903. Geology)
[DAB; DSB v.8]

Marcou, Jules (1824-1898. Geology, Paleontology, Topography)
[DAB; DSB v.9]

Meek, Fielding Bradford (1817-1876. Geology, Paleontology)
[DAB; DSB v.9]

Mitchell, Maria (1818-1889. Astronomy)
[BES; DAB; DSB v.9]

Newberry, John Strong (1822-1892. Geology, Paleontology)
[DAB; DSB v.10]

Norton, John Pitkin (1822-1852. Agricultural chemistry, Agriculture)
[DAB; DSB v.10]

Pourtales, Louis Francois De (1823/24-1880. Oceanography)
[DAB; DSB v.11]

Rutherfurd, Lewis Morris (1816-1892. Astrophysics)
[DAB; DSB v.12]

Silliman, Benjamin, Jr. (1816-1885. Chemistry, Geology)
[DAB; DSB v.12]

Stallo, Johann Bernhard (1823-1900. Philosophy of science)
[DAB; DSB v.12]

Whitney, Josiah Dwight (1819-1896. Geology)
[DAB; DSB v.14]

Winchell, Alexander (1824-1891. Geology, Education)
[DAB; DSB v.14]

1850-1859 Cohort (reached age 25)

Bond, George Phillips (1825-1865. Astronomy)
[BES; DAB; DSB v.2]

Clark, Alvan Graham (1832-1897. Astronomical instrumentation)
[BES; DAB; DSB v.3]

Clark, George Bassett (1827-1891. Astronomical instrumentation)
[DSB v.3]

Cooke, Josiah Parsons, Jr. (1827-1894. Chemistry)
[DAB; DSB v.3]

Dalton, John Call (1825-1889. Medicine, Physiology)
[DAB; DSB v.15]

Gulick, John Thomas (1832-1923. Evolution, Zoology)
[DAB; DSB v.17]

Hall, Asaph (1829-1907. Astronomy)
[BES; DAB; DSB v.6]

Hayden, Ferdinand Vandiveer (1829-1887. Geology)
[DAB; DSB v.6]

Hunt, Thomas Sterry (1826-1892. Chemistry, Geology)
[DAB; DSB v.6]

Langley, Samuel Pierpont (1834-1906. Astrophysics)
[BES; CBDS; DAB; DSB v.8]

Marsh, Othniel Charles (1831-1899. Vertebrate paleontology)
[BES; DAB; DSB v.9]

Mitchell, Silas Weir (1829-1914. Medicine, Neurology)
[DAB; DSB v.9]

Newton, Hubert Anson (1830-1896. Astronomy, Mathematics)
[DAB; DSB v.10]

Orton, James (1830-1877. Exploration, Natural history)
[DAB; DSB v.10]

Palmer, Edward (1831-1911. Natural history)
[DSB v.10]

Powell, John Wesley (1834-1902. Ethnology, Geology)
[DAB; DSB v.11]

Rood, Ogden Nicholas (1831-1902. Physics)
[DAB; DSB v.11]

Schott, Charles Anthony (1826-1901. Geophysics)
[DAB; DSB v.12]

Stimpson, William (1832-1872. Marine zoology)
[DAB; DSB v.13]

Watson, Sereno (1826-1892. Botany)
[DAB; DSB v.14]

Whitfield, Robert Parr (1828-1910. Invertebrate paleontology, Stratigraphy)
[DAB; DSB v.14]

Winlock, Joseph (1826-1875. Astronomy, Mathematics)
[DAB; DSB v.14]

Young, Charles Augustus (1834-1908. Astronomy)
[DAB; DSB v.14]

1860-1869 Cohort (reached age 25)

Abbe, Cleveland (1838-1916. Meteorology)
[BES; DAB; DSB v.1]

Agassiz, Alexander (1835-1910. Engineering, Oceanography, Zoology)
[DAB; DSB v.1]

Allen, Joel Asaph (1838-1921. Zoology)
[DAB; DSB v.17]

Atwater, Wilbur Olin (1844-1907. Agricultural chemistry, Physiology, Scientific administration)
[DAB; DSB v.1]

Babcock, Stephen Moulton (1843-1931. Agricultural chemistry)
[BES; CBDS; DABs1; DSB v.1]

Bailey, Loring Woart (1839-1925. Geology)
[DSB v.1]

Bowditch, Henry Pickering (1840-1911. Physiology)
[DAB; DSB v.2]

Brashear, John Alfred (1840-1920. Astrophysical instrumentation)
[DAB; DSB v.2]

Brooks, William Robert (1844-1921. Astronomy)
[DAB; DSB v.2]

Burnham, Sherburne Wesley (1838-1921. Astronomy)
[DAB; DSB v.2]

Chamberlin, Thomas Chrowder (1843-1928. Cosmology, Geology)
[BES; DAB; DSB v.3; ID20thCB; NTCS]

Cope, Edward Drinker (1840-1897. Vertebrate paleontology, Zoology)
[BES; DAB; DSB v.15]

Coues, Elliott (1842-1899. Ornithology)
[DAB; DSB v.3]

Draper, Henry (1837-1882. Astronomy)
[BES; DAB; DSB v.4]

Dutton, Clarence Edward (1841-1912. Geology)
[BES; DAB; DSB v.4]

Emerson, Benjamin Kendall (1843-1932. Geology)
[DABs1; DSB v.4]

Emmons, Samuel Franklin (1841-1911. Geology, Mining)
[DAB; DSB v.4]

Forbes, Stephen Alfred (1844-1930. Biology)
[DAB; DSB v.5]

Gabb, William More (1839-1878. Geology)
[DAB; DSB v.5]

Gibbs, Josiah Willard (1839-1903. Theoretical physics)
[BES; CBDS; DAB; DSB v.5]

Gilbert, Grove Karl (1843-1918. Geology, Geomorphology)
[DAB; DSB v.5]

Gill, Theodore Nicholas (1837-1914. Ichthyology)
[DAB; DSB v.5]

Grote, Augustus Radcliffe (1841-1903. Entomology)
[DAB; DSB v.5]

Hague, Arnold (1840-1917. Geology)
[DAB; DSB v.6]

Harkness, William (1837-1903. Astronomy)
[DAB; DSB v.6]

Hill, George William (1838-1914. Mathematical astronomy)
[DAB; DSB v.6]

Horn, George Henry (1840-1897. Coleopterology)
[DAB; DSB v.6]

Hough, George Washington (1836-1909. Astronomy, Meteorology)
[DAB; DSB v.6]

Hyatt, Alpheus (1838-1902. Invertebrate paleontology, Zoology)
[DAB; DSB v.6]

James, William (1842-1910. Psychology, Philosophy)
[DAB; DSB v.7; ID20thCB]

King, Clarence Rivers (1842-1901. Geology)
[DAB; DSB v.7]

Lyman, Benjamin Smith (1835-1920. Geology)
[DAB; DSB v.8]

Mayer, Alfred Marshall (1836-1897. Physics)
[DAB; DSB v.9]

Montgomery, Edmund Duncan (1835-1911. Cell biology, Philosophy)
[DAB; DSB v.9]

Morley, Edward Williams (1838-1923. Chemistry, Physics)
[BES; CBDS; DAB; DSB v.9; NTCS]

Morse, Edward Sylvester (1838-1925. Zoology)
[DAB; DSB v.9]

Newcomb, Simon (1835-1909. Astronomy)
[BES; CBDS; DAB; DSB v.10]

Packard, Alpheus Spring, Jr. (1839-1905. Entomology)
[DAB; DSB v.10]

Peirce, Charles Sanders (1839-1914. Geodesy, History of science, Logic, Mathematics, Philosophy)
[DAB; DSB v.10]

Proctor, Richard Anthony (1837-1888. Astronomy. Came to U.S. in ca.1881)
[BES; DSB v.11]

Pumpelly, Raphael (1837-1923. Ethnology, Geography, Geology)
[DAB; DSB v.11]

Putnam, Frederic Ward (1839-1915. Anthropology, Archaeology)
[DAB; DSB v.11]

Scudder, Samuel Hubbard (1837-1911. Systematic entomology)
[DAB; DSB v.12]

Shaler, Nathaniel Southgate (1841-1906. Geology)
[DAB; DSB v.12]

Smith, Sidney Irving (1843-1926. Zoology)
[DSB v.12]

Thurston, Robert Henry (1839-1903. Steam engineering, Testing of materials, Engineering education)
[DAB; DSB v.13]

Trowbridge, John (1843-1923. Physics)
[DAB; DSB v.13]

Verrill, Addison Emery (1839-1926. Zoology)
[DAB; DSB v.14]

Whitman, Charles Otis (1842-1910. Zoology)
[DAB; DSB v.14]

Wiley, Harvey Washington (1844-1930. Chemistry)
[DAB; DSB v.14; ID20thCB]

Winchell, Newton Horace (1839-1914. Geology, Archaeology)
[DAB; DSB v.14]

Wood, Horatio Charles (1841-1920. Pharmacology, Therapeutics)
[DAB; DSB v.14]

Wright, George Frederick (1838-1921. Geology)
[DAB; DSB v.14]

1870-1879 Cohort (reached age 25)

Bailey, Solon Irving (1854-1931. Astronomy, Meteorology)
[DABs1; DSB v.1]

Becker, George Ferdinand (1847-1919. Geology)
[DAB; DSB v.1]

Bell, Alexander Graham (1847-1922. Technology)
[BES; CBDS; DAB; DSB v.1]

Bessey, Charles Edwin (1845-1915. Botany, Education)
[DAB; DSB v.2]

Birge, Edward Asahel (1851-1950. Limnology)
[DABs4; DSB v.2]

Boss, Lewis (1846-1912. Positional astronomy)
[BES; DAB; DSB v.2]

Brooks, William Keith (1848-1908. Embryology, Zoology)
[DAB; DSB v.2]

Carroll, James (1854-1907. Bacteriology)
[DAB; DSB v.3]

Carus, Paul (1852-1919. Philosophy of science)
[DAB; DSB v.15]

Chandler, Seth Carlo (1846-1913. Astronomy)
[BES; DAB; DSB v.3]

Clarke, Frank Wigglesworth (1847-1931. Geochemistry)
[DABs1; DSB v.3]

Councilman, William Thomas (1854-1933. Pathology)
[DABs1; DSB v.3]

Cross, Charles Whitman (1854-1949. Petrology)
[DABs4; DSB v.3]

Dall, William Healey (1845-1927. Malacology, Paleontology)
[DAB; DSB v.3]

Davis, William Morris (1850-1934. Geography, Geology, Geomorphology, Meteorology)
[BES; DABs1; DSB v.3; ID20thCB]

Edison, Thomas Alva (1847-1931. Technology)
[BES; CBDS; DABs1; DSB v.4; ID20thCB; NTCS]

Hall, Granville Stanley (1846 [or 1844]-1924. Psychology, Education)
[DAB; DSB v.6]

Halsted, George Bruce (1853-1922. Mathematics, Education)
[DAB; DSB v.6]

Halsted, William Stewart (1852-1922. Surgery)
[BES; DAB; DSB v.6]

Holden, Edward Singleton (1846-1914. Astronomy)
[DAB; DSB v.6]

Jordan, David Starr (1851-1931. Ichthyology, Education)
[DAB; DSB v.7; ID20thCB]

Lloyd, John Uri (1849-1936. Chemistry, Pharmacy)
[DABs2; DSB v.8]

Martin, Henry Newell (1848-1896. Physiology. In U.S. ca.1876-ca.1893)
[DAB; DSB v.9]

Meltzer, Samuel James (1851-1920. Pharmacology, Physiology. Came to U.S. in 1883)
[DAB; DSB v.9]

Merrill, George Perkins (1854-1929. Geology, Meteoritics)
[DAB; DSB v.9]

Michael, Arthur (1853-1942. Chemistry)
[DABs3; DSB v.9]

Michelson, Albert Abraham (1852-1931. Metrology, Optics, Physics)
[BES; CBDS; DAB; DSB v.9; NTCS]

Minot, Charles Sedgwick (1852-1914. Anatomy, Embryology)
[DAB; DSB v.9]

Ott, Isaac (1847-1916. Physiology)
[DAB; DSB v.10]

Peirce, Benjamin Osgood, II (1854-1914. Mathematics, Physics)
[DAB; DSB v.10]

Pickering, Edward Charles (1846-1919. Astronomy)
[BES; CBDS; DAB; DSB v.10]

Power, Frederick Belding (1853-1927. Chemistry, Pharmacy)
[DAB; DSB v.11]

Prudden, Theophil Mitchell (1849-1924. Bacteriology, Pathology, Public health, Archaeology)
[DAB; DSB v.11]

Reed, Walter (1851-1902. Medicine)
[BES; CBDS; DAB; DSB v.11; NTCS]

Remsen, Ira (1846-1927. Chemistry, Education)
[BES; CBDS; DAB; DSB v.11]

Ridgway, Robert (1850-1929. Ornithology)
[DAB; DSB v.11]

Rowland, Henry Augustus (1848-1901. Physics)
[BES; CBDS; DAB; DSB v.11]

Schaeberle, John Martin (1853-1924. Practical astronomy)
[DAB; DSB v.12]

Smith, Edgar Fahs (1854-1928. Chemistry)
[DAB; DSB v.12]

Smith, Erwin Frink (1854-1927. Bacteriology, Plant pathology)
[DAB; DSB v.12]

Stejneger, Leonhard Hess (1851-1943. Herpetology, Ornithology. Came to U.S. in 1881)
[DABs3; DSB v.13]

Takamine, Jokichi (1854-1922. Chemistry. Came to U.S. in 1890)
[BES; CBDS; DAB; ID20thCB]

Thomson, Elihu (1853-1937. Electrical engineering)
[BES; DABs2; DSB v.13]

Walcott, Charles Doolittle (1850-1927. Paleontology)
[BES; DAB; DSB v.14]

Welch, William Henry (1850-1934. Pathology, Public health, Medical education)
[BES; CBDS; DAB; DSB v.14]

White, Israel Charles (1848-1927. Geology)
[DAB; DSB v.14]

Williams, Henry Shaler (1847-1918. Paleontology, Stratigraphy)
[DAB; DSB v.14]

Williston, Samuel Wendell (1851-1918. Entomology, Medicine, Vertebrate paleontology)
[DAB; DSB v.14]

Woodward, Robert Simpson (1849-1924. Applied mathematics, Geophysics)
[DAB; DSB v.14]

1880-1889 Cohort (reached age 25)

Abel, John Jacob (1857-1938. Biochemistry, Pharmacology)
[BES; DABs2; DSB v.1]

Aitken, Robert Grant (1864-1951. Astronomy)
[BES; DABs5; DSB v.1]

Akeley, Carl Ethan (1864-1926. Natural history, Sculpture)
[DAB; ID20thCB]

Allen, Eugene Thomas (1864-1964. Geochemistry)
[DSB v.17]

Ames, Joseph Sweetman (1864-1943. Physics)
[DABs3; DSB v.1]

Baekeland, Leo Hendrik (1863-1944. Chemistry. Came to U.S. in 1889)
[BES; CBDS; DABs3; DSB v.1; DTCWB; ID20thCB]

Bailey, Liberty Hyde, Jr. (1858-1954. Agriculture, Botany, Horticulture)
[DABs5; DSB v.1; ID20thCB]

Barnard, Edward Emerson (1857-1923. Astronomy)
[BES; DAB; DSB v.1; ID20thCB]

Barus, Carl (1856-1935. Physics)
[DABs1; DSB v.1]

Boas, Franz (1858-1942. Anthropology. Came to U.S. in ca.1887)
[DABs3; DSB v.2; DTCWB; ID20thCB]

Bolza, Oskar (1857-1942. Mathematics. Came to U.S. in 1888 and remained until 1910)
[DABs3; DSB v.2]

Brace, DeWitt Bristol (1859-1905. Optics)
[DAB; DSB v.2]

Britton, Nathaniel Lord (1859-1934. Botany)
[DABs1; DSB v.2]

Bumpus, Hermon Carey (1862-1943. Biometry, Evolution, Natural history, Zoology, Education)
[DSB v.17]

Campbell, Douglas Houghton (1859-1953. Botany)
[DABs5; DSB v.3]

Campbell, William Wallace (1862-1938. Astronomy, Education)
[DABs2; DSB v.3; ID20thCB]

Cannon, Annie Jump (1863-1941. Astronomy)
[BES; DABs3; DSB v.3; ID20thCB; NTCS]

Cattell, James McKeen (1860-1944. Psychology, Scientific publishing)
[DABs3; DSB v.3]

Chamberlain, Charles Joseph (1863-1943. Botany)
[DABs3; DSB v.3]

Chapman, Frank Michler (1864-1945. Conservation, Ornithology)
[DABs3; DSB v.17]

Chittenden, Russell Henry (1856-1943. Physiological chemistry)
[BES; CBDS; DABs3; DSB v.3]

Cole, Frank Nelson (1861-1926. Mathematics)
[DAB; DSB v.3]

Comstock, George Cary (1855-1934. Astronomy)
[DABs1; DSB v.3]

Conklin, Edwin Grant (1863-1952. Biology)
[DABs5; DSB v.3]

Crew, Henry (1859-1953. Astrophysics, Physics, Education)
[DSB v.17]

Day, David Talbot (1859-1925. Chemical geology)
[BES; CBDS; DAB; DSB v.3]

Donaldson, Henry Herbert (1857-1938. Neurology)
[DABs2; DSB v.4]

Eigenmann, Carl H. (1863-1927. Ichthyology)
[DAB; DSB v.4]

Elkin, William Lewis (1855-1933. Meteoritics, Positional astronomy)
[DABs1; DSB v.4]

Fine, Henry Burchard (1858-1928. Mathematics)
[DAB; DSB v.4]

Fleming, Williamina Paton Stevens (1857-1911. Astronomy)
[BES; DAB; DSB v.5]

Flexner, Simon (1863-1946. Bacteriology, Pathology)
[DABs4; DSB v.5; ID20thCB; NTCS]

Foerste, August Frederick (1862-1936. Invertebrate paleontology, Stratigraphy)
[DSB v.5]

Hall, Charles Martin (1863-1914. Commercial chemistry)
[BES; CBDS; DAB; DSB v.6]

Hall, Edwin Herbert (1855-1938. Physics)
[BES; DABs2; DSB v.6]

Harper, Robert Almer (1862-1946. Botany)
[DSB v.6]

Hektoen, Ludvig (1863-1951. Microbiology, Pathology)
[DSB v.6]

Herrick, Clarence Luther (1858-1904. Comparative neurology, Psychobiology)
[DSB v.6]

Howard, Leland Ossian (1857-1950. Applied entomology)
[DABs4; DSB v.6]

Howe, James Lewis (1859-1955. Chemistry)
[DSB v.6]

Howell, William Henry (1860-1945. Physiology)
[DABs3; DSB v.6]

Hussey, William Joseph (1862-1926. Astronomy)
[DAB; DSB v.6]

Iddings, Joseph Paxson (1857-1920. Petrology)
[DAB; DSB v.7]

Johnson, Willard Drake (1859-1917. Geomorphology)
[DSB v.7]

Keeler, James Edward (1857-1900. Astronomy)
[BES; CBDS; DAB; DSB v.7]

Kennelly, Arthur Edwin (1861-1939. Electrical engineering. Came to U.S. in 1887)
[BES; CBDS; DABs2; DSB v.7; DTCWB]

Lane, Alfred Church (1863-1948. Geology)
[DSB v.17]

Lawson, Andrew Cowper (1861-1952. Geology)
[DABs5; DSB v.8]

Leverett, Frank (1859-1943. Glacial geology)
[DABs3; DSB v.18]

Lindgren, Waldemar (1860-1939. Geology)
[DABs2; DSB v.8; ID20thCB]

Loeb, Jacques (1859-1924. Biology, Physiology. Came to U.S. in 1891)
[BES; DAB; DSB v.8; NTCS]

Lovell, John Harvey (1860-1939. Botany, Entomology)
[DSB v.8]

Lowell, Percival (1855-1916. Astronomy)
[BES; CBDS; DAB; DSB v.8; ID20thCB]

Mall, Franklin Paine (1862-1917. Anatomy, Embryology, Physiology)
[DAB; DSB v.9]

Merriam, Clinton Hart (1855-1942. Biology)
[DABs3; DSB v.9]

Miller, George Abram (1863-1951. Mathematics)
[DABs5; DSB v.9]

Moore, Eliakim Hastings (1862-1932. Mathematics)
[DABs1; DSB v.9]

Nef, John Ulric (1862-1915. Chemistry)
[BES; DAB; DSB v.10]

Novy, Frederick George (1864-1957. Microbiology)
[DABs6; DSB v.10]

Noyes, William Albert (1857-1941. Chemistry)
[BES; DABs3; DSB v.10]

Osborn, Henry Fairfield (1857-1935. Vertebrate paleontology)
[BES; DABs1; DSB v.10; ID20thCB]

Osborne, Thomas Burr (1859-1929. Protein chemistry)
[DAB; DSB v.10]

Osgood, William Fogg (1864-1943. Mathematics)
[DABs3; DSB v.10]

Park, William Hallock (1863-1939. Bacteriology, Public health)
[DABs2; ID20thCB]

Parkhurst, John Adelbert (1861-1925. Astronomy)
[DAB; DSB v.10]

Pickering, William Henry (1858-1938. Astronomy)
[BES; CBDS; DSB v.10]

Pupin, Michael Idvorsky (1858-1935. Applied physics)
[DABs1; DSB v.11]

Reid, Harry Fielding (1859-1944. Geophysics)
[BES; DSB v.11; ID20thCB]

Ritchey, George Willis (1864-1945. Astronomy)
[BES; DABs3; DSB v.11]

Ritter, William Emerson (1856-1944. Biology)
[DABs3; DSB v.18]

Ruedemann, Rudolf (1864-1956. Geology, Paleontology. Came to U.S. in 1892)
[DSB v.11]

St. John, Charles Edward (1857-1935. Astronomy)
[DABs1; DSB v.12]

Salisbury, Rollin Daniel (1858-1922. Geology, Physical geography)
[DAB; DSB v.12]

Sauveur, Albert (1863-1939. Metallography, Metallurgy)
[DABs2; DSB v.12]

Schuchert, Charles (1858-1942. Paleontology)
[DABs3; DSB v.12]

Scott, William Berryman (1858-1947. Geology, Vertebrate paleontology)
[DABs4; DSB v.12]

Setchell, William Albert (1864-1943. Botany, Geography)
[DABs3; DSB v.12]

Smith, Theobald (1859-1934. Comparative pathology, Microbiology)
[BES; CBDS; DABs1; DSB v.12]

Sperry, Elmer Ambrose (1860-1930. Engineering, Technology)
[BES; CBDS; DAB; DSB v.12; ID20thCB; NTCS]

Stevens, Nettie Maria (1861-1912. Cytology, Heredity)
[DSB v.18; NTCS]

Stewart, George Neil (1860-1930. Physiology. Came to U.S. in 1893)
[DAB; DSB v.13]

Stratton, Samuel Wesley (1861-1931. Metrology, Physics, Science and engineering education)
[DAB; DSB v.18; ID20thCB]

Taylor, Frank Bursley (1860-1938. Geology)
[DABs2; DSB v.13]

Taylor, Frederick Winslow (1856-1915. Engineering)
[DAB; DSB v.13; NTCS]

Tesla, Nikola (1856-1943. Electrical engineering, Physics. Came to U.S. in 1884)
[BES; CBDS; DABs3; DSB v.13; NTCS]

Thaxter, Roland (1858-1932. Cryptogamic botany)
[DAB; DSB v.13]

Thayer, William Sydney (1864-1932. Medicine)
[DAB; DSB v.13]

Trelease, William (1857-1945. Botany)
[DABs3; DSB v.13]

Ulrich, Edward Oscar (1857-1944. Paleontology, Stratigraphy)
[DABs3; DSB v.13]

Van Hise, Charles Edward (1857-1918. Geology)
[DAB; DSB v.13]

Webster, Arthur Gordon (1863-1923. Physics)
[DAB; DSB v.18]

White, (Charles) David (1862-1935. Geology)
[DABs1 ; DSB v.14; ID20thCB]

Whitehead, Alfred North (1861-1947. Mathematical logic, Mathematics, Philosophy, Theoretical physics. Came to U.S. in 1924)
[BES; CBDS; DABs4 ; DSB v.14; DTCWB; ID20thCB; NTCS]

Willis, Bailey (1857-1949. Geology)
[DABs4; DSB v.14]

Wilson, Edmund Beecher (1856-1939. Cytology, Embryology, Heredity)
[BES; DABs2 ; DSB v.14; NTCS]

1890-1899 Cohort (reached age 25)

Abbot, Charles Greeley (1872-1973. Astrophysics)
[ID20thCB]

Angell, James Rowland (1869-1949. Psychology)
[DABs4; ID20thCB]

Atwood, Wallace Walter (1872-1949. Geography, Geology, Geomorphology)
[DABs4; DSB v.17]

Austin, Louis Winslow (1867-1932. Radio physics)
[DSB v.1]

Bancroft, Wilder Dwight (1867-1953. Chemistry)
[DABs5; DSB v.1]

Barrell, Joseph (1869-1919. Geology)
[DAB; DSB v.1; ID20thCB]

Bauer, Louis Agricola (1865-1932. Geophysics)
[DABs1; DSB v.1]

Benedict, Francis Gano (1870-1957. Chemistry, Physiology)
[DSB v.1]

Bensley, Robert Russell (1867-1956. Anatomy)
[DABs6; DSB v.1]

Bocher, Maxime (1867-1918. Mathematics)
[DAB; DSB v.2]

Boltwood, Bertram Borden (1870-1927. Radiochemistry)
[BES; CBDS; DAB; DSB v.2]

Bowie, William (1872-1940. Geology)
[DABs2; DSB v.2; NTCS]

Briggs, Lyman James (1874-1963. Metrology, Physics)
[DABs7; DSB v.17]

Brodel, Max (1870-1941. Anatomy, Medical illustration)
[DABs3; DSB v.15]

Brooks, Alfred Hulse (1871-1924. Geology)
[DAB; DSB v.2]

Brown, Ernest William (1866-1938. Celestial mechanics)
[DAB; DSB v.2; ID20thCB]

Buckingham, Edgar (1867-1940. Physics)
[DSB v.2]

Burton, William Merriam (1865-1954. Chemistry)
[ID20thCB]

Calkins, Gary Nathan (1869-1943. Zoology)
[DABs3; DSB v.3]

Cannon, Walter Bradford (1871-1945. Physiology)
[BES; DABs3; DSB v.15; ID20thCB]

Carrel, Alexis (1873-1944. Experimental biology, Surgery. Came to U.S. in 1904)
[BES; CBDS; DABs3; DSB v.3; ID20thCB; NTCS]

Castle, William Ernest (1867-1962. Biology)
[DSB v.3]

Child, Charles Manning (1869-1954. Zoology)
[DABs5; DSB v.3; NTCS]

Clements, Frederic Edward (1874-1945. Botany)
[DABs3; DSB v.3]

Coblentz, William Weber (1873-1962. Infrared spectroscopy)
[BES; DSB v.3]

Coghill, George Ellett (1872-1941. Anatomy, Embryology, Psychobiology)
[DABs3; DSB v.3]

Coolidge, Julian Lowell (1873-1954. Mathematics)
[DABs5; DSB v.3]

Coolidge, William David (1873-1975. Physics)
[CBDS; DSB v.17; ID20thCB; NTCS]

Cowles, Henry Chandler (1869-1939. Botany, Ecology)
[DABs2; ID20thCB]

Curtis, Heber Doust (1872-1942. Astronomy)
[BES; DABs3; DSB v.3]

Cushing, Harvey Williams (1869-1939. Neurophysiology, Neurosurgery)
[BES; CBDS; DABs2; DSB v.3; DTCWB; ID20thCB]

Daly, Reginald Aldworth (1871-1957. Geology, Geophysics)
[DABs6; DSB v.3; ID20thCB; NTCS]

Davenport, Charles Benedict (1866-1944. Eugenics, Zoology)
[BES; DABs3; DSB v.3]

Dean, Bashford (1867-1928. Ichthyology)
[DAB; DSB v.3]

DeForest, Lee (1873-1961. Electronics)
[BES; CBDS; DABs7; DSB v.4; DTCWB; ID20thCB; NTCS]

Dickson, Leonard Eugene (1874-1954. Mathematics)
[DABs5; DSB v.4]

Douglass, Andrew Ellicott (1867-1962. Astronomy)
[BES; DABs7; ID20thCB]

Duane, William (1872-1935. Physics, Radiology)
[DABs1; DSB v.4]

Duggar, Benjamin Minge (1872-1956. Plant pathology)
[BES; DABs6; DSB v.4]

Erlanger, Joseph (1874-1965. Physiology)
[BES; CBDS; DABs7; DSB v.4; DTCWB; ID20thCB; M-H:MS&E; NTCS]

Evans, Alexander William (1868-1959. Botany)
[DSB v.15]

Ewing, James (1866-1943. Pathology)
[DABs3; DSB v.4]

Fairchild, David Grandison (1869-1954. Agricultural exploration, Botany)
[DABs5; ID20thCB]

Fenneman, Nevin Melancthon (1865-1945. Geology)
[DABs3; DSB v.4]

Fenner, Clarence Norman (1870-1949. Petrology, Volcanology)
[DSB v.4]

Fernald, Merritt Lyndon (1873-1950. Botany)
[DABs4; DSB v.4]

Fessenden, Reginald Aubrey (1866-1932. Radio engineering)
[BES; CBDS; DABs1; DSB v.4]

Folin, Otto Knut Olof (1867-1934. Biochemistry)
[DABs1; DSB v.5]

Frost, Edwin Brant (1866-1935. Astronomy)
[DABs1; DSB v.5]

Gay, Frederick Parker (1874-1939. Bacteriology, Pathology)
[DABs2; DSB v.5]

Goldberger, Joseph (1874-1929. Epidemiology)
[BES; DAB; DSB v.5; ID20thCB]

Gomberg, Moses (1866-1947. Organic chemistry)
[BES; DABs4; DSB v.5]

Grabau, Amadeus William (1870-1946. Geology, Paleontology)
[DABs4; DSB v.5]

Granger, Walter Willis (1872-1941. Paleontology)
[DABs3; DSB v.5]

Guyer, Michael Frederic (1874-1959. Zoology)
[DSB v.5]

Hale, George Ellery (1868-1938. Astrophysics)
[BES; CBDS; DABs2; DSB v.6; ID20thCB; NTCS]

Harkins, William Draper (1873-1951. Physical chemistry)
[BES; DABs5; DSB v.6]

Harrison, Ross Granville (1870-1959. Biology)
[BES; CBDS; DABs6; DSB v.6; ID20thCB]

Hart, Edwin Bret (1874-1953. Biochemistry, Nutrition)
[DABs5; DSB v.6]

Hayford, John Fillmore (1868-1925. Geodesy)
[DAB; DSB v.6]

Henderson, Yandell (1873-1944. Physiology)
[DABs3; DSB v.6]

Herrick, Charles Judson (1868-1960. Comparative neurology, Psychobiology)
[DSB v.6]

Hitchcock, Albert Spear (1865-1935. Botany)
[DSB v.15]

Hober, Rudolf Otto Anselm (1873-1953. Physiology. Came to U.S. in ca.1936)
[DSB v.17]

Hrdlicka, Ales (1869-1943. Physical anthropology)
[DABs3; DSB v.6; ID20thCB; NTCS]

Hunt, Reid (1870-1948. Pharmacology)
[DABs4; DSB v.17]

Huntington, Edward Vermilye (1874-1952. Mathematics)
[DABs5; DSB v.6]

Jaggar, Thomas Augustus, Jr. (1871-1953. Geology, Volcanology)
[DABs5; DSB v.7]

Jeffrey, Edward Charles (1866-1952. Botany)
[DABs5; DSB v.7]

Jennings, Herbert Spencer (1868-1947. Zoology)
[DABs4; DSB v.7]

Jepson, Willis Linn (1867-1946. Botany)
[DABs4; DSB v.7]

Johannsen, Albert (1871-1962. Petrography, Petrology)
[DSB v.7]

Jones, Harry Clary (1865-1916. Physical chemistry)
[DAB; DSB v.7]

Jones, Walter Jennings (1865-1935. Biochemistry)
[DSB v.17]

Jordan, Edwin Oakes (1866-1936. Bacteriology)
[DABs2; DSB v.7]

Juday, Chancey (1871-1944. Limnology)
[DABs3; DSB v.7]

Kahlenberg, Louis Albrecht (1870-1941. Chemistry)
[DSB v.7]

Kay, George Frederick (1873-1943. Geology)
[DSB v.7]

Kellogg, Vernon Lyman (1867-1937. Entomology, Zoology)
[DSB v.7]

King, Helen Dean (1869-1955. Genetics)
[DSB v.17]

Kofoid, Charles Atwood (1865-1947. Zoology)
[DABs4; DSB v.7]

Landsteiner, Karl (1868-1943. Immunology, Medicine, Serology. Came to U.S. in 1922)
[BES; CBDS; DABs3; DSB v.7; DTCWB; ID20thCB; NTCS]

Leavitt, Henrietta Swan (1868-1921. Astronomy)
[BES; CBDS; DAB; DSB v.8; ID20thCB; NTCS]

Levene, Phoebus Aaron Theodor (1869-1940. Biochemistry)
[BES; CBDS; DABs2; DSB v.8; ID20thCB]

Lillie, Frank Rattray (1870-1947. Embryology, Zoology)
[DABs4; DSB v.8; NTCS]

Loeb, Leo (1869-1959. Medicine, Pathology. Came to U.S. in ca.1897)
[DABs6; DSB v.8]

Loewi, Otto (1873-1961. Pharmacology, Physiology. Came to U.S. in 1940)
[BES; CBDS; DABs7; DSB v.8; DTCWB; ID20thCB; NTCS]

Lovejoy, Arthur Oncken (1873-1962. Epistemology, History of ideas)
[DABs7; DSB v.8; ID20thCB]

Lusk, Graham (1866-1932. Nutrition, Physiology)
[BES; CBDS; DABs1; DSB v.8]

Lyman, Theodore (1874-1954. Physics)
[BES; CBDS; DABs5; DSB v.8]

Macmillan, William Duncan (1871-1948. Astronomy, Mathematics)
[DSB v.8]

Magnus-Levy, Adolf (1865-1955. Medicine, Physiology. Came to U.S. in 1940)
[DSB v.18]

Mast, Samuel Ottmar (1871-1947. Botany)
[DSB v.9]

Maury, Antonia Caetana DePaiva Pereira (1866-1952. Astronomy)
[BES; DSB v.9; NTCS]

McClung, Clarence Erwin (1870-1946. Cytology, Zoology)
[DABs4; DSB v.8]

Mendel, Lafayette Benedict (1872-1935. Physiological chemistry)
[DABs1; DSB v.9]

Merriam, John Campbell (1869-1945. Paleontology)
[DABs3; DSB v.9]

Miller, Dayton Clarence (1866-1941. Physics)
[BES; DABs3; DSB v.9]

Millikan, Robert Andrews (1868-1953. Physics)
[BES; CBDS; DABs5; DSB v.9; DTCWB; ID20thCB; NTCS]

Montgomery, Thomas Harrison, Jr. (1873-1912. Zoology)
[DAB; DSB v.9]

Morgan, Thomas Hunt (1866-1945. Embryology, Genetics)
[BES; CBDS; DABs3; DSB v.9; DTCWB; ID20thCB; NTCS]

Moulton, Forest Ray (1872-1952. Astronomy, Mathematics)
[BES; DABs5; DSB v.9; ID20thCB; NTCS]

Nichols, Ernest Fox (1869-1924. Physics)
[DAB; DSB v.10]

Noyes, Arthur Amos (1866-1936. Chemistry)
[DABs2; DSB v.10]

Osterhout, Winthrop John Vanleuven (1871-1964. Botany, General physiology)
[DABs7; DSB v.18]

Patterson, John Thomas (1873-1960. Embryology, Evolution, Genetics)
[DSB v.18]

Perrine, Charles Dillon (1867-1951. Astronomy)
[DABs5; DSB v.10]

Pierce, George Washington (1872-1956. Applied physics)
[DSB v.10]

Poor, Charles Lane (1866-1951. Astronomy)
[DSB v.11]

Pratt, Frederick Haven (1873-1958. Physiology)
[DSB v.11]

Richards, Theodore William (1868-1928. Chemistry)
[BES; CBDS; DAB; DSB v.11; ID20thCB; NTCS]

Ricketts, Howard Taylor (1871-1910. Pathology)
[BES; CBDS; DABs1; DSB v.11]

Ross, Frank Elmore (1874-1966. Astronomy)
[DSB v.11]

Sabin, Florence Rena (1871-1953. Anatomy, Immunology)
[DABs5; DSB v.12; ID20thCB; NTCS]

Sabine, Wallace Clement Ware (1868-1919. Physics)
[BES; DAB; DSB v.12]

Schlesinger, Frank (1871-1943. Astronomy)
[DABs3; DSB v.12]

Seares, Frederick Hanley (1873-1964. Astronomy)
[DSB v.12]

See, Thomas Jefferson Jackson (1866-1962. Astronomy)
[DSB v.12]

Steinmetz, Charles Proteus (1865-1923. Engineering)
[DAB; DSB v.13; ID20thCB; NTCS]

Stiles, Charles Wardell (1867-1941. Public health, Zoology)
[DABs3; DSB v.13]

Streeter, George Linius (1873-1948. Embryology)
[DABs4; DSB v.13]

Sumner, Francis Bertody (1874-1945. Biology)
[DABs3; DSB v.13]

Tennent, David Hilt (1873-1941. Biology)
[DSB v.13]

Washington, Henry Stephens (1867-1934. Geology)
[DAB; DSB v.14; ID20thCB]

Wheeler, William Morton (1865-1937. Entomology)
[DABs2; DSB v.14]

Wilson, Harold Albert (1874-1964. Physics. Came to U.S. in 1912)
[DSB v.18]

Winchell, Alexander Newton (1874-1958. Mineralogy, Petrology)
[DSB v.14]

Winchell, Horace Vaughn (1865-1923. Geology, Mining engineering)
[DAB; DSB v.14]

Wood, Robert Williams (1868-1955. Experimental physics)
[BES; CBDS; DSB v.14]

Wright, Orville (1871-1948. Aeronautics)
[BES; CBDS; DABs4 ; DSB v.14; DTCWB; ID20thCB; NTCS]

Wright, Wilbur (1867-1912. Aeronautics)
[BES; CBDS; DAB; DSB v.14; DTCWB; ID20thCB; NTCS]

Wright, William Hammond (1871-1959. Astronomy)
[DSB v.14]

1900-1909 Cohort (reached age 25)

Adams, Walter Sydney (1876-1956. Astrophysics)
[BES; CBDS; DABs6; DSB v.1; ID20thCB; NTCS]

Alexanderson, Ernst Frederik Werner (1878-1975. Electrical engineering)
[ID20thCB; NTCS]

Andrews, Roy Chapman (1884-1960. Exploration, Natural history)
[BES; DABs6; ID20thCB]

Arnold, Harold De Forest (1883-1933. Electronics)
[DABs1; DSB v.1]

Avery, Oswald T. (1877-1955. Biology)
[BES; CBDS; DABs5; DSB v.1; DTCWB; ID20thCB; M-H:MS&E; NTCS]

Babcock, Ernest Brown (1877-1954. Botany, Genetics)
[DSB v.17]

Bassler, Raymond Smith (1878-1961. Paleontology)
[DSB v.1]

Bateman, Harry (1882-1946. Mathematics, Mathematical physics. Came to U.S. in 1910)
[DABs4; DSB v.1]

Beebe, (Charles) William (1877-1962. Exploration, Natural history)
[BES; CBDS; DABs7; ID20thCB]

Bell, Eric Temple (1883-1960. Mathematics)
[DABs6; DSB v.1]

Bennett, Hugh Hammond (1881-1960. Soil science)
[DABs6; ID20thCB]

Bernstein, Felix (1878-1956. Mathematics. Came to U.S. in 1934 and stayed until 1948)
[DSB v.2]

Birkhoff, George David (1884-1944. Mathematics)
[BES; DABs3; DSB v.2; DTCWB; ID20thCB; NTCS]

Bliss, Gilbert Ames (1876-1951. Mathematics)
[DABs5; DSB v.2]

Bowman, Isaiah (1878-1950. Geography)
[DABs4; DSB v.2; ID20thCB]

Bridgman, Percy Williams (1882-1961. Physics, Philosophy of science)
[BES; CBDS; DABs7; DSB v.2; ID20thCB; M-H:MS&E; NTCS]

Carlson, Anton Julius (1875-1956. Physiology)
[DABs6; DSB v.3]

Clark, William Mansfield (1884-1964. Chemistry)
[DSB v.3; M-H:MS&E]

Cohen, Morris Raphael (1880-1947. Philosophy of science)
[DABs4; DSB v.3; ID20thCB]

Cole, Leon Jacob (1877-1948. Genetics)
[DSB v.17]

Cottrell, Frederick Gardner (1877-1948. Chemistry, Engineering)
[DABs4; DSB v.3]

Crampton, Henry Edward (1875-1956. Evolutionary biology)
[DSB v.17]

Curtiss, Ralph Hamilton (1880-1929. Astronomy)
[DSB v.3]

Cushman, Joseph Augustine (1881-1949. Micropaleontology)
[DABs4; DSB v.3]

Dakin, Henry Drysdale (1880-1952. Biochemistry. Came to U.S. in 1905)
[DABs5; DSB v.17]

Danforth, Charles Haskell (1883-1969. Anatomy, Genetics)
[DSB v.3]

Davisson, Clinton Joseph (1881-1958. Physics)
[BES; DABs6; DSB v.3; ID20thCB; NTCS]

Debye, Peter Joseph William (1884-1966. Chemical physics. Came to U.S. in ca.1940)
[BES; CBDS; DABs8; DSB v.3; ID20thCB; M-H:MS&E; NTCS]

Dehn, Max (1878-1952. Mathematics. Came to U.S. in 1940)
[DSB v.4]

Dugan, Raymond Smith (1878-1940. Astronomy)
[DSB v.4]

Duncan, John Charles (1882-1967. Astronomy)
[DSB v.4]

East, Edward Murray (1879-1938. Genetics)
[DABs2; DSB v.4; ID20thCB]

Einstein, Albert (1879-1955. Physics. Came to U.S. ca.1933)
[BES; CBDS; DABs5; DSB v.4; DTCWB; ID20thCB; NTCS]

Eisenhart, Luther Pfahler (1876-1965. Mathematics)
[DABs7; DSB v.4]

Epstein, Paul Sophus (1883-1966. Theoretical physics. Came to U.S. in 1921)
[DSB v.17]

Fletcher, Harvey (1884-1981. Physics)
[ID20thCB]

Forbes, Alexander (1882-1965. Physiology)
[DSB v.5]

Franck, James (1882-1964. Physics. Came to U.S. in 1935)
[BES; CBDS; DABs7; DSB v.5; DTCWB; ID20thCB; M-H:MS&E; NTCS]

Frank, Philipp G. (1884-1966. Mathematics, Physics, Education, Philosophy of science. Came to U.S. in 1938)
[DABs8; DSB v.5]

Funk, Casimir (1884-1967. Biochemistry)
[BES; DABs8; DSB v.5; ID20thCB]

Gaines, Walter Lee (1881-1950. Dairy science)
[DSB v.5]

Gesell, Arnold Lucius (1880-1961. Psychology)
[DABs7; DSB v.5; ID20thCB]

Goddard, Robert Hutchings (1882-1945. Physics, Rocket engineering)
[BES; CBDS; DABs3; DSB v.5; DTCWB; ID20thCB; NTCS]

Goldschmidt, Richard Benedict (1878-1958. General biology, Zoology. Came to U.S. in 1936)
[DSB v.5; NTCS]

Gregory, William King (1876-1970. Comparative anatomy, Vertebrate paleontology)
[DSB v.17]

Grinnell, Joseph (1877-1939. Zoology)
[DSB v.5]

Harper, Roland McMillan (1878-1966. Botany, Demography, Geography)
[DSB v.6]

Hellinger, Ernest (1883-1950. Mathematics. Came to U.S. in 1939)
[DSB v.6]

Helly, Eduard (1884-1943. Mathematics. Came to U.S. in 1938)
[DSB v.17]

Henderson, Lawrence Joseph (1878-1942. Biochemistry, Physiology)
[DABs3; DSB v.6; ID20thCB]

Hess, Victor Franz (Francis) (1883-1964. Physics. Came to U.S. in ca.1920-1923, ca.1938-)
[BES; CBDS; DABs7; DSB v.6; DTCWB; ID20thCB; NTCS]

Hoagland, Dennis Robert (1884-1949. Plant physiology)
[DABs4; DSB v.6]

Hudson, Claude Silbert (1881-1952. Chemistry)
[DABs5; DSB v.6]

Hull, Albert Wallace (1880-1966. Electron physics)
[CBDS; DSB v.6]

Hull, Clark Leonard (1884-1952. Psychology)
[DABs5; ID20thCB]

Huntington, Ellsworth (1876-1947. Geography)
[DABs4; ID20thCB]

Ives, Herbert Eugene (1882-1953. Physics)
[DSB v.7]

Jacobs, Walter Abraham (1883-1967. Organic chemistry)
[DSB v.15]

Jewett, Frank Baldwin (1879-1949. Telecommunications)
[DABs4; DSB v.7; NTCS]

Johnson, Douglas Wilson (1878-1944. Geomorphology)
[DABs3; DSB v.7]

Karman, Theodore von (1881-1963. Aerodynamics. Came to U.S. in 1929)
[BES; DABs7; DSB v.7; ID20thCB; M-H:MS&E; NTCS]

Karpinski, Louis Charles (1878-1956. Cartography, History of mathematics)
[DSB v.15]

Kettering, Charles Franklin (1876-1958. Engineering, Invention)
[DABs6; DSB v.7; NTCS]

Knopf, Adolph (1882-1966. Economic geology, Petrology)
[DSB v.17]

Kraus, Charles August (1875-1967. Chemistry)
[BES; DSB v.7]

Ladenburg, Rudolf Walther (1882-1952. Physics. Came to U.S. in 1931)
[DSB v.7]

Langmuir, Irving (1881-1957. Chemistry, Physics)
[BES; CBDS; DABs6; DSB v.8; DTCWB; ID20thCB; NTCS]

Larsen, Esper Signius, Jr. (1879-1961. Geology)
[DSB v.8; M-H:MS&E]

Lefschetz, Solomon (1884-1972. Mathematics)
[DSB v.18; M-H:MS&E]

Lewis, Gilbert Newton (1875-1946. Physical chemistry)
[BES; CBDS; DABs4; DSB v.8; NTCS]

Livingston, Burton Edward (1875-1948. Physiological ecology, Plant physiology)
[DABs4; DSB v.8]

Lotka, Alfred James (1880-1949. Demography, Statistics)
[DABs4; DSB v.8]

Maanen, Andriaan Van (1884-1946. Astronomy. Came to U.S. in 1911)
[BES; DSB v.8]

Mason, Max (1877-1961. Invention, Mathematical physics)
[DABs7; ID20thCB]

McAtee, Waldo Lee (1883-1962. Ecology, Ornithology)
[DSB v.18]

McCollum, Elmer Verner (1879-1967. Nutrition, Organic chemistry)
[BES; DSB v.8; NTCS]

Meinzer, Oscar Edward (1876-1948. Groundwater hydrology)
[DABs4; DSB v.9]

Merrill, Elmer Drew (1876-1956. Botany)
[DABs6; DSB v.15]

Meyerhof, Otto (1884-1951. Biochemistry. Came to U.S. in 1940)
[BES; CBDS; DABs5; DSB v.9; ID20thCB; NTCS]

Michaelis, Leonor (1875-1949. Biochemistry. Came to U.S. in 1926)
[BES; DABs4; DSB v.18]

Mises, Richard Von (1883-1953. Mathematics, Mechanics, Probability. Came to U.S. in 1939)
[DSB v.9; ID20thCB; NTCS]

Moore, Joseph Haines (1878-1949. Astronomy)
[DABs4; DSB v.9]

Moore, Robert Lee (1882-1974. Mathematics)
[DSB v.18]

Morgan, Herbert Rollo (1875-1957. Astronomy)
[DSB v.9]

Moss, William Lorenzo (1876-1957. Medicine, Pathology)
[DSB v.9]

Murphy, James Bumgardner (1884-1950. Biology)
[DABs4; DSB v.9]

Newman, Horatio Hackett (1875-1957. Genetics)
[DSB v.18]

Nieuwland, Julius Arthur (1878-1936. Organic chemistry)
[BES; CBDS; DABs2; DSB v.10; ID20thCB]

Noguchi, (Seisaku) Hideyo (1876-1928. Microbiology)
[BES; DAB; DSB v.10; ID20thCB; NTCS]

Papanicolaou, George Nicholas (1883-1962. Anatomy, Cytology. Came to U.S. in ca.1913)
[DABs7; DSB v.10; ID20thCB; NTCS]

Payne, Fernandus (1881-1977. Genetics)
[DSB v.18]

Pearl, Raymond (1879-1940. Biology, Genetics)
[DABs2; DSB v.10; ID20thCB]

Pease, Francis Gladhelm (1881-1938. Astronomy)
[DSB v.10]

Pegram, George Braxton (1876-1958. Physics, Academic administration)
[DABs6; DSB v.18]

Raymond, Percy Edward (1879-1952. Geology, Paleontology)
[DSB v.11]

Richards, Alfred Newton (1876-1976. Pharmacology, Physiology)
[DSB v.18]

Richtmyer, Floyd Karker (1881-1939. Physics)
[DABs2; DSB v.11]

Riddle, Oscar (1877-1968. Endocrinology, Physiology)
[BES; DSB v.18]

Rous, (Francis) Peyton (1879-1970. Pathology, Virology)
[BES; CBDS; DABs8; DTCWB; ID20thCB; M-H:MS&E; NTCS]

Rowe, Allan Winter (1879-1934. Physiological chemistry)
[DSB v.11]

Rudenberg, Reinhold (1883-1961. Electrical engineering. Came to U.S. in 1939)
[DSB v.11]

Russell, Henry Norris (1877-1957. Astrophysics, Spectroscopy)
[BES; DABs6; DSB v.12; DTCWB; ID20thCB; NTCS]

Sabine, Paul Earls (1879-1958. Acoustics)
[DSB v.12]

Sanderson, Ezra Dwight (1878-1944. Entomology, Rural sociology)
[DABs3; DSB v.12]

Sarton, George Alfred Leon (1884-1956. History of science. Came to U.S. in 1915)
[DABs6; DSB v.12; ID20thCB]

Saunders, Frederick Albert (1875-1963. Acoustics, Spectroscopy)
[DSB v.18]

Shelford, Victor Ernest (1877-1968. Ecology)
[DSB v.18]

Shull, Aaron Franklin (1881-1961. Evolution, Genetics)
[DSB v.12]

Slipher, Earl C. (1883-1964. Planetary astronomy)
[DSB v.12]

Slipher, Vesto Melvin (1875-1969. Astronomy)
[BES; DABs8; DSB v.12; ID20thCB; NTCS]

Smith, Philip Edward (1884-1970. Anatomy, Endocrinology)
[DSB v.12]

Stine, Charles Milton Altland (1882-1954. Organic chemistry)
[DABs5; DSB v.13]

Stromberg, Gustaf Benjamin (1882-1962. Astronomy. Came to U.S. in 1916)
[DSB v.13]

Sutton, Walter Stanborough (1877-1916. Biology, Medicine)
[DSB v.13; DTCWB; ID20thCB; NTCS]

Swann, William Francis Gray (1884-1962. Experimental physics, Theoretical physics)
[DSB v.13]

Tolman, Richard Chace (1881-1948. Mathematical physics, Physical chemistry)
[DABs4; DSB v.13]

Twenhofel, William Henry (1875-1957. Geology)
[DSB v.13]

Van Slyke, Donald Dexter (1883-1971. Biochemistry)
[DSB v.13; M-H:MS&E]

Vandiver, Harry Schultz (1882-1973. Mathematics)
[DSB v.18]

Veblen, Oswald (1880-1960. Mathematics)
[DABs6; DSB v.13; ID20thCB]

Washburn, Edward Wight (1881-1934. Physical chemistry)
[DAB; DSB v.14]

Watson, John Broadus (1878-1958. Psychology)
[DABs6; DTCWB; ID20thCB]

Wedderburn, Joseph Henry Maclagan (1882-1948. Mathematics. In U.S. as early as 1904 but came permanently in ca.1918)
[DSB v.14]

Wells, Harry Gideon (1875-1943. Pathology)
[DABs3; DSB v.14]

Wilson, Edwin Bidwell (1879-1964. Mathematics, Physics)
[DSB v.14]

Wright, Frederick Eugene (1877-1953. Petrology)
[DSB v.14]

Yerkes, Robert Mearns (1876-1956. Comparative psychology)
[DABs6 ; DSB v.14; ID20thCB]

Young, John Wesley (1879-1932. Mathematics, Education)
[DAB; DSB v.14]

Zinsser, Hans (1878-1940. Bacteriology, Immunology)
[DABs2 ; DSB v.14; ID20thCB; NTCS]

1910-1919 Cohort (reached age 25)

Adams, Leason Heberling (1887-1969. Geophysics)
[DSB v.17; M-H:MS&E]

Adams, Roger (1889-1971. Organic chemistry)
[BES; DSB v.15; M-H:MS&E; NTCS]

Allee, Warder Clyde (1885-1955. Ecology)
[DSB v.17]

Allen, Edgar (1892-1943. Endocrinology)
[BES; DABs3; DSB v.1]

Armstrong, Edwin Howard (1890-1954. Radio engineering)
[CBDS; DABs5; DSB v.1; DTCWB; ID20thCB; NTCS]

Baade, (Wilhelm Heinrich) Walter (1893-1960. Astronomy. Came to U.S. in 1931)
[BES; CBDS; DABs6; DSB v.1; DTCWB; ID20thCB; M-H:MS&E; NTCS]

Barbour, Henry Gray (1885/1886-1943. Pharmacology, Physiology)
[DABs3; DSB v.1]

Bateman, Alan Mara (1889-1971. Geology)
[DSB v.17]

Benedict, Ruth (Fulton) (1887-1948. Anthropology)
[DABs4; ID20thCB]

Bergmann, Max (1886-1944. Biochemistry, Organic chemistry. Came to U.S. in ca.1934)
[BES; DABs3; DSB v.2]

Birge, Raymond Thayer (1887-1980. Physics)
[DSB v.17]

Boring, Edwin Garrigues (1886-1968. Psychology)
[DABs8; ID20thCB]

Bowen, Norman Levi (1887-1956. Geology)
[DABs6; DSB v.2; ID20thCB; M-H:MS&E]

Bridges, Calvin Blackman (1889-1938. Genetics)
[DABs2; DSB v.2]

Brillouin, Leon Nicolas (1889-1969. Electronics, Information theory, Quantum theory, Solid-state physics. Came to U.S. in 1941)
[DSB v.17]

Bronk, Detlev Wulf (1897-1975. Biophysics)
[BES; DABs9; DSB v.17; ID20thCB; M-H:MS&E; NTCS]

Bryan, Kirk (1888-1950. Geology, Geomorphology)
[DABs4; DSB v.2]

Bucher, Walter Herman (1889-1965. Geology)
[DSB v.2; M-H:MS&E; NTCS]

Buddington, Arthur Francis (1890-1980. Geology)
[DSB v.17; M-H:MS&E]

Bush, Vannevar (1890-1974. Electrical engineering, Science and government)
[CBDS; DABs9; DSB v.17; ID20thCB; M-H:MS&E; NTCS]

Carnap, Rudolf (1891-1970. Logical positivism, Philosophy. Came to U.S. in 1936)
[BES; DABs8; DTCWB; ID20thCB]

Chaney, Ralph Works (1890-1971. Conservation, Paleobotany, Paleoecology)
[DSB v.17; M-H:MS&E]

Chapman, Sydney (1888-1970. Geophysics, Natural philosophy. Came to U.S. in 1953)
[BES; DSB v.17; ID20thCB; M-H:MS&E]

Clausen, Jens Christen (1891-1969. Plant genetics. Came to U.S. in 1931)
[DSB v.17; M-H:MS&E]

Cleland, Ralph Erskine (1892-1971. Botany, Genetics)
[DSB v.17; M-H:MS&E]

Cohn, Edwin Joseph (1892-1953. Biochemistry)
[DABs5; DSB v.3; ID20thCB]

Compton, Arthur Holly (1892-1962. Physics)
[BES; CBDS; DABs7; DSB v.3; DTCWB; ID20thCB; NTCS]

Compton, Karl Taylor (1887-1954. Physics)
[DABs5; DSB v.3]

Conant, James Bryant (1893-1978. Organic chemistry, Science policy)
[BES; DABs10; DSB v.17; ID20thCB]

Courant, Richard (1888-1972. Mathematics. Came to U.S. in 1934)
[DABs9; ID20thCB; M-H:MS&E; NTCS]

Daniels, Farrington (1889-1972. Physical chemistry)
[BES; DABs9; DSB v.17; M-H:MS&E]

DeGolyer, Everette Lee (1886-1956. Geophysics)
[DSB v.4]

Dickinson, Roscoe Gilkey (1894-1945. Crystal structure, Physical chemistry, X rays)
[DSB v.4]

Doisy, Edward Adelbert (1893-1986. Biochemistry)
[BES; CBDS; ID20thCB; M-H:MS&E; NTCS]

Dumond, Jesse William Monroe (1892-1976. Experimental physics)
[DSB v.17]

Dunbar, Carl Owen (1891-1979. Geology, Paleontology)
[DSB v.17; M-H:MS&E]

Dunn, Leslie Clarence (1893-1974. Genetics)
[DSB v.17; M-H:MS&E]

Evans, Griffith Conrad (1887-1973. Mathematics)
[DSB v.17; M-H:MS&E]

Ewald, Paul Peter (1888-1985. Crystallography, Physics. Came to U.S. in ca.1949)
[DSB v.17]

Fajans, Kasimir (1887-1975. Chemistry. Came to U.S. in ca.1936)
[BES; DSB v.17]

Fenn, Wallace Osgood (1893-1971. Physiology)
[DSB v.17; M-H:MS&E]

Fischer, Hermann Otto Lorenz (1888-1960. Biochemistry. Came to U.S. in 1948)
[DSB v.5]

Foshag, William Frederick (1894-1956. Geology, Mineralogy)
[DABs6; DSB v.17]

Gasser, Herbert Spencer (1888-1963. Physiology)
[BES; CBDS; DABs7; DSB v.5; ID20thCB; M-H:MS&E; NTCS]

Goldring, Winifred (1888-1971. Geology, Paleontology)
[DSB v.17; NTCS]

Goodpasture, Ernest William (1886-1960. Pathology)
[BES; ID20thCB; M-H:MS&E]

Goodspeed, Thomas Harper (1887-1966. Botany, Plant genetics)
[DSB v.17]

Gutenberg, Beno (1889-1960. Seismology. Came to U.S. in 1930)
[BES; DSB v.5; ID20thCB; NTCS]

Guthrie, Edwin Ray, Jr. (1886-1959. Psychology)
[DABs6; ID20thCB]

Harvey, Edmund Newton (1887-1959. Physiology)
[DSB v.17]

Hill, Lester Sanders (1890-1961. Mathematics)
[DSB v.6]

Hooton, Earnest Albert (1887-1954. Anthropology)
[DABs5; ID20thCB]

Hubble, Edwin Powell (1889-1953. Cosmology, Observational astronomy)
[BES; CBDS; DABs5; DSB v.6; DTCWB; ID20thCB; NTCS]

Hubbs, Carl Leavitt (1894-1979. Ichthyology)
[DSB v.17; M-H:MS&E]

Hyman, Libbie Henrietta (1888-1969. Invertebrate zoology)
[BES; DSB v.17; ID20thCB; NTCS]

Jeffries, Zay (1888-1965. Industrial metallurgy)
[DSB v.7; NTCS]

Kelser, Raymond Alexander (1892-1952. Microbiology, Veterinary medicine)
[DABs5; DSB v.7]

Kendall, Edward Calvin (1886-1972. Biochemistry, Endocrinology)
[BES; CBDS; DABs9; DSB v.15; DTCWB; ID20thCB; M-H:MS&E; NTCS]

Kinsey, Alfred Charles (1894-1956. Sexology, Zoology)
[BES; DABs6; DTCWB; ID20thCB; NTCS]

Knight, Samuel Howell (1892-1975. Geology)
[DSB v.17]

Knudsen, Vern Oliver (1893-1974. Physics)
[DSB v.17]

Kohler, Wolfgang (1887-1967. Animal behavior, Gestalt psychology. Came to U.S. in 1935)
[DABs8; DSB v.17; DTCWB; ID20thCB; M-H:MS&E]

Kunitz, Moses (1887-1978. Enzymology, Physical biochemistry)
[DABs10; DSB v.17]

Lark-Horovitz, Karl (1892-1958. Physics. Came to U.S. in ca.1926)
[DSB v.17]

Lashley, Karl Spencer (1890-1958. Neurophysiology, Psychology)
[DABs6; DSB v.8; ID20thCB]

Latimer, Wendell Mitchell (1893-1955. Chemistry)
[DABs5; DSB v.8]

Leopold, (Rand) Aldo (1886-1948. Conservation, Ecology)
[DABs4; ID20thCB; NTCS]

Little, Clarence Cook (1888-1971. Mammalian genetics)
[DSB v.18]

Loewner, Charles (Karl) (1893-1968. Mathematics. Came to U.S. in ca.1939)
[DSB v.8]

Marvel, Carl Shipp ("Speed") (1894-1988. Organic chemistry, Polymer science)
[BES; DSB v.18; M-H:MS&E]

Mather, Kirtley Fletcher (1888-1978. Geology)
[DSB v.18; M-H:MS&E]

Meggers, William Frederick (1888-1966. Physics)
[DSB v.9]

Merica, Paul Dyer (1889-1957. Metallurgy)
[DSB v.9]

Midgley, Thomas, Jr. (1889-1944. Chemistry)
[BES; CBDS; DABs3; DSB v.9; ID20thCB; NTCS]

Minot, George Richards (1885-1950. Medicine)
[BES; CBDS; DABs4; DSB v.9; NTCS]

Moore, Raymond Cecil (1892-1974. Geology, Paleontology)
[DSB v.18; M-H:MS&E; NTCS]

Mordell, Louis Joel (1888-1972. Mathematics. Went to England in ca.1907 and became a British citizen in 1929)
[BES; DSB v.18; M-H:MS&E]

Muller, Hermann Joseph (1890-1967. Eugenics, Evolution, Genetics)
[BES; CBDS; DABs8; DSB v.9; ID20thCB; M-H:MS&E; NTCS]

Neyman, Jerzy (1894-1981. Statistics. Came to U.S. in 1938)
[DSB v.18; ID20thCB; M-H:MS&E]

Nicholson, Seth Barnes (1891-1963. Observational astronomy)
[BES; DABs7; DSB v.10]

Noble, Gladwyn Kingsley (1894-1940. Ethology, Experimental biology, Herpetology)
[DABs2; DSB v.18; NTCS]

Northrop, John Howard (1891-1987. Biochemistry)
[BES; CBDS; ID20thCB; M-H:MS&E; NTCS]

Painter, Theophilus Shickel (1889-1969. Cytogenetics, Genetics)
[DSB v.10]

Rademacher, Hans (1892-1969. Mathematics. Came to U.S. in 1933)
[DABs8; DSB v.11]

Reichenbach, Hans (1891-1953. Philosophy of science. Came to U.S. in 1938)
[BES; DABs5; DSB v.11; ID20thCB]

Ritt, Joseph Fels (1893-1951. Mathematics)
[DSB v.11]

Romer, Alfred Sherwood (1894-1973. Paleontology, Vertebrate anatomy)
[BES; DABs9; DSB v.18; ID20thCB; M-H:MS&E; NTCS]

Sax, Karl (1892-1973. Chromosome studies, Horticulture, Demography)
[DSB v.18; M-H:MS&E]

Scatchard, George (1892-1973. Physical chemistry)
[DSB v.18; M-H:MS&E]

Schmidt, Karl Patterson (1890-1957. Herpetology)
[DSB v.18]

Schrader, Franz [Otto Johann Wolfgang] (1891-1962. Cytology)
[DSB v.18]

Shapley, Harlow (1885-1972. Astronomy)
[BES; CBDS; DABs9; DSB v.12; ID20thCB; M-H:MS&E; NTCS]

Shewhart, Walter Andrew (1891-1967. Physics, Research statistics)
[DSB v.18]

Steenbock, Harry (1886-1967. Agricultural chemistry, Biochemistry, Nutrition)
[DSB v.18]

Stern, Otto (1888-1969. Physics. Came to U.S. in ca.1933)
[BES; CBDS; DABs8; DSB v.13; DTCWB; ID20thCB; M-H:MS&E; NTCS]

Stock, Chester (1892-1950. Paleontology)
[DSB v.13]

Sturtevant, Alfred Henry (1891-1970. Genetics)
[BES; CBDS; DSB v.13; M-H:MS&E; NTCS]

Sumner, James Batcheller (1887-1955. Biochemistry)
[BES; CBDS; DABs5; DSB v.13; ID20thCB; M-H:MS&E; NTCS]

Szent-Gyorgyi, Albert von (1893-1986. Biochemistry. Came to U.S. in 1947)
[BES; CBDS; ID20thCB; M-H:MS&E; NTCS]

Taylor, Charles Vincent (1885-1946. Biology)
[DSB v.13]

Taylor, Hugh Stott (1890-1974. Physical chemistry. Came to U.S. in 1914)
[BES; DSB v.18; M-H:MS&E]

Trumpler, Robert Julius (1886-1956. Astronomy. Came to U.S. in 1915)
[BES; DSB v.13]

Urey, Harold Clayton (1893-1981. Chemistry, Geophysics, Planetary science)
[BES; CBDS; DSB v.18; DTCWB; ID20thCB; M-H:MS&E; NTCS]

Vickery, Hubert Bradford (1893-1978. Biochemistry)
[DSB v.18; M-H:MS&E]

Waksman, Selman Abraham (1888-1973. Microbiology, Pharmacology, Soil science)
[BES; CBDS; DABs9; DSB v.18; DTCWB; ID20thCB; M-H:MS&E; NTCS]

Webster, David Locke (1888-1976. Physics)
[DSB v.18]

Wiener, Norbert (1894-1964. Mathematics)
[BES; CBDS; DABs7; DSB v.14; DTCWB; ID20thCB; M-H:MS&E; NTCS]

Williams, Robert Runnels (1886-1965. Chemistry, Nutrition)
[BES; DSB v.14; M-H:MS&E]

Wright, Sewall (1889-1988. Genetics, Mathematics)
[BES; DTCWB; M-H:MS&E; NTCS]

Wu, Hsien (1893-1959. Biochemistry, Nutrition. In the U.S. ca.1911 to 1920, ca.1947 to 1959)
[BES; DSB v.14]

1920-1929 Cohort (reached age 25)

Aiken, Howard Hathaway (1900-1973. Computers, Invention, Mathematics)
[BES; DTCWB; ID20thCB; NTCS]

Albright, Fuller (1900-1969. Endocrinology, Medicine)
[DSB v.17]

Allison, Samuel King (1900-1965. Physics)
[DABs7; DSB v.17]

Allport, Gordon Willard (1897-1967. Psychology)
[DABs8; ID20thCB]

Anderson, Edgar (1897-1969. Plant genetics)
[DSB v.17; M-H:MS&E]

Badger, Richard McLean (1896-1974. Physical chemistry, Spectroscopy)
[DSB v.17; M-H:MS&E]

Bateson, Gregory (1904-1980. Anthropology. Came to U.S. in 1940)
[DABs10; ID20thCB]

Beadle, George Wells (1903-1989. Genetics)
[BES; CBDS; ID20thCB; M-H:MS&E; NTCS]

Bekesy, Georg Von (1899-1972. Biophysics, Psychology. Came to U.S. in ca.1949)
[BES; DABs9; DSB v.17; ID20thCB; M-H:MS&E; NTCS]

Bergman, Stefan (1895-1977. Mathematics. Came to U.S. in ca.1939)
[DSB v.17]

Bjerknes, Jacob Aall Bonnevie (1897-1975. Meteorology, Oceanography. Came to U.S. in 1940)
[BES; CBDS; DSB v.17; ID20thCB; M-H:MS&E; NTCS]

Black, Harold Stephen (1898-1983. Electrical engineering)
[ID20thCB; M-H:MS&E]

Blalock, Alfred (1899-1964. Physiology, Surgery)
[DABs7; ID20thCB; M-H:MS&E]

Bochner, Salomon (1899-1982. Mathematics. Came to U.S. in 1933)
[DSB v.17; M-H:MS&E]

Bowen, Ira Sprague (1898-1973. Astronomy, Physics)
[BES; DABs9; DSB v.17; M-H:MS&E]

Brattain, Walter Houser (1902-1987. Physics)
[BES; DTCWB; ID20thCB; M-H:MS&E; NTCS]

Brauer, Richard Dagobert (1901-1977. Mathematics. Came to U.S. in 1933; in Canada 1935-1948)
[DSB v.17; M-H:MS&E]

Brode, Wallace Reed (1900-1974. Organic chemistry)
[DSB v.17; M-H:MS&E]

Brouwer, Dirk (1902-1966. Astronomy. Came to U.S. in 1927)
[BES; DSB v.2; M-H:MS&E]

Campbell, Ian (1899-1978. Geology)
[DSB v.17]

Carothers, Wallace Hume (1896-1937. Chemistry)
[BES; CBDS; DABs2; DSB v.3; ID20thCB; NTCS]

Condon, Edward Uhler (1902-1974. Physics)
[CBDS; DABs9; ID20thCB; M-H:MS&E]

Coon, Carleton Stevens (1904-1981. Anthropology)
[ID20thCB; M-H:MS&E]

Cori, Carl Ferdinand (1896-1984. Biochemistry. Came to U.S. in 1922)
[BES; ID20thCB; M-H:MS&E; NTCS]

Cori, Gerty Theresa Radnitz (1896-1957. Biochemistry. Came to U.S. in 1922)
[BES; CBDS; DABs6; DSB v.3; ID20thCB; M-H:MS&E; NTCS]

Cournand, Andre Frederic (1895-1988. Medicine, Physiology. Came to U.S. in ca.1930)
[BES; ID20thCB; M-H:MS&E; NTCS]

Demerec, Milislav (1895-1966. Genetics)
[BES; DSB v.17; M-H:MS&E]

Dennison, David Mathias (1900-1976. Physics)
[DSB v.17]

Dirac, Paul Adrien Maurice (1902-1984. Cosmology, Quantum mechanics, Relativity. Came to U.S. in 1971)
[BES; CBDS; DSB v.17; DTCWB; ID20thCB; M-H:MS&E; NTCS]

Dobzhansky, Theodosius G. (1900-1975. Evolution, Genetics. Came to U.S. in 1927)
[BES; CBDS; DABs9; DSB v.17; DTCWB; ID20thCB; M-H:MS&E; NTCS]

Douglas, Jesse (1897-1965. Mathematics)
[DSB v.4]

Dryden, Hugh Latimer (1898-1965. Aerodynamics, Physics)
[DSB v.17; ID20thCB; M-H:MS&E]

Dubos, Rene Jules (1901-1982. Microbiology)
[BES; ID20thCB; M-H:MS&E; NTCS]

DuVigneaud, Vincent (1901-1978. Biochemistry)
[BES; CBDS; DABs10; DTCWB; ID20thCB; M-H:MS&E; NTCS]

Eckart, Carl Henry (1902-1973. Acoustics, Geophysics, Physics)
[DSB v.17; M-H:MS&E]

Eckert, Wallace John (1902-1971. Celestial mechanics, Computation)
[DSB v.15]

Edgerton, Harold Eugene (1903-1990. Electrical engineering)
[ID20thCB; M-H:MS&E; NTCS]

Edinger, Johanna Gabrielle Ottilie (Tilly) (1897-1967. Vertebrate paleontology. Came to U.S. in 1940)
[DSB v.17; NTCS]

Elvehjem, Conrad Arnold (1901-1962. Biochemistry)
[BES; DABs7; DSB v.4; M-H:MS&E]

Emerson, Robert (1903-1959. Plant physiology)
[DSB v.4]

Enders, John Franklin (1897-1985. Virology)
[BES; ID20thCB; M-H:MS&E; NTCS]

Esau, Katherine (1898-. Botany)
[ID20thCB; M-H:MS&E; NTCS]

Eyring, Henry (1901-1981. Physical chemistry)
[BES; DSB v.17; M-H:MS&E]

Fermi, Enrico (1901-1954. Physics. Came to U.S. in ca.1938)
[BES; CBDS; DABs5; DSB v.4; DTCWB; ID20thCB; M-H:MS&E; NTCS]

Fieser, Louis Frederick (1899-1977. Organic chemistry)
[DABs10; DSB v.17; M-H:MS&E; NTCS]

Fulton, John Farquhar (1899-1960. History of medicine, Physiology)
[DABs6; DSB v.5]

Gamow, George (1904-1968. Physics. Came to U.S. in ca.1932)
[BES; DABs8; DSB v.5; DTCWB; ID20thCB; M-H:MS&E; NTCS]

Giauque, William Francis (1895-1982. Chemical physics, Physical chemistry)
[BES; CBDS; DSB v.17; ID20thCB; M-H:MS&E; NTCS]

Gilluly, James (1896-1980. Geology)
[DSB v.17; M-H:MS&E]

Goudsmit, Samuel Abraham (1902-1978. Physics. Came to U.S in 1927)
[BES; CBDS; DABs10; DSB v.17; ID20thCB; M-H:MS&E; NTCS]

Hartline, Haldan Keffer (1903-1983. Neurophysiology)
[BES; ID20thCB; M-H:MS&E; NTCS]

Iselin, Columbus O'Donnell (1904-1971. Oceanography)
[ID20thCB; M-H:MS&E]

Jepsen, Glenn Lowell (1903-1974. Paleontology)
[DSB v.17]

Julian, Percy Lavon (1899-1975. Organic chemistry)
[DABs9; DSB v.17; ID20thCB; M-H:MS&E; NTCS]

Kay, Marshall (1904-1975. Geology)
[DSB v.17]

Kharasch, Morris Selig (1895-1957. Organic chemistry)
[DABs6; DSB v.7]

Lancefield, Rebecca Craighill (1895-1981. Bacteriology)
[DSB v.17; ID20thCB; NTCS]

Lawrence, Ernest Orlando (1901-1958. Physics)
[BES; CBDS; DABs6; DSB v.8; DTCWB; ID20thCB; M-H:MS&E; NTCS]

London, Fritz (1900-1954. Physics, Theoretical chemistry. Came to U.S. in ca.1939)
[BES; DSB v.8; NTCS]

Long, Cyril Norman Hugh (1901-1970. Biochemistry, Endocrinology. Came to U.S. in 1932)
[DSB v.18]

Mayr, Ernst (1904-. Biology. Came to U.S. in ca.1932)
[BES; ID20thCB; M-H:MS&E; NTCS]

McClintock, Barbara (1902-1992. Genetics)
[BES; CBDS; ID20thCB; NTCS]

Mead, Margaret (1901-1978. Anthropology)
[DABs10; DTCWB; ID20thCB; M-H:MS&E]

Mehl, Robert Franklin (1898-1976. Chemistry, Metallurgy)
[DSB v.18; M-H:MS&E]

Mirsky, Alfred Ezra (1900-1974. Biochemistry, Cell biology)
[DABs9; DSB v.18]

Mulliken, Robert Sanderson (1896-1986. Physical chemistry)
[BES; DTCWB; ID20thCB; M-H:MS&E; NTCS]

Nicholas, John Spangler (1895-1963. Biology)
[DSB v.10]

Onsager, Lars (1903-1976. Chemistry, Physics. Came to U.S. in 1928)
[BES; CBDS; DABs10: DSB v.18; ID20thCB; M-H:MS&E; NTCS]

Oppenheimer, Julius Robert (1904-1967. Theoretical physics)
[BES; CBDS; DABs8; DSB v.10; DTCWB; ID20thCB; M-H:MS&E; NTCS]

Pauling, Linus Carl (1901-1994. Chemistry)
[BES; CBDS; DTCWB; ID20thCB; M-H:MS&E; NTCS]

Pincus, Gregory Goodwin (1903-1967. Endocrinology)
[BES; CBDS; DABs8; DSB v.10; ID20thCB; M-H:MS&E; NTCS]

Post, Emil Leon (1897-1954. Logic, Mathematics)
[DSB v.11]

Rabi, Isidor Isaac (1898-1988. Physics)
[BES; CBDS; DTCWB; ID20thCB; M-H:MS&E; NTCS]

Rado, Tibor (1895-1965. Mathematics. Came to U.S. in ca.1929)
[DSB v.11]

Ramsdell, Lewis Stephen (1895-1975. Crystallography, Mineralogy)
[DSB v.18]

Rhine, Joseph Banks (1895-1980. Parapsychology)
[DABs10; DTCWB; ID20thCB]

Richter, Charles Francis (1900-1985. Geophysics, Seismology)
[BES; ID20thCB; NTCS]

Rossby, Carl-Gustaf Arvid (1898-1957. Meteorology, Oceanography. Came to U.S. in 1926 and returned to Sweden in 1950)
[BES; CBDS; DSB v.11; ID20thCB; M-H:MS&E; NTCS]

Russell, Richard Joel (1895-1971. Climatology, Geomorphology)
[DSB v.18; M-H:MS&E]

Schairer, John Frank (1904-1970. Mineralogy, Physical chemistry)
[DSB v.18]

Schoenheimer, Rudolf (1898-1941. Biochemistry. Came to U.S. in ca.1933)
[BES; DABs3; DSB v.18]

Shepard, Francis Parker (1897-1985. Geology)
[BES; ID20thCB; M-H:MS&E]

Simpson, George Gaylord (1902-1984. Paleontology)
[BES; DTCWB; ID20thCB; M-H:MS&E; NTCS]

Skinner, Burrhus Frederic (1904-1990. Psychology)
[DTCWB; ID20thCB; M-H:MS&E]

Slater, John Clarke (1900-1976. Quantum mechanics, Theoretical physics)
[DABs10; DSB v.18; M-H:MS&E; NTCS]

Slichter, Louis Byrne (1896-1978. Geophysics)
[DABs10; DSB v.18; M-H:MS&E]

Smith, Homer William (1895-1962. Evolutionary biology, Physiology)
[DABs7; DSB v.12]

Snell, George Davis (1903-. Immunogenetics)
[BES; ID20thCB; M-H:MS&E; NTCS]

Stanley, Wendell Meredith (1904-1971. Chemistry, Virology, Education)
[BES; DABs9; DSB v.18; ID20thCB; M-H:MS&E; NTCS]

Stern, Curt (1902-1981. Genetics. Came to U.S. in ca.1932)
[BES; DSB v.18; M-H:MS&E]

Struve, Otto (1897-1963. Astronomy. Came to U.S. in 1921)
[BES; DABs7; DSB v.13; ID20thCB; M-H:MS&E]

Szilard, Leo (1898-1964. Biology, Physics. Came to U.S. in ca.1938)
[BES; CBDS; DABs7; DSB v.13; DTCWB; ID20thCB; M-H:MS&E; NTCS]

Tarski, Alfred (1901-1983. Algebra, Mathematical logic, Set theory. Came to U.S. in 1939)
[BES; DSB v.18; ID20thCB; NTCS]

Theiler, Max (1899-1972. Microbiology)
[BES; ID20thCB; M-H:MS&E; NTCS]

Tuve, Merle Antony (1901-1982. Physics)
[BES; DSB v.18; M-H:MS&E; NTCS]

Van de Graaff, Robert Jemison (1901-1967. Physics)
[BES; CBDS; DABs8; DSB v.13; ID20thCB; M-H:MS&E; NTCS]

Van Vleck, John Hasbrouck (1899-1980. Chemical physics, Magnetism, Quantum theory of solids, Spectroscopy)
[BES; DABs10; DSB v.18; ID20thCB; M-H:MS&E; NTCS]

Vishniac, Roman (1897-1990. Biology. Came to U.S. in 1940)
[ID20thCB]

Von Neumann, John (1903-1957. Mathematical physics, Mathematics. Came to U.S. in 1930)
[BES; CBDS; DABs6; DSB v.14; DTCWB; ID20thCB; M-H:MS&E; NTCS]

Wald, Abraham (1902-1950. Mathematics, Statistics, Economics. Came to U.S. in 1938)
[DSB v.14]

Wigner, Eugene Paul (1902-1995. Mathematical physics. Came to U.S. in 1930)
[BES; CBDS; DTCWB; ID20thCB; M-H:MS&E; NTCS]

Wintner, Aurel (1903-1958. Mathematics. Came to U.S. in 1930)
[DSB v.14]

Wolfrom, Melville Lawrence (1900-1969. Carbohydrate chemistry)
[DSB v.18; M-H:MS&E]

Youden, William John (1900-1971. Mathematical statistics)
[DSB v.14]

Zwicky, Fritz (1898-1974. Astrophysics, Physics, Rocketry. Came to U.S. in 1925)
[BES; DABs9 ; DSB v.18; ID20thCB]

1930-1939 Cohort (reached age 25)

Abelson, Philip Hauge (1913-. Physical chemistry)
[BES; ID20thCB; M-H:MS&E; NTCS]

Albert, Abraham Adrian (1905-1972. Mathematics)
[DSB v.17; M-H:MS&E]

Alvarez, Luis Walter (1911-1988. Physics)
[BES; CBDS; ID20thCB; M-H:MS&E; NTCS]

Anderson, Carl David (1905-1991. Physics)
[BES; CBDS; ID20thCB; NTCS]

Axelrod, Julius (1912-. Pharmacology)
[BES; ID20thCB; M-H:MS&E; NTCS]

Bardeen, John (1908-1991. Physics)
[BES; CBDS; DTCWB; ID20thCB; M-H:MS&E; NTCS]

Bates, Marston (1906-1974. Ecology, Epidemiology)
[DSB v.17]

Begle, Edward Griffith (1914-1978. Mathematics)
[ID20thCB]

Berkner, Lloyd Viel (1905-1967. Geophysics, Radio engineering, Science administration)
[DSB v.17; ID20thCB; M-H:MS&E]

Bethe, Hans Albrecht (1906-. Physics. Came to U.S. in 1935)
[BES; CBDS; DTCWB; ID20thCB; M-H:MS&E; NTCS]

Bloch, Felix (1905-1983. Physics. Came to U.S. in ca.1933)
[BES; CBDS; DTCWB; ID20thCB; M-H:MS&E; NTCS]

Bloch, Konrad Emil (1912-. Biochemistry. Came to U.S. in 1936)
[BES; ID20thCB; M-H:MS&E; NTCS]

Borlaug, Norman Ernest (1914-. Plant pathology, Genetics)
[BES; DTCWB; ID20thCB; NTCS]

Brown, George Harold (1908-1987. Electrical engineering)
[ID20thCB; M-H:MS&E]

Brown, Herbert Charles (1912-. Organic chemistry)
[BES; ID20thCB; M-H:MS&E; NTCS]

Calvin, Melvin (1911-. Chemistry)
[BES; CBDS; ID20thCB; M-H:MS&E; NTCS]

Cameron, Angus Ewan (1906-1981. Mass spectroscopy, Physical chemistry)
[DSB v.17]

Carlson, Chester Floyd (1906-1968. Invention, Physics)
[BES; CBDS; DABs8; DTCWB; ID20thCB; NTCS]

Carson, Rachel Louise (1907-1964. Ecology, Marine biology, Natural history)
[DABs7; DSB v.17; ID20thCB; NTCS]

Chandrasekhar, Subrahmanyan (1910-1995. Astronomy, Astrophysics. Came to U.S. in 1936)
[BES; CBDS; DTCWB; ID20thCB; M-H:MS&E; NTCS]

Clark, Kenneth Bancroft (1914-. Psychology)
[ID20thCB]

Cochran, William Gemmell (1909-1980. Statistics. Came to U.S. in 1939)
[DSB v.17]

Colbert, Edwin Harris (1905-. Paleontology)
[ID20thCB]

Dantzig, George Bernard (1914-. Mathematics)
[ID20thCB; M-H:MS&E; NTCS]

Davis, D(elbert) Dwight (1908-1965. Evolutionary morphology)
[DSB v.17]

Delbruck, Max (1906-1981. Biology. Came to U.S. in 1937)
[BES; CBDS; ID20thCB; M-H:MS&E; NTCS]

Dulbecco, Renato (1914-. Biology. Came to U.S. in 1947)
[BES; CBDS; ID20thCB; M-H:MS&E; NTCS]

Ewing, (William) Maurice (1906-1974. Geophysics, Oceanography, Seismology)
[BES; CBDS; DABs9; DSB v.17; ID20thCB; M-H:MS&E]

Fankuchen, Isidor (1905-1964. Crystallography)
[DSB v.4]

Feller, William (1906-1970. Mathematics. Came to U.S. in 1939)
[DSB v.17; M-H:MS&E]

Flory, Paul John (1910-1985. Organic chemistry)
[BES; CBDS; ID20thCB; M-H:MS&E; NTCS]

Fowler, William Alfred (1911-. Physics)
[BES; ID20thCB; M-H:MS&E; NTCS]

Godel, Kurt Friedrich (1906-1978. General relativity, Mathematical logic, Set theory. Came permanently to U.S. in 1940)
[BES; DABs10; DSB v.17; DTCWB; ID20thCB; M-H:MS&E; NTCS]

Hansen, William Webster (1909-1949. Microwave electronics, Physics)
[DABs4; DSB v.6]

Hershey, Alfred Day (1908-. Microbiology)
[BES; CBDS; ID20thCB; M-H:MS&E; NTCS]

Hess, Harry Hammond (1906-1969. Geological geophysics, Geology, Mineralogy, Oceanography)
[BES; CBDS; DSB v.17; NTCS]

Hunt, Frederick Vinton (1905-1972. Acoustics)
[DSB v.17; M-H:MS&E]

Jansky, Karl Guthe (1905-1950. Radio astronomy, Radio engineering)
[BES; CBDS; DABs4; DTCWB; ID20thCB; NTCS]

Kerst, Donald William (1911-1993. Physics)
[BES; ID20thCB]

Kirkwood, John Gamble (1907-1959. Theoretical chemistry)
[DABs6; DSB v.7]

Knipling, Edward Fred (1909-. Entomology)
[ID20thCB; M-H:MS&E]

Kuiper, Gerard Peter (1905-1973. Astronomy. Came to U.S. in ca.1933)
[BES; DABs9; ID20thCB; NTCS]

Kusch, Polykarp (1911-1993. Physics)
[BES; ID20thCB; M-H:MS&E; NTCS]

Lamb, Willis Eugene, Jr. (1913-. Physics)
[BES; DTCWB; ID20thCB; M-H:MS&E; NTCS]

Levinson, Norman (1912-1975. Mathematics)
[DSB v.18]

Libby, Willard Frank (1908-1980. Chemistry)
[BES; CBDS; DABs10; DTCWB; ID20thCB; M-H:MS&E; NTCS]

Luria, Salvador Edward (1912-1991. Biology. Came to U.S. in 1940)
[BES; ID20thCB; M-H:MS&E; NTCS]

MacLeod, Colin Munro (1909-1972. Medicine, Microbiology)
[BES; DABs9; DSB v.18; M-H:MS&E; NTCS]

Mauchly, John William (1907-1980. Engineering)
[BES; DABs10; ID20thCB; NTCS]

Mayer, Maria Goeppert (1906-1972. Physics. Came to U.S. in ca.1930)
[BES; CBDS; DABs9; DSB v.18; ID20thCB; M-H:MS&E; NTCS]

McMillan, Edwin Mattison (1907-1991. Physics)
[BES; CBDS; ID20thCB; M-H:MS&E; NTCS]

Moore, Stanford (1913-1982. Biochemistry)
[BES; CBDS; ID20thCB; M-H:MS&E; NTCS]

Ochoa, Severo (1905-1993. Biochemistry. Came to U.S. in 1941)
[BES; ID20thCB; M-H:MS&E; NTCS]

Palade, George Emil (1912-. Cell biology. Came to U.S. in ca.1946)
[BES; CBDS; ID20thCB; M-H:MS&E; NTCS]

Peterson, Roger Tory (1908-. Natural history, Ornithology)
[ID20thCB]

Pickering, William Hayward (1910-. Engineering, Physics)
[ID20thCB; M-H:MS&E]

Purcell, Edward Mills (1912-. Physics)
[BES; CBDS; ID20thCB; M-H:MS&E; NTCS]

Reber, Grote (1911-. Radio astronomy, Engineering)
[BES; CBDS; ID20thCB; NTCS]

Revelle, Roger (1909-1991. Oceanography, Geology)
[BES; ID20thCB; M-H:MS&E; NTCS]

Sabin, Albert Bruce (1906-1993. Microbiology, Virology)
[BES; DTCWB; ID20thCB; M-H:MS&E; NTCS]

Salk, Jonas Edward (1914-1995. Microbiology, Virology)
[BES; DTCWB; ID20thCB; M-H:MS&E; NTCS]

Scheffe, Henry (1907-1977. Mathematics, Statistics)
[DSB v.18]

Seaborg, Glenn Theodore (1912-. Nuclear chemistry)
[BES; CBDS; DTCWB; ID20thCB; M-H:MS&E; NTCS]

Segre, Emilio Gino (1905-1989. Physics. Came to U.S. in ca.1938)
[BES; CBDS; ID20thCB; M-H:MS&E; NTCS]

Shockley, William Bradford (1910-1989. Physics)
[BES; CBDS; DTCWB; ID20thCB; M-H:MS&E; NTCS]

Sonneborn, Tracy Morton (1905-1981. Genetics, Protozoology)
[BES; DSB v.18; M-H:MS&E]

Sperry, Roger Wolcott (1913-1994. Psychobiology)
[BES; CBDS; ID20thCB; M-H:MS&E; NTCS]

Stebbins, George Ledyard (1906-. Botany)
[BES; ID20thCB; M-H:MS&E]

Stein, William Howard (1911-1980. Biochemistry)
[BES; DSB v.18; ID20thCB; M-H:MS&E; NTCS]

Stevens, Stanley Smith (1906-1973. Experimental psychology, Psychophysics)
[DSB v.18; M-H:MS&E]

Tatum, Edward Lawrie (1909-1975. Biochemistry, Genetics)
[BES; DABs9; DTCWB; ID20thCB; M-H:MS&E; NTCS]

Teller, Edward (1908-. Physics. Came to U.S. in ca.1935)
[BES; CBDS; DTCWB; ID20thCB; M-H:MS&E; NTCS]

Tombaugh, Clyde William (1906-. Astronomy)
[BES; ID20thCB; NTCS]

Ulam, Stanislaw Marcin (1909-1984. Mathematics)
[BES; ID20thCB; M-H:MS&E]

Van Allen, James Alfred (1914-. Physics)
[BES; CBDS; DTCWB; ID20thCB; M-H:MS&E; NTCS]

Vinograd, Jerome Ruben (1913-1976. Molecular biology, Physical biochemistry)
[DSB v.18; M-H:MS&E]

von Braun, Werner Magnus Maximilian (1912-1977. Rocket engineering. Came to U.S. in 1945)
[BES; CBDS; DABs10; DTCWB; ID20thCB; M-H:MS&E; NTCS]

Wald, George (1906-. Biochemistry)
[BES; ID20thCB; M-H:MS&E; NTCS]

Weisskopf, Victor Frederick (1908-. Physics. Came to U.S. in ca.1937)
[ID20thCB; M-H:MS&E]

Wheeler, John Archibald (1911-. Theoretical physics)
[BES; DTCWB; M-H:MS&E; NTCS]

Wilhelm, Richard Herman (1909-1968. Chemical engineering)
[DSB v.18; M-H:MS&E]

Wilks, Samuel Stanley (1906-1964. Mathematical statistics)
[DSB v.14]

Winstein, Saul (1912-1969. Chemistry)
[DSB v.18; M-H:MS&E]

Wolfowitz, Jacob (1910-1981. Information theory, Mathematical statistics)
[DSB v.18]

Wu, Chien-Shiung (1912-. Physics. Came to U.S. in 1936)
[BES; CBDS; ID20thCB; M-H:MS&E; NTCS]

Zachariasen, (Fredrik) William Houlder (1906-1979. Chemical crystallography, X-ray diffraction)
[DSB v.18]

1940-1949 Cohort (reached age 25)

Alpher, Ralph Asher (1921-. Physics)
[BES; CBDS; DTCWB]

Anderson, Philip Warren (1923-. Physics)
[BES; CBDS; ID20thCB; M-H:MS&E; NTCS]

Anfinsen, Christian Boehmer (1916-. Biochemistry)
[BES; ID20thCB; M-H:MS&E; NTCS]

Bandy, Orville Lee (1917-1973. Geology, Micropaleontology, Paleontology, Stratigraphy)
[DSB v.17]

Benacerraf, Baruj (1920-. Immunology)
[BES; ID20thCB; M-H:MS&E; NTCS]

Bloembergen, Nicolaas (1920-. Physics. Came to U.S. in 1946)
[BES; ID20thCB; M-H:MS&E; NTCS]

Bruner, Jerome Seymour (1915-. Psychology)
[ID20thCB]

Chamberlain, Owen (1920-. Physics)
[BES; DTCWB; ID20thCB; M-H:MS&E; NTCS]

Christofilos, Nicholas C. (1916-1972. Engineering, Magnetic fusion, Nuclear physics, Particle accelerator design. Born in U.S. but resident in Greece ca.1923-1953)
[DSB v.17]

Commoner, Barry (1917-. Biology, Ecology)
[ID20thCB; NTCS]

Eckert, John Presper, Jr. (1919-1995. Computers, Electrical engineering)
[BES; CBDS; DTCWB; ID20thCB; M-H:MS&E; NTCS]

Chronology: Scientist Cohorts, 1940-1949   437

Feynman, Richard Phillips (1918-1988. Physics)
[BES; CBDS; DTCWB; ID20thCB; M-H:MS&E; NTCS]

Fitch, Val Logsdon (1923-. Physics)
[BES; ID20thCB; M-H:MS&E; NTCS]

Gajdusek, Daniel Carleton (1923-. Medical science, Virology)
[BES; CBDS; ID20thCB; M-H:MS&E; NTCS]

Gold, Thomas (1920-. Astronomy. Came to U.S. in 1956)
[BES; CBDS; ID20thCB; NTCS]

Good, Robert Alan (1922-. Immunology)
[BES; ID20thCB; M-H:MS&E]

Guillemin, Roger Charles Louis (1924-. Medical science. Came to U.S. in 1953)
[BES; ID20thCB; M-H:MS&E; NTCS]

Hauptman, Herbert Aaron (1917-. Chemistry)
[BES; ID20thCB; NTCS]

Heezen, Bruce Charles (1924-1977. Oceanography)
[BES; DABs10; DSB v.17; ID20thCB; M-H:MS&E]

Hofstadter, Robert (1915-1990. Physics)
[BES; ID20thCB; M-H:MS&E; NTCS]

Holley, Robert William (1922-1993. Biochemistry)
[BES; ID20thCB; M-H:MS&E; NTCS]

Karle, Jerome (1918-. Physical chemistry)
[BES; ID20thCB; NTCS]

Khorana, Har Gobind (1922-. Biochemistry. Came to U.S. in 1960)
[BES; CBDS; ID20thCB; M-H:MS&E; NTCS]

Kline, Nathan Schellenberg (1916-1983. Psychology)
[ID20thCB]

Kornberg, Arthur (1918-. Biochemistry)
[BES; CBDS; DTCWB; ID20thCB; M-H:MS&E; NTCS]

Kuhn, Thomas Samuel (1922-. History of science)
[BES; ID20thCB]

Lipscomb, William Nunn, Jr. (1919-. Chemistry)
[BES; ID20thCB; M-H:MS&E; NTCS]

Merrifield, Robert Bruce (1921-. Chemistry)
[BES; ID20thCB; NTCS]

Rainwater, (Leo) James (1917-1986. Physics)
[BES; CBDS; ID20thCB; M-H:MS&E; NTCS]

Robbins, Frederick Chapman (1916-. Pediatrics, Physiology, Microbiology)
[BES; ID20thCB; M-H:MS&E; NTCS]

Robinson, Abraham (1918-1974. Aerodynamics, Logic, Mathematics. Came to U.S. in ca.1962)
[DSB v.18]

Sager, Ruth (1918-. Genetics)
[ID20thCB; NTCS]

Savage, Leonard Jimmie (1917-1971. Mathematics, Philosophy, Statistics)
[DSB v.18]

Schawlow, Arthur Leonard (1921-. Physics)
[BES; ID20thCB; NTCS]

Schiff, Leonard Isaac (1915-1971. Physics)
[DSB v.18]

Schwinger, Julian Seymour (1918-1994. Physics)
[BES; ID20thCB; M-H:MS&E; NTCS]

Shannon, Claude Elwood (1916-. Mathematics)
[BES; CBDS; ID20thCB; M-H:MS&E; NTCS]

Stommel, Henry Melson (1920-1992. Oceanography)
[ID20thCB; M-H:MS&E; NTCS]

Sutherland, Earl Wilbur, Jr. (1915-1974. Pharmacology, Physiology)
[BES; DABs9; ID20thCB; M-H:MS&E; NTCS]

Taube, Henry (1915-. Chemistry. Came to U.S. in 1937)
[BES; ID20thCB; M-H:MS&E; NTCS]

Townes, Charles Hard (1915-. Physics)
[BES; CBDS; ID20thCB; M-H:MS&E; NTCS]

Wang, Hsien Chung (1918-1978. Mathematics. Came to U.S. in 1949)
[DSB v.18]

Weller, Thomas Huckle (1915-. Parasitology, Physiology)
[BES; ID20thCB; M-H:MS&E; NTCS]

Woodward, Robert Burns (1917-1979. Organic chemistry)
[BES; CBDS; DABs10; ID20thCB; M-H:MS&E; NTCS]

Yalow, Rosalyn Sussman (1921-. Medical science)
[BES; CBDS; ID20thCB; M-H:MS&E; NTCS]

Yang, Chen Ning (1922-. Physics. Came to U.S. in ca.1945)
[BES; CBDS; DTCWB; ID20thCB; M-H:MS&E; NTCS]

1950-1959 Cohort (reached age 25)

Berg, Paul (1926-. Biochemistry)
[BES; CBDS; ID20thCB; M-H:MS&E; NTCS]

Blumberg, Baruch Samuel (1925-. Medical science)
[BES; ID20thCB; M-H:MS&E; NTCS]

Cooper, Leon Neil (1930-. Physics)
[BES; CBDS; ID20thCB; M-H:MS&E; NTCS]

Cronin, James Watson (1931-. Physics)
[BES; ID20thCB; NTCS]

Drake, Frank Donald (1930-. Astronomy)
[BES; ID20thCB]

Edelman, Gerald Maurice (1929-. Biochemistry)
[BES; ID20thCB; M-H:MS&E; NTCS]

Gell-Mann, Murray (1929-. Physics)
[BES; DTCWB; ID20thCB; M-H:MS&E; NTCS]

Giaever, Ivar (1929-. Physics. Came to U.S. in 1956)
[BES; ID20thCB; M-H:MS&E; NTCS]

Gilbert, Walter (1932-. Molecular biology)
[BES; CBDS; ID20thCB; NTCS]

Glaser, Donald Arthur (1926-. Physics)
[BES; CBDS; ID20thCB; M-H:MS&E; NTCS]

Glashow, Sheldon Lee (1932-. Physics)
[BES; CBDS; DTCWB; ID20thCB; NTCS]

Hubel, David Hunter (1926-. Neurobiology. Came to U.S. in 1954)
[BES; CBDS; ID20thCB; NTCS]

Lederberg, Joshua (1925-. Biology, Genetics)
[BES; DTCWB; ID20thCB; M-H:MS&E; NTCS]

Lee, Tsung-Dao (1926-. Physics)
[BES; CBDS; DTCWB; ID20thCB; M-H:MS&E; NTCS]

MacArthur, Robert Helmer (1930-1972. Ecology)
[BES; DSB v.18; NTCS]

Maiman, Theodore Harold (1927-. Physics)
[BES; CBDS; ID20thCB; M-H:MS&E; NTCS]

Miller, Stanley Lloyd (1930-. Biochemistry)
[BES; ID20thCB; NTCS]

Mottelson, Ben Roy (1926-. Physics. Born in U.S. and in 1973 became Danish citizen)
[BES; ID20thCB; M-H:MS&E; NTCS]

Nathans, Daniel (1928-. Microbiology)
[BES; ID20thCB; M-H:MS&E; NTCS]

Nirenberg, Marshall Warren (1927-. Biochemistry)
[BES; ID20thCB; M-H:MS&E; NTCS]

Penzias, Arno (1933-. Physics)
[BES; CBDS; ID20thCB; M-H:MS&E; NTCS]

Richter, Burton (1931-. Physics)
[BES; CBDS; ID20thCB; M-H:MS&E; NTCS]

Sagan, Carl Edward (1934-. Astronomy)
[BES; ID20thCB; NTCS]

Schally, Andrew Victor (1926-. Medical science, Endocrinology. Came to U.S. in ca.1957)
[BES; ID20thCB; M-H:MS&E; NTCS]

Schmidt, Maarten (1929-. Astronomy. Came to U.S. in ca.1959)
[BES; ID20thCB]

Schrieffer, John Robert (1931-. Physics)
[BES; ID20thCB; M-H:MS&E; NTCS]

Smith, Hamilton Othanel (1931-. Microbiology)
[BES; CBDS; ID20thCB; M-H:MS&E; NTCS]

Temin, Howard Martin (1934-1994. Virology)
[BES; CBDS; ID20thCB; M-H:MS&E; NTCS]

Wasserburg, Gerald Joseph (1927-. Geophysics)
[ID20thCB; M-H:MS&E]

Watson, James Dewey (1928-. Biochemistry, Biology)
[BES; CBDS; DTCWB; ID20thCB; M-H:MS&E; NTCS]

Weinberg, Steven (1933-. Physics)
[BES; CBDS; DTCWB; ID20thCB; M-H:MS&E; NTCS]

Wilson, Edward Osborne (1929-. Biology)
[BES; ID20thCB; M-H:MS&E; NTCS]

1960-1969 Cohort (reached age 25)

Baltimore, David (1938-. Microbiology)
[BES; ID20thCB; M-H:MS&E; NTCS]

Brown, Michael Stuart (1941-. Molecular genetics)
[BES; ID20thCB; NTCS]

Goldstein, Joseph Leonard (1940-. Molecular genetics)
[BES; ID20thCB; NTCS]

Gould, Stephen Jay (1941-. Paleontology)
[BES; ID20thCB; NTCS]

Hoffmann, Roald (1937-. Chemistry)
[BES; ID20thCB; M-H:MS&E; NTCS]

Johanson, Donald Carl (1943-. Physical anthropology)
[BES; ID20thCB]

Ting, Samuel Chao Chung (1936-. Physics)
[BES; CBDS; ID20thCB; M-H:MS&E; NTCS]

Wilson, Kenneth Geddes (1936-. Physics)
[BES; CBDS; ID20thCB; NTCS]

Wilson, Robert Woodrow (1936-. Physics, Radio astronomy)
[BES; ID20thCB; M-H:MS&E; NTCS]

1970-1979 Cohort (reached age 25)

Bowen, Rufus (Robert) (1947-1978. Mathematics)
[DSB v.17]

3

Research Guide

Introduction

The chronology of the history of science in the United States is a map over time and a very generalized view at that. For some, it may suffice to answer a particular question or to give a glimpse of salient features during a certain time period. Others will want to go further in exploring the topic of science in the American context and this research guide is intended for those readers.

The secondary literature and primary sources for the subject are large and constantly growing. Neither this nor any other guide can achieve complete coverage; this one does not set its sight on that goal. The aim, as with the chronology, is to give a view of some important features and to invite the reader to enter and to explore.

What This Guide Does

This is a selective guide to bibliographies, reference works, and other resources that facilitate access to literature, information, and research materials on the history of science in America. That such a guide is possible, as well as necessary, is a gauge of the growth and vitality of the field. An active community of scholars has contributed to the description, evaluation, and control of the literature and to improved access to information and to archival and other primary sources. This guide attempts to coordinate these efforts. Most citations in the guide are to relatively recent publications. However, some older reference works are included when appropriate.

Reference works, bibliographies, or indexes relating to general United States history are *not* directly cited here. Also excluded are similar sources for access to the scientific literature. Obviously these are two categories of great significance for the history of American science, but they are dealt with well in other reference works. Any serious search for reference materials in all categories will begin with the American Library Association's *Guide to Reference Books* and its British counterpart, *Walford's Guide to Reference Material*, which has a separate volume for science and technology.[1] There are various guides to sources on United States history; of the general works, two of the more recent and best are Prucha's *Handbook* and Blazek and Perrault, *United States History: A Selective Guide*.[2] Some of the bibliographies cited in this research guide will direct the user to reference sources on the primary literature of science; in certain instances the earlier literature (especially for the period before 1800) will be the object of the bibliography cited. The nineteenth century is well served by the Royal Society *Catalogue*,[3] but the growth of scientific specialization in the twentieth century requires a discipline approach to the primary literature. These are represented by such massive sources as *Biological Abstracts* and *Chemical Abstracts* as well as innovative interdisciplinary reference works such as *Science Citation Index*.

Organization of the Guide

The research guide is two interrelated parts. The first is several short bibliographic essays, generally organized around type of material. This part refers to the second, which is an alphabetically arranged bibliography of reference works. The works mentioned or listed are selective and not exhaustive.

Part one also has sections on electronic access and on manuscript and archival sources. Some of the information in these sections do not lend themselves to the usual bibliographic description and therefore may not appear in the bibliography.

The essays in this first part of the guide relate to:
1) Guides and bibliographies
2) Reference works
3) Electronic access
4) Primary sources, organizations, and projects

1. Guides and Bibliographies

As with sources for the history of science in America overall, guides and bibliographies vary in their content and emphasis. Those featured in this work generally are specific to the history of science and its collateral areas rather than for history in general or for science *per se*. Nonetheless, these other categories of works should not be overlooked and for particular types of historical investigation are indispensable.

Some publications, for example bibliographies relating to the history of particular disciplines, have sections on reference works relevant to that field, sometimes broadly defined. Jayawardene, *Reference Books for the Historian of Science* (1982)[4] is a general work in this area and some of its entries will be useful for historians of American science. Gordon Miller, *The History of Science: An Annotated Bibliography* (1992) is a general and selective work that includes some reference tools. Of continuing value, despite its age, is the classic Sarton, *A Guide to the History of Science* (1952). There is much useful bibliographic and other information in this work, including some relevant specifically to the United States; it has the virtue of presentation by the individual who is largely credited with the institutional founding of the history of science as a field of study.

Basic to all bibliographic approaches in the history of science is the "Isis Critical Bibliography" (beginning in 1989, called "Current Bibliography") published annually since 1913 in *Isis*, which is the journal of the History of Science Society. The annual issues have been gathered in several series as the *Isis Cumulative Bibliography* (now covering the period to 1985). The *Isis* bibliographies include books, articles, and other works, in a classified arrangement. The "Current Bibliography" now is available on-line, beginning with 1976, through the Research Libraries Group (RLG) as one of its CitaDel databases. Early announcements of newly published books and dissertations appear in *History of Science in America: News and Views*, and in the newsletters of the History of Science Society and other groups.

The history of technology is well served by a similar "Current Bibliography in the History of Technology," which has appeared annually in the journal *Technology and Culture* since 1962; it now is issued as a separate volume (see bibliography under Technology and Culture ...). It also is available as part of the same RLG on-line database with the "Isis Current Bibliography," but coverage for technology begins with 1987.

Earlier works in the history of medicine can be found in Genevieve Miller, editor, *Bibliography of the History of Medicine* (1964), which covers the period 1939-1960, with annual issues for 1961-1964. In 1965, the *Bibliography of the History of Medicine* began issuance annually from the U.S. Public Health Service. There are five-year cumulations.

In addition to these master bibliographies, in recent years several useful overall guides to research issues and bibliography in the history of science have appeared. Corsi and Weindling, editors, *Information Sources in the History of Science and Medicine* (1983) and Durbin, editor, *A Guide to the Culture of Science, Technology and Medicine* (1980) are good overall reviews of the field and its sources. Corsi and Weindling has a separate chapter on American science and medicine. Also helpful for the history of science generally, and incidentally for the United States, is Brush, *The History of Modern Science: A Guide* (1988). Concentrating on the period 1800-1950 and works published in the previous ten years, Brush has aimed to produce a work useful for teachers; the book is generally organized around major topics, such as evolution. Mitcham and Williams, *The Reader's Adviser* (1994) is the science, technology, and medicine volume of a standard reference work that gives historical and biographical information as well as up-to-date guidance on the structure and literature of the field.

The state of the history of science in America and its associated literature was the subject of Kohlstedt and Rossiter, editors, "Historical Writing on American Science," a collection of review essays by various historians that inaugurated the revival of the journal *Osiris* (1985); it subsequently was reprinted as a paperback volume. The basic selected and annotated bibliography for works on American science is Rothenberg, *The History of Science and Technology in the United States* (1982, 1993). Two older works that combined a concern for the historiography of the field with some degree of bibliographic guidance are Bell, *Early American Science: Needs and Opportunities for Study* (1955) and Hindle, *Technology in Early America* (1966), with the same subtitle. Bell has a selected bibliography on fifty scientists, while Hindle's bibliographic orientation is more general. The latter includes a useful directory of artifact collections, by Lucius F. Ellsworth. The University of Toronto, *Science in Society: An Annotated Guide to Resources* (1989) is a short but useful work relating to current science issues as well as to historical considerations.

Specialized guides and bibliographies are characterized by different types of coverage. Black, *American Science and Technology: A Bicentennial Bibliography* (1979) includes only works published in the national anniversary year of 1976. Harkanyi, *The Natural Sciences and American Scientists in the Revolutionary Era* (1990) purports to focus on science during a particular time period. In fact this work has a great deal of other bibliographic information on American history and American science beyond what the title would suggest. Selection seems to have been every source encountered by the compiler that in actuality did contain, or potentially might include, something relevant to the period from 1760 to 1789. Elliott, "Collective Lives of American Scientists" (1990) includes a bibliography of collective biographies. Weisbard and Apple, editors, *The History of Women and Science, Health, and Technology: A Bibliographic Guide* (1993) approaches the topic from the view of a particular group, both as participants and as subjects of science and its collateral fields. A related bibliographic work is Bindocci and Ochs, *Women and Technology: An Annotated Research Guide* (1993)[5]. Eastwood, *Directory of Audio-Visual Sources: History of Science, Medicine, and Technology* (1979), while dated both as to content and technology, stands as a general guide to sources in a particular format. Like some other general works, however, there is a question here of ease of use specifically for American science, when the work itself has not been organized from that perspective.

The most common bibliographic guides are by subject or discipline. Many, perhaps most, of these, in the very nature of their orientation, do not feature national boundaries. Nonetheless, they contain much of interest for the history of science in the United States and are essential tools. In some cases, American aspects can be garnered from the index, from sections on biography that identify by nationality, through the names of institutions, and the like. In other cases, browsing is the best if not the only option. When such an approach is required, knowledgeable users will, of course, recognize more than the novice, but titles of works and the place of their publication often reveal national focus or content.

Among the subject-oriented guides and bibliographies those in the biological and medical sciences appear to be particularly numerous.[6] Two general books in this field are Overmier, *The History of Biology* (1989) and Bridson, *The History of Natural History* (1994).[7] The latter is a masterly work; in addition to references to the secondary literature there is a great deal of information on other bibliographies, reference works, societies and organizations for the history of natural history, and periodicals. In the area of ecology and the environment are Egerton, "The History of Ecology" (1983, 1985); Elbers, *Changing Wilderness Values, 1930-1990* (1991), with both historical and contemporary content; and Owings, *Environmental Values, 1860-1972* (1976). Aldrich (1974), in a bibliographic review essay, discusses historians' analyses of the reception of Darwinism in the United States. McIver, *Anti-Evolution* (1988) lists what really are primary sources for the evolution debate, from the view of the opposition.

The history of biology easily melds with that of medicine in certain respects. Grainger, *A Guide to the History of Bacteriology* (1958) is an older annotated bibliography on a topic that inherently spans science and medicine. Erlen, *The History of the Health Care Sciences and Health Care, 1700-1980* (1984) interprets its topic broadly. The Wellcome Institute for the History of Medicine (London), *Subject Catalogue* (1980) is a reproduction of the Institute's card catalogue of secondary literature. The 18 volumes are in three parts: subjects, biographies, and geographic areas. American medicine is represented in each series and the geographic volumes include large sections on the United States and its subdivisions. B. Bullough, V.L. Bullough, and Elcano, *Nursing: A Historical Bibliography* (1981) features one aspect of the health care field. The recently published Higby and Stroud, *The History of Pharmacy* (1995) will assist historians of medicine and should be of interest to historians of chemistry as well.

Blain and Barton, *History of American Psychiatry* (1979) includes bibliographic essays on the topic and a chronology of psychiatry in the United States. In the field of history of psychology, several bibliographic and research guides are available. There are two works by Watson, *Eminent Contributors to Psychology* (1974-1976) and *The History of Psychology and the Behavioral Sciences* (1978). The latter is a topical approach while the former is biographical and includes bibliographies of works by and about the psychologist under study. Viney, Wertheimer, and Wertheimer, *History of Psychology* (1979) is another available bibliography, partly annotated. Fried, *American Popular Psychology* (1994) is an annotated bibliography that blends a historical and present perspective. Kemper and Phinney, *The History of Anthropology* (1977) is an unannotated guide to its topic.

Stannard, "Early American Botany and Its Sources" (1966) is a bibliographic essay for the period before the Civil War that includes within its scope published sources as well as botanists, manuscripts, herbaria, botanic gardens, and other topics, assessing both accomplishments and needs. Fahl, *North American Forest and Conservation History* (1977) is another bibliographic source for the botanical field.

Edwards, *Bibliography of the History of Agriculture in the United States* (1930) is a basic source for this area. Not all of agriculture is science related, of course, and this bibliography will include a wide range of topics. Several works are available that help to update Edwards. These include Ogden, *The United States Forest Service* (1976); Sharrer, *1001 References for the History of American Food Technology* (1978); and Rossiter, *A List of References for the History of Agricultural Science in America* (1980). Schlebecker, *Bibliography of Books and Pamphlets on the History of Agriculture in the United States, 1607-1967* (1969) attempts comprehensiveness for the type of publication and time period covered. Most up-to-date but only related in part to the United States is Hurt and Hurt, *The History of Agricultural Science and Technology* (1994); in addition to other types of works, there is a useful section on scientific bibliographies and on-line databases.

Porter, *The Earth Sciences* (1983) is means of access to publications on the history of geology and related topics; the compiler admits to an emphasis on items relating to the period from the mid-eighteenth to the latter part of the nineteenth century. Sarjeant, *Geologists and the History of Geology* (1980, 1987) is a monument to bibliography and historical data gathering. It includes general historical entries but is especially strong for biographical references. Though not annotated, it includes, in addition to bibliographic citations, succinct information on subjects of biographies. Arrangement or indexing by nationality will facilitate access to entries on the United States. Brown and Wheeler, *A Bibliography of Geographic Thought* (1989) attempts comprehensiveness for English language books and articles but does not include annotations. The arrangement, as in some other works delineated here, will require scanning or close study by the user to extract references to United States history topics. Also related to this area of study is Dunbar, *History of Modern Geography* (1985).

As might be expected, the general areas of astronomy, space exploration, and related topics are fairly well served by bibliographers. These include DeVorkin, *History of Modern Astronomy and Astrophysics* (1982); Dickson, *History of Aeronautics and Astronautics* (1968); Hallion, *The Literature of Aeronautics, Astronautics, and Air Power* (1984), which is an extended bibliographic essay;[8] and Pisano and

Lewis, editors, *Air and Space History* (1988). For access to a special topic in this general area, see Collea and Aveni, *A Selected Bibliography of Native American Astronomy* (1978), which covers both North and South America during the pre-Columbian era. Also see Brush and Landsberg, *History of Geophysics and Meteorology* (1985).

The general history of physics since the 1890s is shown in Brush and Belloni, *The History of Modern Physics* (1983). This work is complemented by Heilbron and Wheaton, *Literature on the History of Physics in the 20th Century* (1981). About half of the pages of this unannotated work are devoted to biographies. Some of the larger implications of modern physics are bibliographically explored in Burns' two works, *The Atomic Papers: A Citizen's Guide* (1984) and *The Nuclear Present* (1992). Much of the content of these two publications are references to what amount to primary documents for the history of the atomic bomb, national policy, the peace movement, etc. The 1992 volume includes a chronology of "nuclear events," with its own index. Sturchio, editor, *Corporate History and the Chemical Industries* (1985) includes sections on historical works as well as on issues relating to archives, records management, and oral history. Higham, *A Guide to the Sources for United States Military History* (1975, 1981-1993) includes sections on science, technology, and medicine and the chapters sometimes include sections on manuscript sources as well as publications.

May, *Bibliography and Research Manual on the History of Mathematics* (1973) includes a number of biographical references as well as those related to other aspects of the subject. Its value is limited, however, by the omission of titles of articles. This work is complemented and updated by Dauben, *The History of Mathematics from Antiquity to the Present* (1985).

Cortada, *A Bibliographic Guide to the History of Computing, Computers, and the Information Processing Industry* (1990) is international in coverage but includes much that relates to the United States. This is to be expected for a field of relatively recent emergence and with such a significant presence by Americans.

The fields of engineering and technology have several bibliographic guides. Channell, *The History of Engineering Science* (1989) is a starting point. It does not include chemical or electrical technologies. For those subjects, see Multhauf, *The History of Chemical Technology* (1983) and Finn, *The History of Electrical Technology* (1991). Also of interest is Shiers, *Bibliography of the History of Electronics* (1972). Other bibliographic works in the area of engineering include Molloy, *The History of Metal Mining and Metallurgy* (1986) and Stapleton, *The History of Civil Engineering Since 1600* (1986). Also relevant and dealing specifically with American topics is Hoy and Robinson, *Public Works History in the United States* (1982). The perspective of Cutcliffe and others, *Technology and Values in American Civilization* (1980) will place the foregoing topics in a wider historical and contemporary context.

Business and industry relate to the history of science and technology in various ways, through their products, processes, and research. Geahigan, *U.S. and Canadian Businesses, 1955 to 1987: A Bibliography* (1988) lists about 4,000 titles relating to the history and operations of individual companies and also includes biographical references. The unannotated entries are arranged by a classified scheme based on type of industry, with headings such as agricultural production, metal mining, chemicals and allied products, electronic and other electrical equipment, health services. Nasrallah, *United States Corporation Histories* (1991) also offers an approach through histories of individual firms.

The topic of science and public policy is of great importance in the area of science studies but its recency as a social and political phenomenon diminishes the historical perspective. Caldwell, *Science, Technology and Public Policy* (1969-1972) has some historical entries but others may be useful as primary sources. Coverage is for English-language works appearing 1945-1970. U.S. Congress [...], *Bibliography of Studies and Reports on Science Policy and Related Topics, 1945-1985* (1987) includes especially references to government documents. Most are contemporary in focus, but there is a final section for specifically historical works.

The history of science is a very broad area with relations to various social and cultural sectors. Schatzberg, Waite, and Johnson, editors, *The Relations of Literature and Science* (1987) explores its topic internationally. Scholnick, "Bibliography: American Literature and Science through 1989" (1992) is specific to the United States; it is not annotated. Hall, editor, *Science Fiction and Fantasy Reference Index* (1987, 1993) is an index to secondary works relating to a focal point of contact between science and literature. Melton and Poggi, *Magic, Witchcraft, and Paganism in America* (1992) may prove more useful as a means to view a community parallel to science than one that is interactive with it.

2. Reference Works

Reference information for the history of science in America is found in many places. Certain secondary monographic works in history emphasize factual content and in conjunction with an effective index can function as works of reference. For example, Ralph Bates, *Scientific Societies in the United States*[9] is an overall historical account of an important category of scientific institution but also useful to have nearby on a shelf; it can be consulted for data on individual societies (especially on their founding but also for some limited additional information as well). Likewise, Arnold Mallis's *American Entomologists*[10] can be read through, but because of its organization also can be used as a biographical dictionary of entomology. Raymond Stearns, *Science in the British Colonies of America*,[11] because of its scholarship and sweep of the colonial period, has a great deal of factual information on that time period. The usefulness of the foregoing category of works notwithstanding, the reference publications discussed in this section of the research guide generally have the primary intent to be consulted and only secondarily to be read through.

Reference works, together with bibliographies, have the function of giving not only access but structure to a subject. Almost inevitably, they are selective and in selecting the compiler circumscribes and helps to define the field of inquiry. It is well, therefore, that users be aware of this fact, to be clear about an author's intent and definition.

This compiler is no exception to the foregoing point, in arguing, at least by action, that the underlying character and structure of the history of science can be shown in a selective chronology of events. The chronology that forms the main body of this reference work, however, is, in part, derivative from and embedded in other chronologies that cast the history of science in its general and international form, or focus on the unfolding of a particular area of study. Gascoigne, *A Chronology of the History of Science, 1450-1900* (1987) is international but limited in time span, as is true also of Parkinson, *Breakthroughs* (1985) which ends in 1930 and concentrates very largely on scientific thought *per se*. Chronologies that take the subject to the time of their publication are Asimov, *Asimov's Chronology of Science and Discovery* (1994); Hellemans and Bunch, *The Timetables of Science* (1988); Bunch and Hellemans, *The Timetables of Technology* (1993); and *History of Science and Technology: A Narrative Chronology* (1988), originally published in Italian in 1975 and covering the subject to 1970.[12]

Among the available chronologies that concentrate on a particular subject and which include American topics are: H. Davis, *Electrical and Electronic Technologies* (1981-1985); Thompson, *A Chronology of Geological Thinking from Antiquity to 1899* (1988), which really is a chronological bibliography with notes on the significance of each work; and U.S. National Aeronautics and Space Administration, *Aeronautics and Astronautics* (1961) and its succeeding volumes to 1985.

Some reference works, systematic in coverage and ordinarily alphabetical in arrangement, are listed in the bibliography in the second part of this research guide. These encyclopedias, handbooks, and the like may be general or relate to particular topics.[13] Bynum, Browne, and Porter, *Dictionary of the History of Science* (1981) is organized by scientific topic; American contributions can be gauged by consulting the index, where the nationality of scientists mentioned in the text are listed with the name. Mount and List, *Milestones in Science and Technology* (1994) is arranged topically; the indexes include chronological and geographical access. Two works edited by Travers, *World of Invention* (1994) and *World of Scientific Discovery* (1994), are alphabetically arranged and include biographical and topical entries. Teich and Pace, *Science and Technology in the USA* (1986) is not, strictly, a reference nor a historical work but rather one to be read for current purposes. Nonetheless, its wealth of information on the contemporary organizational state of American science makes it a valuable source to be consulted also.

Some other historical encyclopedias or handbooks are available for certain subject areas. These include the international coverage of McGrew and McGrew, *Encyclopedia of Medical History* (1985), with a topical organization and exclusion of biography. Howells and Osborn, *A Reference Companion to the History of Abnormal Psychology* (1984) is an alphabetically arranged work that includes entries on individuals, institutions, medical terms, etc. There also is Schapsmeier and Schapsmeier, *Encyclopedia of American Agricultural History* (1975) and R. Davis, *Encyclopedia of American Forest and Conservation History* (1983).

Cortada, *Historical Dictionary of Data Processing* (1987) is in three volumes, relating respectively to biographies, organizations, and technology. Another area related to business is explored in Downard, *Dictionary of the History of the American Brewing and Distilling*

Industries (1980). Also of interest is Gunn, editor, *The New Encyclopedia of Science Fiction* (1988).

A work of a somewhat different character in terms of content and structure is Bali, *Space Exploration: A Reference Handbook* (1990); it incorporates a chronology, and sections that include biographies, fact and data tables, directory of organizations, an annotated bibliography of publications, and films and videocassettes. Firth, editor, *Handbook of Scientific and Technical Awards in the United States and Canada, 1900-1952* (1956) has information not easily found elsewhere.

Other than bibliographies of various descriptions, biographical reference works probably are the most common type of reference work. Some of the bibliographies of secondary works, discussed earlier in this research guide, include references to biographies. There also are some other works which, to varying degrees, are bio-bibliographic in nature, giving some personal information but having the primary purpose of leading to more substantial information sources. Of these, one of the most significant, for brief biographical information as well as for references to publications by and about individuals contributing to mathematics, astronomy, chemistry, physics, geophysics, and related areas is Poggendorff, *Biographisch-literarisches Handworterbuch der exakten Naturwissenschaften* (1863-). Barr, *An Index to Biographical Fragments in Unspecialized Scientific Journals* (1973) and Elliott, *Biographical Index to American Science* (1990) cover the period to 1920, while Ireland, *Index to Scientists of the World* (1962) and Pelletier, *Prominent Scientists* (1994) take coverage near to the time of their publication. These works give little or no biographical information directly; with the exception of Elliott they are international in coverage.

There also are biographical bibliographies or indexes that are limited to particular subjects. These include Gilbert, *A Compendium of the Biographical Literature on Deceased Entomologists* (1977); Barnhart, *Biographical Notes upon Botanists* (1965); Fruton, *A Bio-Bibliography for the History of Biochemical Sciences since 1800* (1982, 1985); Holloway, *Medical Obituaries* (1981); Morton and Moore, *A Bibliography of Medical and Biomedical Biography* (1994); and Roysdon and Khatri, *American Engineers of the Nineteenth Century* (1978). Of these, Holloway and Roysdon-Khatri are specific to the United States.

The relatively large number of biographical dictionaries and directories in the bibliographic part of this research guide precludes listing them in full here. However, none of the works listed will obviate the need to consult the *Dictionary of American Biography*[14] and other standard works of biographical reference. The *Dictionary of Scientific Biography* (1970-1990) is the major up-to-date source for the most important, deceased scientists. The bibliography lists other general international biographical sources, some of which include living persons; see, for example, under: Asimov (1982); Daintith, Mitchell, Tootill, and Gjerstem (1994); *McGraw-Hill Modern Scientists and Engineers* (1980); *Notable Twentieth Century Scientists* (1995); Porter (1994); Williams (1994); and *World Who's Who in Science* (1968). Also useful are the biographical works on Nobel prize winners, by Schlessinger and Schlessinger (1990), and by Wasson (1987).

For American scientists specifically, see Elliott (1979) for the period up to the early twentieth century, and for the years thereafter, *American Men and Women of Science* (1906-), including the cumulative index for issues through 1979. *Who Was Who in American History: Science and Technology* (1976) is a handy compilation from Who's Who, prepared at the time of the American bicentennial. The U.S. National Academy of Sciences, *Biographical Memoirs* (1877-) is a significant source both for biographical information and for lists of publications by deceased members of the Academy.

In recent years, several useful biographical reference works on women in science have been published. Siegel and Finley (1985) is a combination biographical source and annotated bibliography, while Bailey (1994) is a more conventional biographical dictionary. Ogilvie (1986) is international but includes information on American women through the end of the nineteenth century.

Sammons (1990) is a useful directory and guide for research on blacks in American science. Geiser (1959) is limited to scientists who were associated with a geographical area, namely Texas.

Other biographical reference works are directed toward researchers in a particular field. These include the works by Ewan and Ewan (1981) on Rocky Mountain naturalists; Abbott (1973) on malacologists; The Auk (1954) for ornithologists; Axelrod and Phillips (1993) and Stroud (1985) in relation to the environment and conservation; Humphrey (1961) for botanists; Kelly and Burrage (1928) and Kaufman, Galishoff, and Savitt (1984) in medicine; and nursing in the work of Kaufman (1988) and in that of V.L. Bullough, Church, and Stein (1988 and 1992). In chemistry, there is the work by Miles (1976),[15] while publications by the American Society of Civil Engineers (1972) and American Society of Mechanical Engineers (1980) are sources for two engineering groups. Cowart and Wymer (1981) have produced a work on American science fiction writers; C. Smith (1986) also has entries for American contributors to that genre.

The bibliography also includes subject-related biographical reference works that are international, including Cortada, as a volume in his *Historical Dictionary of Data Processing* (1987); Hawthorne (1992) on space exploration; Mann (1988) for anthropologists; Larkin (1993) in geography; and Zusne (1984) for psychology. Grinstein (with others) has edited biobibliographic works on women in chemistry and physics (1993) and in mathematics (1987). Gacs and others (1988) have edited a work on women in anthropology.

Kohlstedt, "Biographical Directory of the AAAS, 1848-60" (1976) is useful in its own right, as means of access to biographical data on the early American scientific community. It also is a reminder that biographical and other reference information may show up in unexpected places.

In addition to persons, institutions of various types are a fundamental part of the history of science in America. Not nearly as much has been done on this topic compared to available biographical reference works. Kiger, *Research Institutions and Learned Societies* (1982),[16] while not specific to science has a number of entries for science and technology institutions. Danilov, *America's Science Museums* (1990) has brief descriptions for close to 500 museums and some historical information is included. The industrial sector of American science, among the weakest areas in terms of historical investigation and available reference works, is of increasing importance for an understanding of the workings of science and technology. The *International Directory of Company Histories* (1988-) is arranged by type of industry, with categories including chemicals, electronics, technology and mainframe computers, and others. Though the emphasis in the entries is on business aspects of the company, there is some science-related content of an institutional character. Science and technology are encountered in another institutional context in Findling and Pelle, editors, *Historical Dictionary of World's Fairs and Expositions, 1851-1988* (1990); the bibliographic sections may be especially useful but the short historical essays on individual fairs give general orientation, even though much is about aspects other than science and technology.

Research Guide 459

3. Electronic Access

The topic of this section not only is difficult to define but even more difficult to delineate in any timely manner. Major libraries and other organizations are devoting significant resources to promote the availability of information electronically, through locally accessible catalogues as well as by the Internet. At the same time, indexes, text, and images and sound also are available on CD-ROM, often in periodically updated versions accessible in library reference rooms.

There also is a widespread and in many respects *ad hoc* activity by individuals and groups who have created means for electronic discussion on a multitude of topics, through mailing lists (e.g., the Listserv system on Bitnet), in electronic newsletters, and the like. Data of a wide-ranging character are accessible by Internet, some of it mediated through traditional means of peer or professional review, at other times the product of enthusiasm, dedication, or interest but without means on the part of the ultimate user to gauge its character or reliability. Dependable bibliographic reference to many electronic sources also is difficult. Published reference works have a physical existence in terms of which they can be described. Not only can the content of sources on the Internet be changed (one of its virtues) but they also can be moved from one address to another, or presumably can be removed all together. In a research guide such as this, description and location can only dependably hold for the moment of writing. In accessing the electronic medium, the technical or procedural questions can be hard enough, but a consciousness of underlying assumptions, structures, and practices also has to be kept constantly in mind.

Among the library-related organizations, two are of special note. The Research Libraries Group (RLG) provides some special bibliographic databases through its CitaDel files (chiefly journal articles)—including the history of science and technology, as indicated elsewhere in this research guide. The RLG bibliographic file, Research Libraries Information Network (RLIN), also is a major electronic union catalogue for a number of research libraries; in 1995 it included more than 22 million titles for material in a number of formats. Use of the RLG files is facilitated by their on-line search system, Eureka. The second major bibliographic utility is the Online Computer Library Center, Inc. (OCLC) and its FirstSearch service. The OCLC facility includes its WorldCat, which facilitates bibliographic access to more than 28 million items in thousands of libraries worldwide (a limited number of manuscript resources also are available). FirstSearch also makes available searching of various other databases for journal articles and similar material. Therefore, although Internet access to a number of individual library catalogues now is possible, researchers may want to rely in the first instance on RLIN or OCLC and revert to other catalogues when that search fails to locate wanted items.[17]

This section of the research guide has to be modest, both because of the compiler's relative lack of experience with the medium, and because the medium itself changes so rapidly. There is no denying that the universe of scholarly—as well as social—communication is undergoing an enormous change, and electronic discussion groups and newsletters are a leading manifestation as well as promoter of that phenomenon. Individuals will have to decide on their own whether to enter this sea of exchange, in which access to substantive discussion about a topic, news of events, and information about resources can emerge. This is part of the infrastructure of the community and combines elements of the telephone, mail, newsletters and journals in a more instantaneous and widespread expression of those established communication channels.

Not to deny their importance, discussion groups and their other interactive relatives are not the focus here, which is on access to bibliographic and similar information of a somewhat more structured kind. There are, of course, books that one can and should consult about the use of the Internet in all its many forms. These publications undoubtedly will proliferate in the future and currency will be important.[18] Familiarity with the Internet and its services, or access to published or other information about it, is a presupposition of this section of the research guide.

Access to resources for the history of science in America through the Internet has the characteristics of access through more traditional means. That is, the universe of concern includes secondary sources in history generally, and those specific to the fields of history of science and to American science, the scientific research literature, and manuscript and archival materials. In fact, there currently is little among the Internet resources categorically devoted to the history of American science and technology. The strategy, therefore, is to search from the subject perspective within those electronic resources that encompass the various aspects of the history of science in the United States.

The Clearinghouse of Subject-Oriented Internet Resources Guides, maintained at the University of Michigan by the University Library and the School of Information and Library Studies,[19] is one place to go for specially prepared information on subject access. Use of the Gopher search facility, through its menu structure, allows access to much that is on the Internet. Some gophers are particularly useful for a subject approach, including that at the Library of Congress.[20] Two tools are especially designed to search gopher menus by word(s) in the title. These are Veronica and Jughead, the first a facility to search all gophers, the latter confined to a specified universe of gopher menus.[21] On Veronica, for example, one can request a search for "history and science," either limited to directory titles or to include the titles of menu items themselves. In spring 1995, the foregoing search words retrieved nearly 150 directories, and many more menu items.[22]

HNSource is a central information source for history, maintained by the history department at the University of Kansas.[23] Through it, one can access a variety of history-related information, news, guides, historical texts, and other Internet sites that provide access to additional bibliographic and other sources. HNSource solicits additional material from users.

The World Wide Web is a means of access to the Internet that uses hypertext. Highlighted words or phrases can be activated to display linked text. They also can display images or sound if the user has the proper software (such as Mosaic or Netscape).[24] One of the resources available on the Web is the so-called Virtual Library, which arranges sources by subject.[25] When the user enters the Virtual Library, there is an alphabetical list of subjects (which can be searched by keyword). Among the subject areas, in addition to history general, is History of Science, Technology and Medicine; this resource was established in September 1994 and is maintained at the Australian National University (the Australian Science Archives Project, or ASAP). Among the categories of resources accessible by this facility are announcements, discussion groups, electronic journals, primary and secondary historical texts, bibliographies and library catalogues, directory and other information on research facilities, reference information, and more.[26]

There are other tools for subject access to the World Wide Web. For example, the Yahoo Search guide arranges items in broad categories, but also has word searching of addresses, titles, and comments.[27] Yahoo also will connect to other Web search facilities, including WebCrawler, which features keyword retrieval, searching contents of documents as well as addresses and titles.[28]

Access to information from the science side also is a possibility although one would expect this to be largely current rather than historical. Nonetheless, it is not a resource to overlook. It is likely that researchers will be most successful if they approach the topic from the point of view of a particular discipline rather than science in general. For example, the Smithsonian Institution's Natural History gopher provides access to A Biologist's Guide to Internet Resources.[29] The Biologist's Guide has a great deal of well-organized information that may be of interest to historians. As with all searches on the Internet, much will depend on the particular interest of the individual user. In fact much of the incentive for the Internet itself would appear to be a sense that "This is bound to be of interest to someone, somewhere. Let's share it." Between that motive of the provider and the need of the user lies much serendipity, persistence, frustration, help from your friends, but a great adventure.

The same uncertainties that relate to subject searching as outlined above pertain to the search for archival and manuscript sources, and other materials by format. Elsewhere in this guide, there is reference to the History of Science and Technology bibliographic database available on-line through the Research Libraries Group's CitaDel service. As discussed in the section that follows, RLG and, to a lesser extent, OCLC, also have become focal points for locating archival and manuscript sources in all fields. Increasingly, repositories will make available finding aids to their collections and these more detailed descriptions or inventories often will be available through a locally constructed gopher, World Wide Web page, or an on-line library catalogue.[30] Along with other Internet developments, this area is volatile and evolving. The Clearinghouse of Subject-Oriented Internet Resource Guides (mentioned above) has an on-line "Guide to Archives on the Internet" (the most recent dated May 27, 1994). At present, Johns Hopkins University makes available finding aids from various repositories.[31] These resources, and gopher and World Wide Web sites, have become available through an Archives and Archivists Web page at Miami University (Ohio).[32]

In addition to on-line resources, electronic access also is possible through use of local technologies such as floppy disk or CD-ROM. Many of these are in the category of reference tools purchased ordinarily only by libraries. For example, *America: History and Life*, a major abstracting source for articles on American history, now is available on disk, as is the more general *Historical Abstracts*. The standard directory of current biographical information, *American Men and Women of Science*, is available as part of the R.R. Bowker Company's *SciTech Reference Plus*, a CD-ROM publication (also available on-line through DIALOG Information Services, Inc.). A number of science indexes and abstracting tools also are available electronically, both on disk or on-line (e.g., *Biological Abstracts* and *Chemical Abstracts*), though generally it will take some years of cumulation to give these databases a genuinely historical usefulness.

The future will bring more databases and publications in electronic form, some of them on-line through subscription services (e.g., CD Plus Technologies—formerly BRS Online, and DIALOG) and others in CD-ROM or other format. But this already is a substantial enterprise; one of the leading published guides to electronic databases in 1995 consisted of two substantial volumes with information on more than 9,000 databases.[33] In spite of this output, however, there is relatively little directly on the history of science or science in America. Nevertheless, the interdisciplinary character of the field, as indicated earlier, should make a number of sources useful to individuals who are prepared to explore their features and their applicability to a particular topic.

The attractiveness of electronic access is derived from its convenience and the variety and range of sources to which one can have access. But it also entails a substantial change in the quality and nature of access. This is manifest, first, in the greatly expanded flexibility of searching that electronic access makes possible. Secondly, it permits usage that allows direct and simultaneous access not only to text, but to images and sound as well. All of these factors not only facilitate research but formulate or redirect it at the same time. Through an ability to get at and to manipulate sources, new connections and relevancies come to the fore.

Enthusiasm notwithstanding, there is no doubt that learning and using the capabilities of the new electronic media can be hard and frustrating work. And progress comes only by doing. Keeping up with developments can come from a number of sources, including traditional ones such as newsletters and dependence on the help and advice of librarians.[34] It also can come from the electronic media itself, through discussion groups or other such information sources. Every historian will have to seek a personal style. We all will continue to learn and adapt together.

4. Primary Sources, Organizations, and Projects

Historiography, Methodology, and Research Resources: General Concerns

Consideration of the overall character, availability, and patterns of use is prelude to reviewing the topic of primary sources for the history of science in America. Kragh, *An Introduction to the Historiography of Science* (1987) is a good place to start, along with Knight, *Sources for the History of Science* (1975). Both are of somewhat circumscribed value for the topic, however, Kragh because the book de-emphasizes social and institutional aspects of the subject, Knight because of a bias toward British science and termination at 1914. Despite these limitations, both raise issues that are of general interest and application. Though less systematic than the foregoing, Reingold, editor, *Symposium* (1990) also explores a number of topics relating to the use of documentation for the history of science.[35]

The enormous growth in potential documentation for the history of science in modern America has drawn forth concern and response from both historians and archivists. The 1960 Conference on Science Manuscripts (see under Conference, in the bibliography) was a seminal event in bringing attention to the problem. Subsequently, the Joint Committee on Archives of Science and Technology, *Understanding Progress as Process* (1983) reviewed the problem of documentation and outlined an agenda for action. The problems and procedures for documenting academic science, from an archival view, have been addressed by Brichford, *Scientific and Technological Documentation* (1969) and by Haas, Samuels, and Trippel, *Appraising the Records of Modern Science and Technology* (1985). Some of the considerations and procedures for dealing with industry are investigated and delineated in Bruemmer and Hochheiser, *The High Technology Company* (1989). The American Institute of Physics Center for the History of Physics has been active in studies in this area. One product of their work is Warnow and others, *Guidelines for Records Appraisal at Major Research Facilities* (1985), which addresses the situation in government laboratories. Harden and Risse, editors, *AIDS and the Historian* (1991) has a section on problems relating to documentation for the history of one of the major ongoing medical events of modern times. Warnow-Blewett, "Documenting Recent Science: Progress and Needs" (1992) is an excellent review of the topics alluded to above by an individual who has been involved centrally or as a consultant to many of the projects in this area.

Primary Source Materials: Character and Access

Primary sources for the history of science in America, as for all fields, vary widely in format. They include printed material such as broadsides, pamphlets, books, and—most importantly for scientific thought during the modern period—journal articles. There also is a large corpus of technical report literature and the like that results from contract or project research. The latter in certain respects occupies a region between published and unpublished sources. It is characteristic of much of the foregoing that they often will be found not in library special collections, or not only there, but integral to a research library's general holdings. In such a context, the focus is the subject of science—or one of its constituent disciplines—rather than the national, institutional, or personal origins of the material. In accessing the published scientific literature, a researcher's starting point may be knowledge of an item already known, with the goal of locating a copy. Or the search may be for items relating to a particular subject; when works of a specific author are sought, it is from knowledge that the individual contributed to the subject under investigation. In each case, the well-developed infrastructure of library catalogues and printed and other indexes is the chief means of access.

Other primary materials include archival and manuscript sources—minutes, diaries, correspondence, notebooks, lectures, and the like. They also include visual materials such as photographs, and artifacts. These types of primary sources might be accessed in relation to a subject but their particularly distinguishing characteristics are their relations to the context of historical activity within which they were created. In this case, the idea of "science in America" is salient insofar as the materials relate to science as practiced by a particular person or within a particular national (or regional or institutional) setting. An adequate view of the history of science requires access to materials in each of the categories above.

PRINT LITERATURE OF SCIENCE

This guide does not undertake to address access to the vast primary literature of science itself. However, there are occasional bibliographic publications that address the technical literature from an historical perspective. These are focused particularly on the early period of American history, where some semblance of comprehensiveness is possible. Two works by Gascoigne, though not limited to the United States, are useful guides to the scientific literature up to 1900. His *A Historical Catalogue of Scientists and Scientific Books* (1984) is basically a chronological list of scientists and is both a reference work and bibliography. Part three relates to the eighteenth and nineteenth centuries and lists scientists by year of birth within subject areas. In addition to references to biographical sources, Gascoigne also lists their major book publications (although such works are not listed for every person in the catalogue). His *A Historical Catalogue of Scientific Periodicals, 1665-1900* (1985) supplements the previous work; periodicals published in the United States are to be found by browsing through the catalogue.

Batschelet, *Early American Scientific and Technical Literature* (1990) is a general, annotated catalogue of publications 1665-1799. There also are other bibliographies for the early period, with a subject focus. Meisel, *A Bibliography of American Natural History* (1924-1929) covers 1769-1865. In addition to citations to the primary published literature (articles and monographs), this work includes references to biographies and incorporates brief histories of scientific organizations. It is, therefore, much more than a bibliography. Other works for access to the early American scientific and technical literature include: Hazen and Hazen, *American Geological Literature, 1669-1850* (1980); Karpinski, *Bibliography of Mathematical Works Printed in America through 1850* (1940), which also includes facsimiles of the title pages of many works; and Rink, *Technical Americana* (1981), which lists separately published works dealing with technology and engineering and gives locations where copies of the works can be found.

In recent years, the American Philosophical Society Library has produced a series of bibliographies based on its own holdings, the richness of which give the publications a general interest. These works, which—like the APS collections themselves—are not limited to items written or published by Americans, include: Crowther and Fawcett, *Science and Medicine to 1870: Pamphlets* (1968); Guerrini, *Natural History and the New World, 1525-1770* (1986); and Stapleton, *Accounts* of European Science, Technology and Medicine Written by American Travellers Abroad, 1735-1860 (1985), which lists both printed and manuscript items. Two bibliographic works by Tucher, *Agriculture in America, 1622-1860* (1984) and *Natural History in America, 1609-1860* (1985) are based on holdings of the American Philosophical Society and several other Philadelphia libraries.

The literature of early American medicine has been well served by historical bibliographers. Items separately published during the period to 1910 are listed in Austin, *Early American Medical Imprints ... 1668-1820* (1961) and Cordasco, *American Medical Imprints, 1820-1910* (1985). Both of these works give locations of copies of the titles in the bibliography. The colonial period is covered over a larger range of types of material in Guerra, *American Medical Bibliography, 1639-1783* (1962). Volumes 1 and 2 of Rutkow, *The History of Surgery in the United States, 1775-1900* (1988-) is an annotated bibliography of the subject in which the notes may include biographical and other information. Asbell, *A Bibliography of Dentistry in America* (1973) attempts a complete census of books and articles published during the period 1790-1840; it includes notes, information on the location of copies of the works listed, and facsimiles of title pages. Morton and Norman, *Morton's Medical Bibliography* (1991) is international in its coverage and includes the United States; some secondary historical publications are listed in addition to its focus on primary works.

Finally (for this section on access to the primary literature of science), there are two near-definitive subject bibliographies of books published in the United States during the period 1876-1982. These are *Pure and Applied Science Books* (1982) and *Health Science Books* (1982).

The collateral area of science fiction works are represented in Clareson, *Science Fiction in America, 1870s-1930s* (1984) and Reginald, *Science Fiction and Fantasy Literature* (1979, 1992), which is international in coverage. The first part of the latter work, covering 1700-1974, includes a biographical directory of modern authors. Benson, *Vintage Science Films, 1896-1949* (1985) combines narrative history with filmography.

MANUSCRIPT AND ARCHIVAL SOURCES

For the last several decades, the *National Union Catalog of Manuscript Collections* (NUCMC)[36] has been the chief general means of access to manuscript materials in the United States.[37] In the emerging electronic environment, this now has changed. National bibliographic utilities such as the Research Libraries Group's Research Libraries Information Network (RLIN) will replace NUCMC.[38] In fact, NUCMC has announced plans to cease publication of its printed catalogue with the volume for 1993 (expected to appear in 1995). Thereafter, the Library of Congress NUCMC office will serve smaller repositories by assisting them in the input of information on manuscript resources into the national bibliographic databases. At the same time, many research libraries are making information regarding their manuscript and archival holdings increasingly available through their on-line catalogues. And this development involves ever more specific information on those holdings, not only general descriptions but finding aids that delineate the contents of collections.[39]

Although electronic access in various guises certainly is the strong trend, the substantial body of published research guides of various types is not to be overlooked. The bibliography in the second part of this guide lists some of these tools for access to archival and manuscript sources in the history of science.

The guides in the bibliography include some that are for individual repositories and others that are union catalogues.[40] The guides to individual repositories encompass some of the major resources for the history of science in America. These include the Academy of Natural Sciences of Philadelphia, American Institute of Physics, the American Philosophical Society (see under APS, and special subject guides under Glass, Kay, Nebeker, M. Smith, Van Keuren), Eleutherian Mills Historical Library, Hunt Institute for Botanical Documentation, Merrimack Valley Textile Museum, Rockefeller Archive Center, and the Smithsonian Institution (see under the Institution, and National Anthropological Archives, and National Museum of History and Technology).

Academic archives and library special collections departments also are central sources for the history of science. In the bibliography these are represented by Harvard University, under Elliott (1974), and the Massachusetts Institute of Technology. Research or professional organizations also maintain sources, such as those listed under the American Medical Association, College of Physicians of Philadelphia, Lick Observatory, New York Botanical Garden, and Scripps Institution of Oceanography.

Over the years a number of very important projects have taken place that have inventoried sources, in a number of repositories, relating to the history of science. Several of these have been regional in focus, including the upper Midwest (Layton) and northern California (Rider and Lowood). There also are guides to sources within a geographical region, for the history of a particular topic, notably atmospheric science sources in the vicinity of the District of Columbia (Fleming). Cartographic and related material located in the region of the national capital also is accessible in a special guide (Ehrenberg).

The most common form of the union catalogue of archival and manuscript sources are those that relate to a particular subject or discipline, sometimes with an international coverage. Notable among these subject catalogues are those relating to: air and space (Gull and Smith); biochemistry and molecular biology (Bearman and Edsall); chemistry (Tselos and Wickey); computers and data-processing (Bruemmer [1987]; Cortada [*Archives...*, 1990]); electrical engineering (Bedi, Kline, and Semsel); forestry (R. Davis [1977]); physics (Kuhn and others; Warnow [1969]; Warnow-Blewett and Teichman, on solid state physics; Wheaton; Wheaton and Heilbron, which is an inventory of published letters);[41] and psychology (Sokal and Rafail). Though on a smaller scale than the foregoing, sometimes special guides are issued for subject-related sources located in various places. Access to NASA history (Roland) and the history of pharmacy (Parascandola and Keeney) are examples. The latter represents a type of publication that includes not only information on source materials but useful guidance through an overview history of the field, secondary works, and biographical information, and the like, with references to manuscript sources integrated into the overall review. Two examples of publications that are subject oriented, but limited both topically and to one or several repositories, are the guides to sources in the U.S. National Archives for World War II research and development (Fishbein) and to scientific exploration during the early nineteenth century in Philadelphia libraries (Stanton).

The bibliography also includes catalogues relating to some special types of material, including: maps (Cobb); medical photographs (Apple); oral history relating to computers (Aspray and Bruemmer), and space history (Collins, Bailey, and Fredericks);[42] scientific apparatus at the College of Charleston (Hughes); and engineering artifacts (Sackheim; Schodek, which is narrative history as well as a useful reference work).

Current information on sources for the history of science can be found in various periodical publications. Two that have been particularly concerned with this subject and have routinely included articles on archival and manuscript resources are the *Mendel Newsletter* (genetics) and the *History of Anthropology Newsletter*.

It is well to keep in mind that many more guides to archival and manuscript materials are available than can be listed here. Even when science is not the focus, they may be useful to a historian of science. The holdings of the National Archives is an obvious instance.[43] Beers, *Bibliographies in American History* (1942, 1982) is one source for references to archival guides as well as much else of bibliographic interest to historians. Access to repository-level information is available in a 1988 national directory,[44] while a recently published bibliography describes a number of repository and collection guides.[45] The likely value of many general access tools and collection guides should increase as one extends the definition of the history of American science beyond national-level and historically developed professional activities, to include the social impact of science, local events in science, and efforts and interests in science on the part of nonscientists in the population.

Organizations and Projects

The development of means for access to information sources for the history of science in the United States is both an individual and a collective undertaking. The same is true, of course, for the scholarly development of the field. The Forum for the History of Science in America, an interest group affiliated with the History of Science Society, publishes the newsletter, *History of Science in America: News and Views*. The Forum also publishes an occasional *Directory, Historians of American Science*, the most recent of which appeared in 1993 as a special issue of *News and Views* (vol. 10, no. 3). The *Guide to the History of Science* (1992), published by the History of Science Society, is an information-rich source on the organization of the field internationally. In addition to a members' directory, it includes useful sections on graduate programs, research centers, societies and organizations, and an annotated bibliography of journals and newsletters. Though some of the information in such a source soon becomes dated, much also is of long-term reference value.[46] Warnow-Blewett, "Documenting Recent Science: Progress and Needs" (1992), in addition to discussing various projects, includes appended directories of discipline history centers and historical documentation projects in the United States. The Warnow-Blewett paper also includes a list of published repository guides for the history of science (most of which also are named in this research guide).

The concerns of scholars interested in the history of particular disciplines have been reasonably well served by existing newsletters and organizations. The *Mendel Newsletter* and the *History of Anthropology Newsletter* already have been mentioned. Interest group publications such as the Geological Society of America, *Newsletter of the History of Geology Division* also serve to coordinate the interests of scholars and relate the professional historical community with that of scientists.

In recent decades, several active discipline history centers have emerged. The activities of these organizations, while they may differ somewhat from one to another, all effectively complement the archival and research activities in the academic sector and elsewhere. While engaging to some degree in the collection of research materials, the centers ordinarily act as a repository-of-last-resort, working to place collections in other repositories whenever possible. Center activities also may include the collection and maintenance of information on the whereabouts of archival and manuscript material, research on archival appraisal and documentation, and advice to organizations (e.g., industry) on the value of archival and historical programs. Among the major organizations in this category are the American Institute of Physics' Center for the History of Physics (the earliest and model for subsequent discipline centers), Charles Babbage Institute Center for the History of Information Processing, Chemical Heritage Foundation–Beckman Center for the History of Chemistry (including an active interest in the biomolecular sciences), and the Institute of Electrical and Electronics Engineers' Center for the History of Electrical Engineering. Each of the above organizations regularly issues a newsletter.

In addition to the formal centers mentioned above, there are other organizations or locations of activity in the United States that relate to aspects of the history of science from a subject or discipline point of view. Again, the categories and range of their activities differ and may in fact change over time. Following is a list of organizations with a particular subject focus. These organizations engage in activities to promote the documentation of their subject in addition to (or in place of) collecting materials directly.

American Institute of the History of Pharmacy
University of Wisconsin
425 N. Charter Street
Madison, Wisconsin 53706

Archival Center for Radiation Studies
Department of Special Collections
Hoskins Library
University of Tennessee
Knoxville, Tennessee 37996

Archives of American Mathematics
Center for American History
SRH 2.109
University of Texas at Austin
Austin, Texas 78712

Archives of the History of American Psychology
University of Akron
Akron, Ohio 44325

Center for History of Microbiology
Albin O. Kuhn Library
University of Maryland-Baltimore County
5401 Wilkens Avenue
Catonsville, Maryland 21228

Center for History of Physics
American Institute of Physics
One Physics Ellipse
College Park, Maryland 20740

Center for the History of Electrical Engineering
39 Union Street
Rutgers University
News Brunswick, New Jersey 08903

Charles Babbage Institute
Center for the History of Information Processing
103 Walter Library
University of Minnesota
Minneapolis, Minnesota 55455

Chemical Heritage Foundation
Beckman Center for the History of Chemistry
3401 Walnut Street
Philadelphia, Pennsylvania 19104
    Relocating in 1996 to:
    315 Chestnut Street
    Philadelphia, Pennsylvania 19106

Forest History Society
701 Vickers Avenue
Durham, North Carolina 27701

Hunt Institute for Botanical Documentation
Carnegie Mellon University
Pittsburgh, Pennsylvania 15213

Iowa State University
Department of Special Collections
403 Parks Library
Ames, Iowa 50011
    Programs include:
    Archives of American Agriculture
    Archives of American Veterinary Medicine
    Archives of Women in Science and Engineering
    Evolution/Creation Archive
    Statistical Archive

Neuroscience History Archives
Brain Research Institute
Center for Health Sciences
University of California
Los Angeles, California 90024

Public Works Historical Society
1313 East 60th Street
Chicago, Illinois 60637

Notes

1. Eugene P. Sheehy, ed., *Guide to Reference Books*, 10th ed. (Chicago and London: American Library Association, 1986); Robert Balay, ed., *Guide to Reference Books: Covering Materials from 1985-1990*, supplement to the 10th ed. (Chicago: American Library Association, 1992); Marilyn Mullay and Priscilla Schlicke, eds., *Walford's Guide to Reference Materials*, vol. 1: *Science and Technology*, 6th ed. (London: Library Association Publishing, 1993). A new edition of the A.L.A. *Guide to Reference Books*, under the editorship of Robert Balay, is in preparation.

2. Francis Paul Prucha, *Handbook for Research in American History: A Guide to Bibliographies and Other Reference Works* (Lincoln: University of Nebraska Press, 1987); Ron Blazek and Anna H. Perrault, *United States History: A Selective Guide to Information Sources* (Englewood, Colo.: Libraries Unlimited, 1994).

3. Royal Society of London, *Catalogue of Scientific Papers, 1800-1900*, 19 vols. (London: Clay, 1867-1902; Cambridge: University Press, 1914-1925), arranged by author, in successive series, covering journal article publications during the periods 1800-1863, 1864-1873, 1874-1883, 1884-1900. See Clark A. Elliott, "The Royal Society Catalogue as an Index to 19th-Century American Science," *Journal of American Society for Information Science* 21 (1970): 396-401.

4. Full citations for works discussed and not footnoted are to be found in the alphabetically arranged bibliography that follows.

5. Marilyn Bailey Ogilvie, *Women and Science: An Annotated Bibliography* (New York: Garland Publishing) is in preparation.

6. Text references here to bibliographic works ordinarily do not include subtitles; see the bibliography for full information.

7. Pieter Smit, *History of the Life Sciences: An Annotated Bibliography* (New York: Hafner, 1974) is a standard work that covers the subject from ancient times but appears to have relatively little that is identifiable as relating to the history of biology and medicine in the United States.

8. See also Hallion's "A Source Guide to the History of Aeronautics and Astronautics," *American Studies International* 20 (Spring 1982): 3-50.

9. Ralph S. Bates, *Scientific Societies in the United States*, 3rd ed. (Cambridge, Mass.: MIT Press, 1965).

10. Arnold Mallis, *American Entomologists* (New Brunswick, N.J.: Rutgers University Press, 1971).

11. Raymond Phineas Stearns, *Science in the British Colonies of America* (Urbana: University of Illinois Press, 1970).

12. Bryan H. Bunch, *The Henry Holt Handbook of Current Science & Technology: A Sourcebook of Facts and Analysis Covering the Most Important Events in Science and Technology* (New York: Holt, 1992) features events occurring in 1990-1991 but gives brief chronologies and discussion as background to topics or subject areas, as well as up-to-date scientific information. It is a very useful supplement to the chronology that constitutes the main part of this book.

13. Marc Rothenberg currently is editing a general encyclopedia of the history of science in the United States, to be issued by Garland Publishing. In addition, Garland has commissioned several discipline or subject-area encyclopedias in the history of science and related fields that will include information that should interest historians of American science.

14. The American Council of Learned Societies and Oxford University Press currently have in preparation an *American National Biography*; when completed, it will largely supersede the *Dictionary of American Biography* in currency and coverage.

15. There is reference to the following publication, not seen by the compiler: Wyndham D. Miles and Robert F. Gould, *American Chemists and Chemical Engineers*, vol. 2 (Guilford, Conn.: Gould Books, 1994).

16. This work is part of a series, "Greenwood Encyclopedia of American Institutions," but is the only volume listed in this research guide. However, other volumes might be of interest to some users. Among the volumes in the "Encyclopedia," that represent institutional categories not included by Kiger, are those on Foundations, Government Agencies, Private Colleges and Universities, and Public Colleges and Universities.

17. It is necessary to keep in mind that libraries may have differing relations to these utilities, so that sometimes the information in the local catalogue may be more, or less, complete, more timely, or accurate than that in the RLIN and OCLC WorldCat files. Access to the RLG databases requires payment of subscription fees, either by a library or an individual. OCLC's FirstSearch is available only to libraries or institutions, but its related EPIC service is available by individual subscription.

18. Among the available publications are Harley Hahn and Rick Stout, *The Internet Complete Reference* (Berkeley: Osborne McGraw-Hill, 1994); Brendan P. Kehoe, *Zen and the Art of the Internet: A Beginner's Guide*, 3rd ed. (Englewood Cliffs, N.J.: PTR Prentice Hall, 1994); and Ed Krol, *The Whole Internet User's Guide and Catalog*, 2nd ed. (Sebastopol, Calif.: O'Reilly & Associates, 1994). In regard specifically to the topic under discussion here, see *Directory of Electronic Journals, Newsletters and Academic Discussion Lists*, 4th ed. (Washington, D.C.: Association of Research Libraries, Office of Scientific and Academic Publishing, 1994), an annual publication; an electronic version is available at the ARL gopher arl.cni.org, and at other Internet locations.

19. Access at gopher una.hh.lib.umich.edu 70, path=inetdirs, or World Wide Web address http://www.lib.umich.edu/chhome.html. (In the Internet addresses given here, the initial instruction is the command to connect to the source, while the path ordinarily is the choice to be made from the menu after connection is achieved.)

20. Gopher marvel.loc.gov 70, path=global electronic library by subject.

21. Hahn and Stout, *The Internet Complete Reference*, p.455.

22. The yield from such a search is very much mixed and includes duplicate entries where the same item appears in more than one location. A good percentage of the information is of local interest primarily (e.g., academic course descriptions).

23. Telnet history.cc.ukans.edu, login: history, or World Wide Web at http://history.cc.ukans.edu/history/WWW_history_main.html.

24. Hahn and Stout, *The Internet Complete Reference*, pp.496, 510.

25. This is available at the World Wide Web address http://info.cern.ch/hytertext/DataSources/bySubject/Overview.html. Another accessible location is telnet lynx.cc.ukans.edu, login: www; for the Virtual Library, on connection find Web Overview link.

26. ASAP also is available directly at the World Wide Web site http://coombs.anu.edu.au/SpecialProj/ASAP/ASAPHome.html.

27. See at http://www.yahoo.com/.

28. For direct connection to the WebCrawler facility, access the address http://webcrawler.cs.washington.edu/WebCrawler/WebQuery.html.

29. Gopher nmnhgoph.si.edu, menu choice Related Gopher and Information Servers, which leads (under Ecology and Evolution) to A Biologist's Guide to Internet Resources. The guide also is accessible at the Web address http://nmnhwww.si.edu/nmnhweb.html.

30. Appropriately, the Charles Babbage Institute Center for the History of Information Processing (CBI) is one of the active participants in these developments. In spring 1995, they had electronic access to their resources through three means, by telnet to the University of Minnesota library catalogue, by the University of Minnesota Libraries gopher, and by the Center's World Wide Web home page. So long as these parallel access tools are maintained, the CBI example will allow users to examine and to compare the three avenues of approach to searching for research materials, and the differing ways in which a repository can take advantage of the features of each. (To reach the Babbage Institute, gopher cutter.lib.umn.edu; choose Subject List, and then Archives and Special Collections, Charles Babbage. The Babbage gopher gives

information on access to their Web page. The University library catalogue also is accessible through that gopher command, or by telnet pubinfo.ais.umn.edu.)

31. Gopher musicbox.mse.jhu.edu, path=Other Gophers/ Other Archives and Manuscript Repositories.

32. Address http://www.muohio.edu/~ArchivesList/.

33. *Gale Directory of Databases* (Detroit: Gale Research Inc., January 1993-). Formed by the merger of several earlier directories, the January 1995 issue devoted volume 1 to on-line databases and volume 2 to CD-ROM, diskette, magnetic tape, handheld, and batch access databases.

34. This is an appropriate and pleasurable place to state my admiration and to express my indebtedness to the bibliographical and instructional skills of Barbara Burg and Debbie Kelley-Millburn. Their work in the Research and Bibliographic Services office in Harvard University's Widener Library is helping an entire community come to grips with the phenomenon of the Internet and other electronic sources. I hope all of you will have equally talented librarians to call upon in your hours of need.

35. Clark A. Elliott, "Citation Patterns and Documentation for the History of Science: Some Methodological Considerations," *American Archivist* 44 (1981): 131-142 reveals some of the underlying practices in the use of different categories of research material.

36. U.S. Library of Congress, *National Union Catalog of Manuscript Collections, 1959-* (Ann Arbor, Mich.: Edwards, 1962; Hamden, Conn.: Shoe String, 1964; Washington, D.C.: Library of Congress, 1965-). See also: *Index to Personal Names in the National Union Catalog of Manuscript Collections 1959-1984* (Alexandria, Va.: Chadwyck-Healey, 1988), and *Index to Subjects and Corporate Names in the National Union Catalog of Manuscript Collections, 1959-1984* (Alexandria, Va.: Chadwyck-Healey, 1994).

37. *National Inventory of Documentary Sources in the United States* (Teaneck, N.J.: Chadwyck-Healey, 1983-) also has made available more widely an array of manuscript collection finding aids (on microfiche). Previously, these inventories were accessible only at the repositories that hold the material.

38. Although RLG has been a leader in the movement to facilitate access to manuscript and archival resources at the national level, the Online Computer Library Center (OCLC) WorldCat also includes a number of catalogue records for such unpublished research materials.

39. A collateral development is the availability of source documents themselves—both textual and visual—in on-line or CD-ROM systems. All of these developments are so fast paced at present that this guide can do little more than note the phenomenon and urge vigilance upon researchers, through consultation with the information services in their local library.

40. Guides to specific collections within repositories have not been included in the bibliography.

41. The American Institute of Physics Center for the History of Physics has in preparation a "Catalog of Sources for Laser History"; publication, under the direction of Joan Warnow-Blewett, is expected during 1995.

42. There is reference to the following publication, not seen by the compiler: Center for the History of Electrical Engineering, *Sources in Electrical History 2: Oral History Collections in U.S. Repositories* (New Brunswick, N.J.: IEEE Center for the History of Electrical Engineering, 1992).

43. U.S. National Archives and Records Administration, *Guide to the National Archives of the United States* (Washington, D.C.: The Archives, 1987), which is a reprint of a 1974 guide with updated information in the appendix.

44. U.S. National Historical Publications and Records Commission, *Directory of Archives and Manuscript Repositories in the United States*, 2nd ed. (Phoenix: Oryx, 1988).

45. Donald L. DeWitt, *Guides to Archives and Manuscript Collections in the United States: An Annotated Bibliography* (Westport, Conn.: Greenwood Press, 1994).

46. The difficulties of compiling and maintaining such a guide are underscored by recent plans of the History of Science Society to produce hereafter only an updated membership directory.

4

Research Guide: Bibliography

Introduction

The following is a selected bibliography of bibliographies, reference works, guides to manuscript sources, and discussions of the characteristics and problems of research materials. For a topical or contextual discussion of the titles listed, see the preceding research guide.

All of the works in the bibliography relate to the history of science in the United States, but many of them are broader, not limited in coverage to the American national scene. Thus, the bibliography is, to some degree, a general guide to the history of science, although this is not its intent. Scientific organization and practice mean that, in some works, American concerns will be integrated with international coverage of a topic. This can pose difficulties for the user who wants to extract information specific to the United States. The index or structure of an information source may anticipate the interests of users with a national interest in science. In other instances, however, a work can be so utilized only through browsing and may call upon a degree of prior knowledge of the topic (e.g., recognition of the names of American scientists, or of subjects that were of special interest to Americans). It is a salient feature of science that it is an activity both local and global. Inevitably, this is reflected in the approaches and interests of historians themselves and in the construction of the information sources that potentially serve their needs.

Abbott, Robert Tucker, ed. *American Malacologists: A National Register of Professional and Amateur Malacologists and Private Shell Collectors and Biographies of Early American Mollusk Workers Born Between 1618 and 1900*. 494pp. Falls Church, Va.: American Malacologists, 1973.

Academy of Natural Sciences of Philadelphia. *Guide to the Manuscript Collections in the Academy of Natural Sciences of Philadelphia*. Compiled by Venia T. Phillips and Maurice E. Phillips. 553pp. Special Collections, 5. Philadelphia: Academy of Natural Sciences of Philadelphia, 1963.

Aldrich, Michele L. "United States: Bibliographic Essay." In *The Comparative Reception of Darwinism*, edited by Thomas F. Glick, pp.207-226. Austin and London: University of Texas Press, 1974.

American Institute of Physics. *Guide to the Archival Collections in the Niels Bohr Library at the American Institute of Physics*. 574pp. AIP International Catalog of Sources for History of Physics and Allied Sciences, Report 7. College Park, Md.: American Institute of Physics, 1994.

American Medical Association. *Guide to the American Medical Association Historical Health Fraud and Alternative Medicine Collection*. Executive editor, Arthur W. Hafner; senior editor, James G. Carson; contributing editor, John F. Zwick. 215pp. Chicago, Ill.: American Medical Association, 1992.

American Men and Women of Science: A Biographical Directory. New York, N.Y.; Garrison, N.Y.; Lancaster, Pa.: Science Press; New York: Bowker, 1906-. Prior to 12th edition (1971) titled *American Men of Science*.

American Men and Women of Science: Cumulative Index [to] Editions 1-14. Compiled by Jacques Cattell Press. 847pp. Ann Arbor, Mich.: R.R. Bowker Co., 1983.

American Philosophical Society. *Catalog of Manuscripts in the American Philosophical Society Library, Including the Archival Shelflist*. 10 vols. Westport, Conn.: Greenwood Press, 1970.

American Philosophical Society. *A New Guide to the Collections in the Library of the American Philosophical Society.* Edited by J. Stephen Catlett. 414pp. Philadelphia, Penn.: American Philosophical Society, 1987. Catalogue of manuscript and archival collections.

American Society of Civil Engineers. *A Biographical Dictionary of American Civil Engineers.* 163pp. ASCE Historical Publication, 2. New York: Committee on History and Heritage of American Civil Engineering, American Society of Civil Engineers, 1972.

American Society of Mechanical Engineers. *Mechanical Engineers in America Born Prior to 1861: A Biographical Dictionary.* Sponsored by the History and Heritage Committee. 330pp. New York: American Society of Mechanical Engineers, 1980.

Apple, Rima D., comp. *Illustrated Catalogue of the Slide Archive of Historical Medical Photographs at Stony Brook.* By the Center for Photographic Images of Medicine and Health Care. 442pp. Westport, Conn.: Greenwood Press, 1984.

Asbell, M.B. *A Bibliography of Dentistry in America, 1790-1840.* 107pp. [Cherry Hill, N.J.]: Sussex House Publications, 1973.

Asimov, Isaac. *Asimov's Biographical Encyclopedia of Science and Technology: The Lives and Achievements of 1510 Great Scientists from Ancient Times to the Present Chronologically Arranged.* 2nd rev. ed. 941pp. Garden City, N.Y.: Doubleday and Co., Inc., 1982.

Asimov, Isaac. *Asimov's Chronology of Science and Discovery.* Updated and Illustrated. 790pp. New York: HarperCollins, 1994.

Aspray, William, and Bruce Bruemmer, eds. *Guide to the Oral History Collections of the Charles Babbage Institute.* 110pp. Minneapolis: Charles Babbage Institute and the Center for the History of Information Processing of the University of Minnesota, 1986.

Auk. *Biographies of Members of the American Ornithological Union.* By T.S. Palmer, and others. 630pp. Washington, D.C.: 1954. Reprinted from *The Auk*, 1884-1954.

Austin, Robert B. *Early American Medical Imprints: A Guide to Works Printed in the United States, 1668-1820.* 240pp. Washington, D.C.: National Library of Medicine, 1961.

Axelrod, Alan, and Charles Phillips. *The Environmentalists: A Biographical Dictionary from the 18th Century to Today.* 258 pp. New York: Facts on File, 1993.

Bailey, Martha J. *American Women in Science: A Biographical Dictionary.* 463pp. Santa Barbara, Calif.: ABC-Clio, 1994.

Bali, Mrinal. *Space Exploration: A Reference Handbook.* 240pp. Contemporary World Issues. Santa Barbara, Calif.: ABC-Clio, 1990.

Barnhart, John H., comp. *Biographical Notes upon Botanists, Maintained in the New York Botanical Garden Library.* 3 vols. Boston: G.K. Hall, 1965.

Barr, Ernest Scott. *An Index to Biographical Fragments in Unspecialized Scientific Journals.* 294pp. University: University of Alabama Press, 1973.

Batschelet, Margaret W. *Early American Scientific and Technical Literature: An Annotated Bibliography of Books, Pamphlets, and Broadsides.* 136pp. Metuchen, N.J.: Scarecrow Press, 1990.

Bearman, David, and John T. Edsall, eds. *Archival Sources for the History of Biochemistry and Molecular Biology: A Reference Guide and Report.* 338pp., plus 6 microfiches. Boston, Mass.: American Academy of Arts and Sciences; Philadelphia, Penn.: American Philosophical Society; produced and distributed by University Microfilms International, 1980.

Bedi, Joyce E., Ronald R. Kline, and Craig Semsel, comps. *Sources in Electrical History: Archives and Manuscript Collections in U.S. Repositories.* 234pp. New York: Center for the History of Electrical Engineering, 1989.

Beers, Henry P., comp. *Bibliographies in American History: Guide to Materials for Research.* 487pp. New York: Wilson, 1942.

Beers, Henry P., comp. *Bibliographies in American History, 1942-1978: Guide to Materials for Research.* 2 vols. Woodbridge, Conn.: Research Publications, 1982.

Bell, Whitfield J., Jr. *Early American Science: Needs and Opportunities for Study.* 85pp. Williamsburg, Va.: Institute of Early American History and Culture, 1955.

Benson, Michael. *Vintage Science Fiction Films, 1896-1949.* 219pp. Jefferson, N.C.: McFarland and Co., 1985. Filmography of 375 works.

Bibliography of the History of Medicine, no. 1- , 1965-. Bethesda, Md.: U.S. Public Health Service, 1966-.

Bindocci, Cynthia Gay, and Kathleen H. Ochs. *Women and Technology: An Annotated Research Guide.* 229pp. New York: Garland Publishing, 1993.

Black, George W., Jr. *American Science and Technology: A Bicentennial Bibliography.* 170pp. Carbondale: Southern Illinois University Press, 1979.

Blain, Daniel, and Michael Barton. *History of American Psychiatry: A Teaching and Research Guide.* 44pp. Task Force Report 15. Washington, D.C.: American Psychiatric Association, 1979.

Brichford, Maynard J. *Scientific and Technological Documentation: Archival Evaluation and Processing of University Records Relating to Science and Technology.* 35pp. Urbana, Ill.: University of Illinois at Urbana-Champaign, 1969.

Bridson, Gavin D.R. *The History of Natural History: An Annotated Bibliography.* 740pp. Garland Reference Library of the Humanities, 991; Bibliographies on the History of Science and Technology, 24. New York: Garland Publishing, 1994.

Brown, Catherine L., and James O. Wheeler, comps. *A Bibliography of Geographic Thought*. 520pp. Bibliographies and Indexes in Geography, 1. Westport, Conn.: Greenwood Press, 1989.

Bruemmer, Bruce. *Resources for the History of Computing: A Guide to U.S. and Canadian Records*. 187pp. Minneapolis: Charles Babbage Institute, Center for the History of Information Processing, University of Minnesota, 1987.

Bruemmer, Bruce H., and Sheldon Hochheiser. *The High-Technology Company: A Historical Research and Archival Guide*. 131pp. Minneapolis: Charles Babbage Institute, University of Minnesota, 1989.

Brush, Stephen G. *The History of Modern Science: A Guide to the Second Scientific Revolution*. 544pp. Ames: Iowa State University Press, 1988.

Brush, Stephen G., and Lanfranco Belloni. *The History of Modern Physics: An International Bibliography*. 334pp. Bibliographies of the History of Science and Technology, 4. New York: Garland Publishing, 1983.

Brush, Stephen G., and Helmut E. Landsberg. *History of Geophysics and Meteorology: An Annotated Bibliography*. 450pp. Bibliographies of the History of Science and Technology, 7. New York: Garland Publishing, 1985.

Bullough, Bonnie, Vern L. Bullough, and Barrett Elcano. *Nursing: A Historical Bibliography*. 408pp. Reference Library of Social Science, 66. New York: Garland Publishing, 1981.

Bullough, Vern L., Olga Church, and Alice P. Stein. *American Nursing: A Biographical Dictionary*. 2 vols. Garland Reference Library of Social Science, 368, 684. New York: Garland Publishing, 1988-1992.

Bunch, Bryan H., and Alexander Hellemans. *The Timetables of Technology: A Chronology of the Most Important People and Events* in the History of Technology. 490 pp. New York: Simon & Schuster, 1993.

Burns, Grant. *The Atomic Papers: A Citizen's Guide to Selected Books and Articles on the Bomb, the Armsrace, Nuclear Power, the Peace Movement, and Related Issues.* 309pp. Metuchen, N.J.: Scarecrow Press, 1984.

Burns, Grant. *The Nuclear Present: A Guide to Recent Books on Nuclear War, Weapons, the Peace Movement, and Related Issues, with a Chronology of Nuclear Events, 1789-1991.* 633pp. Metuchen, N.J.: Scarecrow Press, 1992.

Bynum, William, E. Janet Browne, and Roy Porter, eds. *Dictionary of the History of Science.* 494pp. Princeton, N.J.: Princeton University Press, 1981.

Caldwell, Lynton K. *Science, Technology and Public Policy: A Selected and Annotated Bibliography.* 3 vols. Bloomington: Program in Public Policy for Science and Technology, Department of Government, Indiana University, 1969-1972.

Channell, David F. *The History of Engineering Science: An Annotated Bibliography.* 311pp. Bibliographies of the History of Science and Technology, 16. New York: Garland Publishing, 1989.

Clareson, Thomas D., comp. *Science Fiction in America, 1870s-1930s: An Annotated Bibliography of Primary Sources.* 305pp. Bibliographies and Indexes in American Literature, 1. Westport, Conn.: Greenwood Press, 1984.

Cobb, David A., comp. *Guide to U.S. Map Resources.* 2nd ed. 495pp. Chicago: American Library Association, 1990.

Collea, Beth A., and Anthony F. Aveni. *A Selected Bibliography of Native American Astronomy.* 148pp. Hamilton, N.Y.: Colgate University, Department of Physics and Astronomy, 1978.

College of Physicians of Philadelphia. *A Catalogue of the Manuscripts and Archives of the Library of the College of Physicians of*

Philadelphia. Prepared by the Francis Clark Wood Institute for the History of Medicine. Edited by Rudolf Hirsch, and others. 259pp. Philadelphia: University of Pennsylvania Press, and the College of Physicians of Philadelphia, 1983.

Collins, Martin J., with Jo Ann Bailey and Patricia Fredericks. *Oral History on Space, Science, and Technology: A Catalog of the Collection of the Department of Space History, National Air and Space Museum*. 267pp. Washington, D.C.: National Air and Space Museum, Smithsonian Institution, 1993.

Conference on Science Manuscripts, 5-6 May 1960, Washington, D.C. [Proceedings]. *Isis* 53 no. 1 (March 1962): issue [pp.1-157].

Cordasco, Francesco. *American Medical Imprints, 1820-1910: A Checklist of Publications Illustrating the History and Progress of Medical Science and Education and the Healing Arts in the United States: A Preliminary Contribution*. 2 vols. Totowa, N.J.: Rowman & Littlefield; Fairview, N.J.: Junius-Vaughn Press, 1985.

Corsi, Pietro, and Paul Weindling, eds. *Information Sources in the History of Science and Medicine*. 531pp. London and Boston: Butterworth Scientific, 1983.

Cortada, James W., ed. *Archives of Data-Processing History: A Guide to Major U.S. Collections*. 181pp. Westport, Conn.: Greenwood Press, 1990.

Cortada, James W. *A Bibliographic Guide to the History of Computing, Computers, and the Information Processing Industry*. 644pp. Bibliographies and Indexes in Science and Technology, 6. Westport, Conn.: Greenwood Press, 1990.

Cortada, James W. *Historical Dictionary of Data Processing: Biographies*. 321pp. Westport, Conn.: Greenwood Press, 1987.

Cortada, James W. *Historical Dictionary of Data Processing: Organizations*. 309pp. Westport, Conn.: Greenwood Press, 1987.

Cortada, James W. *Historical Dictionary of Data Processing: Technology.* 415pp. Westport, Conn.: Greenwood Press, 1987.

Cowart, David, and Thomas L. Wymer, eds. *Twentieth Century American Science Fiction Writers.* 2 vols. Dictionary of Literary Biography, 8. Detroit, Mich.: Gale Research, 1981.

Crowther, Simeon J., and Marion Fawcett, comps. *Science and Medicine to 1870: Pamphlets in the American Philosophical Society Library.* 164pp. Library Publication, 1. Philadelphia: American Philosophical Society, 1968.

Cutliffe, Stephen H., and others. *Technology and Values in American Civilization: A Guide to Information Sources.* 704pp. American Studies Information Guide, 9. Detroit: Gale Research, 1980.

Daintith, John, Sarah Mitchell, Elizabeth Tootill, and Derek Gjersten. *Biographical Encyclopedia of Scientists.* 2nd ed. 2 vols. Bristol and Philadelphia: Institute of Physics Publishing, 1994.

Danilov, Victor J. *America's Science Museums.* 483pp. Westport, Conn.: Greenwood Press, 1990.

Dauben, Joseph W. *The History of Mathematics from Antiquity to the Present: A Selective Bibliography.* 467pp. Bibliographies of the History of Science and Technology, 6. New York: Garland Publishing, 1985.

Davis, Henry B.O. *Electrical and Electronic Technologies: A Chronology of Events and Inventors.* 3 vols. Metuchen, N.J.: Scarecrow, 1981-1985. Vols. cover to 1900, 1900 to 1940, 1940-1980.

Davis, Richard C., editor-in-chief. *Encyclopedia of American Forest and Conservation History.* Forest History Society. 2 vols. Riverside, N.J.: Macmillan Publishing Co., 1983.

Davis, Richard C. *North American Forest History: A Guide to Archives and Manuscripts in the United States and Canada.* 376pp. Santa Barbara, Calif.: Clio Books for the Forest History Society, 1977.

DeVorkin, David H. *History of Modern Astronomy and Astrophysics: A Selected, Annotated Bibliography.* 434pp. Bibliographies of the History of Science and Technology, 1. New York: Garland Publishing, 1982.

Dickson, Katherine Murphy. *History of Aeronautics and Astronautics: A Preliminary Bibliography.* 420pp. Washington, D.C.: National Aeronautics and Space Administration, 1968.

Dictionary of Scientific Biography. Edited by Charles C. Gillispie. 16 vols. New York: Scribner's, 1970-1980. Supplement II, vols. 17-18, edited by Frederic L. Holmes. New York: Scribner's, 1990. Also: *Concise Dictionary of Scientific Biography.* 773pp. New York, 1981.

Directory, Historians of American Science 1993. Compiled and edited by Michele L. Aldrich and Alan E. Leviton. 59pp. History of Science in America: News and Views 10, no. 3, special issue. Washington, D.C.: Forum for the History of Science in America, 1993.

Downard, William L. *Dictionary of the History of the American Brewing and Distilling Industries.* 268pp. Westport, Conn.: Greenwood Press, 1980.

Dunbar, Gary S. *History of Modern Geography: An Annotated Bibliography of Selected Works.* 386pp. Bibliographies of the History of Science and Technology, 9. New York: Garland Publishing, 1985.

Durbin, Paul T., general editor. *A Guide to the Culture of Science, Technology, and Medicine.* 723pp. New York: Macmillan Publishing Co. Free Press, 1980.

Eastwood, Bruce. *Directory of Audio-Visual Sources: History of Science, Medicine, and Technology.* 41, [108] pp. New York: Science History Publications, 1979.

Edwards, Everett E. *Bibliography of the History of Agriculture in the United States.* 307pp. U.S. Department of Agriculture

Miscellaneous Publication, 84. Washington, D.C.: Government Printing Office, 1930. Reprints. Detroit: Gale Research, 1967. New York: Franklin, 1970.

Egerton, Frank N. "The History of Ecology: Achievements and Opportunities." *Journal of the History of Biology* 16 (1983): 259-311 and 18 (1985): 103-143.

Ehrenberg, Ralph E. *Scholar's Guide to Washington, D.C., for Cartography and Remote Sensing Imagery (Maps, Charts, Aerial Photographs, Satellite Images, Cartographic Literature and Geographic Information Systems).* 385pp. Woodrow Wilson International Center for Scholars Series. Washington, D.C.: Smithsonian Institution Press, 1987.

Elbers, Joan S., comp. *Changing Wilderness Values, 1930-1990: An Annotated Bibliography.* 138pp. Bibliographies and Indexes in American History, 18. Westport, Conn.: Greenwood Press, 1991.

Eleutherian Mills Historical Library. *A Guide to the Manuscripts in the Eleutherian Mills Historical Library.* By John Beverley Riggs. 2 vols. Greenville, Del.: Eleutherian Mills Historical Library, 1970, 1978.

Elliott, Clark A. "Collective Lives of American Scientists: An Introductory Essay and a Bibliography." In *Beyond History of Science: Essays in Honor of Robert E. Schofield*, edited by Elizabeth Garber, 81-104. Bethlehem: Lehigh University Press; London and Toronto: Associated University Presses, 1990.

Elliott, Clark A. "Sources for the History of Science in the Harvard University Archives." *Harvard Library Bulletin* 22 (1974): 49-71.

Elliott, Clark A. *Biographical Dictionary of American Science: The Seventeenth Through the Nineteenth Centuries.* 360pp. Westport, Conn.: Greenwood Press, 1979.

Elliott, Clark A. *Biographical Index to American Science: The Seventeenth Century to 1920.* 300pp. Bibliographies and Indexes in American History, 16. Westport, Conn.: Greenwood Press, 1990.

Erlen, Jonathon. *The History of the Health Care Sciences and Health Care, 1700-1980: A Selective Annotated Bibliography*. 1028pp. Garland Reference Library of the Humanities, 398. New York: Garland Publishing, 1984.

Ewan, Joseph, and Nesta Dunn Ewan. *Biographical Dictionary of Rocky Mountain Naturalists: A Guide to the Writings and Collections of Botanists, Zoologists, Geologists, Artists, and Photographers 1682-1932*. 253pp. Regnum Vegetabile 107. Utrecht and Antwerp: Bohn, Scheltema and Holkema; The Hague and Boston: W. Junk; (distributed by Kluwer Boston, Inc., 190 Old Derby St., Hingham, Mass. 02043), 1981.

Fahl, Ronald J. *North American Forest and Conservation History: A Bibliography*. 408pp. Santa Barbara, Calif.: Clio Press for the Forest History Society, 1977.

Findling, John E., and Kimberly D. Pelle, eds. *Historical Dictionary of World's Fairs and Expositions, 1851-1988*. 443pp. Westport, Conn.: Greenwood Press, 1990.

Finn, Bernard S. *The History of Electrical Technology: An Annotated Bibliography*. 342pp. Bibliographies on the History of Science and Technology, 18. New York: Garland Publishing, 1991.

Firth, Margaret A., ed. *Handbook of Scientific and Technical Awards in the United States and Canada, 1900-1952*. 491pp. New York: Special Libraries Association, Science-Technology Division, 1956.

Fishbein, Meyer H. "Archival Remains of Research and Development During the Second World War." In *World War II: An Account of Its Documents*, edited by James E. O'Neill and Robert W. Krauskopf, 163-179. National Archives Conferences, 8. Washington, D.C.: Howard University Press, 1976.

Fleming, James R. *Guide to Historical Resources in the Atmospheric Sciences: Archives, Manuscripts and Special Collections in the Washington, D.C. Area*. 167pp. NCAR-327+1A; NCAR Technical Note. Boulder, Colo.: Climate and Global Dynamics Division, National Center for Atmospheric Research, 1989.

Fried, Stephen B. *American Popular Psychology: An Interdisciplinary Research Guide*. 241pp. New York: Garland Publishing, 1994.

Fruton, Joseph S. *A Bio-Bibliography for the History of Biochemical Sciences Since 1800*. 885pp. Philadelphia, Penn.: American Philosophical Society, 1982.

Fruton, Joseph S. *A Supplement to a Bio-Bibliography for the History of the Biochemical Sciences Since 1800*. 262pp. Philadelphia: American Philosophical Society, 1985.

Gacs, Ute, Aisha Khau, Jerrie McIntire, and Ruth Weinberg, eds. *Women Anthropologists: A Biographical Dictionary*. 428pp. Westport, Conn.: Greenwood Press, 1988.

Gascoigne, Robert M. *A Chronology of the History of Science, 1450-1900*. 585pp. New York: Garland Publishing, 1987.

Gascoigne, Robert M. *A Historical Catalogue of Scientific Periodicals, 1665-1900*. 205pp. Garland Reference Library of the Humanities, 583. New York: Garland Publishing, 1985.

Gascoigne, Robert M. *A Historical Catalogue of Scientists and Scientific Books: From the Earliest Times to the Close of the Nineteenth Century*. 1177pp. New York: Garland Publishing, 1984.

Geahigan, Priscilla C. *U.S. and Canadian Businesses, 1955 to 1987: A Bibliography*. 589pp. Metuchen, N.J.: Scarecrow Press, 1988.

Geiser, Samuel W. *Men of Science in Texas, 1820-1880*. approx. 245pp. Dallas: Southern Methodist University Press, [1959]. Reprinted from *Field and Laboratory* 26 and 27 (1958-1959), and retains that page numbering.

Geological Society of America. *Newsletter of the History of Geology Division*. [Boulder, Col.?]: The Society, 1977-.

Gilbert, Pamela. *A Compendium of the Biographical Literature on Deceased Entomologists*. 455pp. Publications-British Museum

(Natural History), 786. London: British Museum (Natural History), 1977.

Glass, Bentley. *A Guide to the Genetics Collections of the American Philosophical Society*. 122pp. APS Library Publication, 13. Philadelphia, Penn.: American Philosophical Society Library, 1988.

Grainger, Thomas H., Jr. *A Guide to the History of Bacteriology*. 210pp. Chronica Botanica no. 18. New York: Ronald Press, 1958.

Grinstein, Louise S., and Paul J. Campbell, eds. *Women of Mathematics: A Biobibliographic Sourcebook*. 292pp. New York: Greenwood Press, 1987.

Grinstein, Louise S., Rose K. Rose, and Miriam H. Rafailovich, eds. *Women in Chemistry and Physics: A Biobibliographic Sourcebook*. 721 pp. Westport, Conn.: Greenwood Press, 1993.

Guerra, Francisco. *American Medical Bibliography, 1639-1783. A Chronological Catalogue, and Critical and Bibliographical Study of Books, Pamphlets, Broadsides, and Articles in Periodical Publications Relating to the Medical Sciences--Medicine, Surgery, Pharmacy, Dentistry, and Veterinary Medicine—Printed in the Present Territory of the United States of America During the British Dominion and the Revolutionary War*. 885pp. Yale University Department of the History of Science and Medicine Publication, 40. New York: Lathrop C. Harper, 1962.

Guerrini, Anita. *Natural History and the New World, 1524-1770: An Annotated Bibliography of Printed Materials in the Library of the American Philosophical Society*. 83pp. APS Library Publication, 11. Philadelphia: American Philosophical Society, 1986.

Guide to the History of Science. Edited by Thomas P. Carroll; Journals by Roy Goodman. 8th ed. 315pp. Philadelphia: History of Science Society, 1992.

Gull, Cloyd Drake, and Charles Louis Smith, eds. *A Directory of Sources for Air and Space History: Primary Historical Collections in United States Repositories*. Original compiler, E.T. Wooldridge,

Jr.; general editor, Martin J. Collins. A Preliminary edition. 312pp. Washington, D.C.: National Air and Space Museum, Smithsonian Institution, 1989.

Gunn, James, ed. *The New Encyclopedia of Science Fiction*. 524pp. New York: Viking, 1988.

Haas, Joan K., Helen Willa Samuels, and Barbara Trippel Simmons. *Appraising the Records of Modern Science and Technology*. 96pp. Cambridge, Mass.: Massachusetts Institute of Technology Archives, distributed by Society of American Archivists, 1985.

Hall, H.W., ed. *Science Fiction and Fantasy Reference Index, 1878-1985: An International Author and Subject Index to History and Criticism*. 2 vols. Detroit: Gale Research, 1987. Supplement for 1985-1991. 677pp. Englewood, Colo.: Libraries Unlimited, 1993.

Hallion, Richard P. *The Literature of Aeronautics, Astronautics, and Air Power*. 66pp. USAF Warrior Studies. Washington, D.C.: Office of Air Force History, U.S. Air Force, 1984.

Harden, Victoria A., and Guenter B. Risse, eds. *AIDS and the Historian: Proceedings of a Conference at the National Institutes of Health, 20-21 March 1989*. 161pp. NIH Publication, 91-1584. Bethesda, Md.: National Institutes of Health, 1991. Contents include: Peter B. Hirtle, "Documentation in the Federal Government of the History of AIDS." Ramunas Kondratas, "The Artifactual Legacy of AIDS." Nancy W. Zinn, "Documenting AIDS: The Role of the University and Other Agencies."

Harkanyi, Katalin, comp. *The Natural Sciences and American Scientists in the Revolutionary Era: A Bibliography*. 510pp. Bibliographies and Indexes in American History, 17. Westport, Conn.: Greenwood Press, 1990.

Hawthorne, Douglas B. *Men and Women of Space*. 904pp. San Diego, Calif.: Univelt, 1992.

Hazen, Robert M., and Margaret Hindle Hazen. *American Geological Literature, 1669-1850.* 431pp. Stroudsburg, Penn.: Dowden, Hutchinson and Ross, 1980.

Health Science Books 1876-1982. 4 vols. New York: R.R. Bowker, 1982. Arranged by subject, with author and title indexes.

Heilbron, J.L., and Bruce R. Wheaton. *Literature on the History of Physics in the 20th Century.* 485pp. Berkeley Papers in History of Science, 5. Berkeley: Office of History of Science and Technology, University of California, 1981.

Hellemans, Alexander, and Bryan Bunch. *The Timetables of Science: A Chronology of the Most Important People and Events in the History of Science.* 656pp. New York: Simon and Schuster, 1988.

Higby, Gregory J., and Elaine C. Stroud. *The History of Pharmacy: A Selected Annotated Bibliography.* 321pp. Bibliographies on the History of Science and Technology, 25. New York: Garland Publishing, 1995.

Higham, Robin D.S., ed. *A Guide to the Sources of United States Military History.* 559pp. Hamden, Conn.: Archon, 1975. Supplements 1-3. 1981-1993.

Hindle, Brooke. *Technology in Early America: Needs and Opportunities for Study.* 145pp. Chapel Hill, N.C.: University of North Carolina Press, 1966.

History of Anthropology Newsletter. Chicago: University of Chicago, 1973-.

History of Science and Technology: A Narrative Chronology. Editorial director, Edgardo Marcorini. 2 vols. New York and Oxford: Facts on File, 1988.

History of Science in America: News and Views. vol. 1-, 1980-. Since vol. 3, no.3 (May/June 1985) published as newsletter to the Forum for the History of Science in America.

Holloway, Lisabeth M. *Medical Obituaries: American Physicians' Biographical Notices in Selected Medical Journals before 1907.* 513pp. Garland Reference Library of Social Sciences, 104. New York: Garland Publishing, 1981.

Howells, John G., and M. Livia Osborn. *A Reference Companion to the History of Abnormal Psychology.* 2 vols. Westport, Conn.: Greenwood Press, 1984.

Hoy, Suellen M., and Michael C. Robinson. *Public Works History in the United States: A Guide to the Literature.* By the Public Works Historical Society. 477pp. Nashville, Tenn.: American Association for State and Local History, 1982.

Hughes, Barbara. *Catalog of the Scientific Apparatus at the College of Charleston, 1800-1940.* Edited with additional material by Ralph Melnick. 94pp. Charleston, S.C.: College of Charleston Library Associates, 1980.

Humphrey, Harry B. *Makers of North American Botany.* 265pp. New York: Ronald Press Co., 1961.

Hunt Institute for Botanical Documentation. *Guide to the Botanical Records and Papers in the Archives of the Hunt Institute*, Part 1 (A-B), Part 2 (C-F), Part 3 (G-H). Compiled by Michael T. Stieber and Anita L. Karg. Pittsburgh: Hunt Institute for Botanical Documentation, 1981, 1984, 1988. Also published in *Huntia*, vol. 4, no.1 (1981), vol. 5, no.3 (1984), vol. 8, no.1 (1989).

Hurt, R. Douglas, and Mary Ellen Hurt. *The History of Agricultural Science and Technology: An International Annotated Bibliography.* 485pp. Bibliographies of the History of Science and Technology, 20. New York: Garland Publishing, 1994.

International Directory of Company Histories. Edited by Thomas Derdak, and others. Chicago: St. James Press, vol. 1-, 1988-.

Ireland, Norma. *Index to Scientists of the World from Ancient to Modern Times: Biographies and Portraits.* 662pp. Useful Reference Series, 90. Boston: Faxon, 1962.

[Isis] "Critical Bibliography of the History of Science and Its Cultural Influences." *Isis*, 1913-.

Isis Cumulative Bibliography: A Bibliography of the History of Science Formed from Isis Critical Bibliographies ... London: Mansel, in conjunction with the History of Science Society, 1971-. Critical Bibliographies 1-90, 1913-1965, in 6 vols.; 91-100, 1966-1975, in 2 vols.; 101-110, 1976-1985, in 2 vols.

Jayawardene, S. A. *Reference Books for the Historian of Science: A Handlist*. 229pp. Occasional Publications, 2. London: Science Museum, 1982.

Joint Committee on Archives of Science and Technology. *Understanding Progress as Process: Documentation of the History of Post-War Science and Technology in the United States*. Edited by Clark A. Elliott. 64pp. Chicago: Distributed by the Society of American Archivists, 1983. Also: Clark A. Elliott. "Joint Committee on Archives of Science and Technology (JCAST): Summary from the Final Report." *Isis* 75 (1984): 158-162.

Karpinski, Louis Charles. *Bibliography of Mathematical Works Printed in America through 1850*. 697pp. Ann Arbor: University of Michigan Press; London: Milford, Oxford University Press, 1940. Supplements 1 and 2, *Scripta Mathematica* 8 (December 1941): 233-236 and 11 (June 1945): 173-177.

Kaufman, Martin, editor-in-chief. *Dictionary of American Nursing Biography*. 462pp. Westport, Conn.: Greenwood Press, 1988.

Kaufman, Martin, Stuart Galishoff, and Todd L. Savitt, eds. *Dictionary of American Medical Biography*. 2 vols. Westport, Conn.: Greenwood Press, 1984.

Kay, Lily E. *Molecules, Cells and Life: An Annotated Bibliography of Manuscript Sources on Physiology, Biochemistry, and Biophysics, 1900-1960, in the Library of the American Philosophical Society*. 95pp. APS Library Publication, 14. Philadelphia: American Philosophical Society Library, 1989.

Kelly, Howard A., and Walter L. Burrage. *Dictionary of American Medical Biography*. 1364pp. New York: Appleton, 1928. Reprinted. Boston: Milford House, 1971.

Kemper, Robert V., and John F.S. Phinney. *The History of Anthropology: A Research Bibliography*. 212pp. New York: Garland Publishing, 1977.

Kiger, Joseph C., ed. *Research Institutions and Learned Societies*. 551pp. Greenwood Encyclopedia of American Institutions, 5. Westport, Conn.: Greenwood Press, 1982.

Knight, David M. *Sources for the History of Science, 1660-1914*. 223pp. Sources of History: Studies in the Uses of Historical Evidence. Ithaca, N.Y.: Cornell University Press, 1975.

Kohlstedt, Sally Gregory. "Biographical Directory of the AAAS, 1848-60." Appendix to *The Formation of the American Scientific Community: The American Association for the Advancement of Science, 1848-1860*. Urbana: University of Illinois Press, 1976.

Kohlstedt, Sally Gregory, and Margaret W. Rossiter, eds. "Historical Writing on American Science" [issue title]. *Osiris*, 2nd ser., volume 1 (1985): 1-321. Reissued. Philadelphia: History of Science Society, 1985. Baltimore and London: Johns Hopkins University Press, 1986. Content: Rossiter and Kohlstedt, "Introduction." Sally Gregory Kohlstedt, "Institutional History." John Harley Warner, "Science in Medicine." Ronald L. Numbers, "Science and Religion." Sharon Gibbs Thibodeau, "Science in the Federal Government." Mott Greee, "History of Geology." Marc Rothenberg, "History of Astronomy." John Servos, "History of Chemistry." Jane Maienschein, "History of Biology." Albert Moyer, "History of Physics." Hamilton Cravens, "History of the Social Sciences." Clara Sue Kidwell, "Native Knowledge in the Americas." George Wise, "Science and Technology." Alex Roland, "Science and War." Margaret W. Rossiter, "Science and Public Policy since World War II." Clark A. Elliott, "Bibliographies, Reference Works, and Archives."

Kragh, Helge. *An Introduction to the Historiography of Science*. 235pp. Cambridge: Cambridge University Press, 1987.

Kuhn, Thomas S., John L. Heilbron, Paul Forman, and Lini Allen. *Sources for History of Quantum Physics: An Inventory and Report*. 176pp. Memoirs of American Philosophical Society, 68. Philadelphia: American Philosophical Society, 1967.

Larkin, Robert P., and Gary L. Peters. *Biographical Dictionary of Geography*. 361pp. New York: Greenwood Press, 1993.

Layton, Edwin T., Jr. *A Regional Union Catalogue of Manuscripts Relating to the History of Science and Technology Located in Indiana, Michigan, and Ohio*. 215pp. Program in the History of Science and Technology, 1. Cleveland: Case Western Reserve University, 1971.

Lick Observatory. *Preliminary Finding Aid to the Archives of the Lick Observatory, from the Card Catalogs Maintained by the Lick Observatory Archives Staff, University of California at Santa Cruz*. Edited by David DeVorkin. 56pp. AIP Publication R-293; National Catalog of Sources for History of Physics, report 5. New York: American Institute of Physics, 1980.

Mann, Thomas L. ed. *Biographical Directory of Anthropologists before 1920*. Compiled by Library - Anthropology Resource Group. 245pp. Garland Reference Library of the Humanities, 439. New York: Garland Publishing, 1988.

Massachusetts Institute of Technology. Institute Archives and Special Collections. *Selective Guide to the Collections*. 108pp. Cambridge, Mass.: Institute Archives and Special Collections, 1988.

May, Kenneth Ownsworth. *Bibliography and Research Manual on the History of Mathematics*. 818pp. Toronto: University of Toronto Press, 1973.

McGraw-Hill Modern Scientists and Engineers. 3 vols. New York: McGraw-Hill, 1980.

McGrew, Roderick, with Margaret B. McGrew. *Encyclopedia of Medical History*. 400pp. New York: McGraw-Hill, 1985.

McIver, Tom. *Anti-Evolution: An Annotated Bibliography*. 385pp. Jefferson, N.C.: McFarland, 1988. Reprint (with new preface). *Anti-Evolution: A Reader's Guide to Writings Before and After Darwin*. Baltimore: Johns Hopkins University Press, 1992.

Meisel, Max. *A Bibliography of American Natural History: The Pioneer Century 1769-1865; The Role Played by the Scientific Societies; Scientific Journals; Natural History Museums and Botanic Gardens; State Geological and Natural History Surveys; Federal Exploring Expeditions in the Rise and Progress of American Botany, Geology, Mineralogy, Paleontology and Zoology*. 3 vols. Brooklyn, N.Y.: Premier Publishing Co., 1924-1929. Reprint. New York: Hafner Publishing Co., 1967.

Melton, J. Gordon, and Isotta Poggi. *Magic, Witchcraft, and Paganism in America: A Bibliography*. 2nd ed. 408pp. Religious Information Systems, 3. New York: Garland Publishing, 1992.

Mendel Newsletter: Archival Sources for the History of Genetics and Allied Sciences. Philadelphia: American Philosophical Society, 1968-.

Merrimack Valley Textile Museum. *The Merrimack Valley Textile Museum: A Guide to the Manuscript Collections*. Edited by Helena E. Wright. 378pp. New York: Garland Publishing, 1983.

Miles, Wyndham D., ed. *American Chemists and Chemical Engineers*. 544pp. Washington, D.C.: American Chemical Society, 1976.

Miller, Genevieve, ed. *Bibliography of the History of Medicine of the United States and Canada, 1939-1960*. 428pp. Baltimore: Johns Hopkins University Press, 1964. Consolidation of annual bibliographies published in *Bulletin of the History of Medicine*. Annual issues, 1961-1964.

Miller, Gordon L. *The History of Science: An Annotated Bibliography*. 193pp. Magill Bibliographies. Pasadena, Calif.: Salem Press, 1992.

Mitcham, Carl, and William F. Williams, eds. *The Reader's Adviser*: Volume 5: *The Best in Science, Technology, and Medicine*. 14th ed. 975pp. New Providence, N.J.: Bowker, 1994.

Molloy, Peter M. *The History of Metal Mining and Metallurgy: An Annotated Bibliography*. 319pp. Bibliographies of the History of Science and Technology, 12. New York: Garland Publishing, 1986.

Morton, Leslie T., and Robert J. Moore. *A Bibliography of Medical and Biomedical Biography*. 2nd ed. 333pp. Aldershot, Hants, England: Scolar Press; Brookfield, Vt.: Ashgate Publishing, 1994.

Morton, Leslie T., and Jeremy M. Norman, eds. *Morton's Medical Bibliography: An Annotated Check-List of Texts Illustrating the History of Medicine (Garrison and Morton)*. Edition edited by Jeremy M. Norman. 5th ed. 1243pp. Aldershot, Hants, England: Scolar Press; Brookfield, Vt.: Gower, 1991.

Mount, Ellis, and Barbara A. List. *Milestones in Science and Technology: The Ready Reference Guide to Discoveries, Inventions, and Facts*. 2nd ed. 206pp. Phoenix, Ariz.: Oryx, 1994

Multhauf, Robert, comp. *The History of Chemical Technology: An Annotated Bibliography*. 299pp. Bibliographies of the History of Science and Technology, 5. New York: Garland Publishing, 1983.

Nasrallah, Wahib. *United States Corporation Histories: A Bibliography, 1965-1990*. 2nd ed. 511pp. Garland Reference Library of Social Science, 807. New York: Garland Publishing, 1991.

National Anthropological Archives. *Catalog to Manuscripts at the National Anthropological Archives*. Department of Anthropology, Museum of Natural History, Smithsonian Institution. 4 vols. Boston: G.K. Hall, 1975.

National Anthropological Archives. *Guide to the National Anthropological Archives, Smithsonian Institution*. By James R. Glenn. 314pp. Washington, D.C.: National Museum of Natural History, Smithsonian Institution, 1992.

National Museum of History and Technology. *Guide to Manuscript Collections in the National Museum of History and Technology.* 143pp. Archives and Special Collections of the Smithsonian Institution, 3. Washington, D.C.: Smithsonian Institution Press, 1978.

Nebeker, Frederik. *Astronomy and the Geophysical Tradition in the United States in the Nineteenth Century: A Guide to the Manuscript Sources in the Library of the American Philosophical Society.* 114pp. APS Library Publication, 16. Philadelphia: American Philosophical Society Library, 1991.

New York Botanical Garden. Library. *Catalog of the Manuscript and Archival Collections, and an Index to the Correspondence of John Torrey.* Compiled by Sara Lanley, and others. 473pp. Boston: G.K. Hall, 1973.

New York Botanical Garden. Library. *Guide to the Archives and Manuscripts of the New York Botanical Garden.* Prepared by Terry Collins, and Steven P. Johnson. 91pp. New York: New York Botanical Garden Library, 1983.

Notable Twentieth Century Scientists. Edited by Emily J. McMurray. 4 vols. Detroit: Gale Research, 1995.

Ogden, Gerald, comp. *The United States Forest Service: A Historical Bibliography, 1876-1972.* 439pp. Davis, Calif.: Agricultural History Center, University of California-Davis, 1976.

Ogilvie, Marilyn Bailey. *Women in Science: Antiquity Through the Nineteenth Century. A Biographical Dictionary and Annotated Bibliography.* 254pp. Cambridge, Mass.: MIT Press, 1986.

Overmier, Judith A. *The History of Biology: A Selected, Annotated Bibliography.* 157pp. Bibliographies of the History of Science and Technology, 15. New York: Garland Publishing, 1989.

Owings, Loren C. *Environmental Values, 1860-1972: A Guide to Information Sources.* 324pp. Gale Information Guide Library, 4. Detroit: Gale Research, 1976.

Parascandola, John, and Elizabeth Keeney, comps. *Sources in the History of American Pharmacy*. 59pp. Madison, Wis.: American Institute of the History of Pharmacy, 1983.

Parkinson, Claire L. *Breakthroughs: A Chronology of Great Achievements in Science and Mathematics, 1200-1930*. 576pp. Boston: G.K. Hall, 1985.

Pelletier, Paul A. *Prominent Scientists: An Index to Collective Biographies*. 3rd ed. 353pp. New York: Neal-Schumann, 1994.

Pisano, Dominick A., and Cathleen S. Lewis, eds. *Air and Space History: An Annotated Bibliography*. Sponsored by National Air and Space Museum, Smithsonian Institution. 571pp. Garland Reference Library of the Humanities, 834. New York: Garland Publishing, 1988.

Poggendorff, Johann C. [and others]. *Biographisch-literarisches Handworterbuch der exakten Naturwissenschaften*. Leipzig: Barth, 1863-1904; Leipzig: Verlag Chemie, 1925-1940; Berlin: Akademie-Verlag, 1955-. Title varies. Reprint, Band 1-6, to 1931. Ann Arbor, Mich.: Edwards, 1945.

Porter, Roy, consultant editor. *The Biographical Dictionary of Scientists*. 2nd ed. 891pp. New York: Oxford University Press, 1994. First edition published in six volumes, each devoted to one area of science, 1983-1985.

Porter, Roy. *The Earth Sciences: An Annotated Bibliography*. 192pp. Bibliographies of the History of Science and Technology, 3. New York: Garland Publishing, 1983.

Pure and Applied Science Books, 1876-1982. 6 vols. New York: R.R. Bowker, 1982. Arranged by subject, with author and title indexes.

Reginald, R(obert). *Science Fiction and Fantasy Literature: A Checklist 1700-1974*. 2 vols. Detroit: Gale Research, 1979.

Reginald, R(obert). *Science Fiction and Fantasy Literature, 1975-1991: A Bibliography of Science Fiction, Fantasy, and Horror Fiction*

Books and Nonfiction Monographs. 1512pp. Detroit: Gale Research, 1992.

Reingold, Nathan, ed. Symposium on Documents, Interpretations, and the History of the Sciences [Proceedings]. *Proceedings of American Philosophical Society* 134 (1990): 338-441.

Rider, Robin E., and Henry E. Lowood. *Guide to Sources in Northern California for History of Science and Technology*. 194pp. Berkeley Papers in History of Science, 10. Berkeley: Office for History of Science and Technology, University of California, 1985.

Rink, Evald. *Technical Americana: A Checklist of Technical Publications Printed before 1831*. Sponsored by Eleutherian Mills Historical Library. 776pp. Millwood, N.Y.: Kraus International Publications, 1981.

Rockefeller Archive Center. *A Guide to Archives and Manuscripts at the Rockefeller Archive Center*. Compiled by Emily J. Oakhill, and Kenneth W. Rose. 77pp. North Tarrytown, N.Y.: Rockefeller Archive Center, 1989.

Roland, Alex. *A Guide to Research in NASA History*. 7th ed. 59pp. Washington, D.C.: History Office, National Aeronautics and Space Administration, 1984.

Rossiter, Margaret, comp. *A List of References for the History of Agricultural Science in America*. 62pp. Davis, Calif.: Agricultural History Center, University of California-Davis, 1980.

Rothenberg, Marc. *The History of Science and Technology in the United States: A Critical and Selective Bibliography*. 2 vols. Bibliographies of the History of Science and Technology, 2, 17. New York: Garland Publishing, 1982, 1993.

Roysdon, Christine, and Linda A. Khatri. *American Engineers of the Nineteenth Century: A Biographical Index*. 247pp. Garland Library of Social Science, 53. New York: Garland Publishing, 1978.

Rutkow, Ira M. *The History of Surgery in the United States, 1775-1900.* Norman Surgery Series, 2. San Francisco: Norman Pub., 1988-[in progress]

Sackheim, Donald E., comp. *Historic American Engineering Record Catalog 1976.* 193pp. Washington, D.C.: National Park Service; Government Printing Office, 1976.

Sammons, Vivian Ovelton. *Blacks in Science and Medicine.* 293pp. New York: Hemisphere Publishing, 1990.

Sarjeant, William A.S. *Geologists and the History of Geology: An International Bibliography from the Origins to 1978.* 5 vol. New York: Arno Press, 1980. Supplement, 1979-1984 and Additions. 2 vols. Malabar, Fla.: R.E. Krieger, 1987.

Sarton, George. *A Guide to the History of Science: A First Guide for the Study of the History of Science with Introductory Essays on Science and Tradition.* 316pp. Waltham, Mass.: Chronica Botanica Co., 1952.

Schapsmeier, Edward L., and Frederick H. Schapsmeier. *Encyclopedia of American Agricultural History.* 467pp. Westport, Conn.: Greenwood Press, 1975.

Schatzberg, Walter, Ronald A. Waite, and Jonathan K. Johnson, eds. *The Relations of Literature and Science: An Annotated Bibliography of Scholarship, 1880-1980.* 458pp. New York: Modern Language Association of America, 1987.

Schlebecker, John T. *Bibliography of Books and Pamphlets on the History of Agriculture in the United States, 1607-1967.* 183pp. Santa Barbara, Calif.: Clio Press, 1969.

Schlessinger, Bernard S., and June H. Schlessinger, eds. *The Who's Who of Nobel Prize Winners.* 2nd ed. 234pp. Phoenix, Ariz.: Oryx, 1990.

Schodek, Daniel L. *Landmarks in American Civil Engineering.* 383pp. Cambridge, Mass.: MIT Press, 1987.

Scholnick, Robert S. "Bibliography: American Literature and Science through 1989." In *American Literature and Science*, edited by Robert S. Scholnick, pp.251-272. Lexington: University Press of Kentucky, 1992.

Scripps Institution of Oceanography Archives. *Selected List of Collections*. 53pp. La Jolla, Calif.: Archives of the Scripps Institution of Oceanography, University of California, San Diego, 1989.

Sharrer, G. Terry, comp. *1001 References for the History of American Food Technology*. 103pp. Davis, Calif.: Agricultural History Center, University of California-Davis, 1978.

Shiers, George. *Bibliography of the History of Electronics*. 323pp. Metuchen, N.J.: Scarecrow Press, 1972.

Siegel, Patricia Joan, and Kay Thomas Finley. *Women in the Scientific Search: An American Bio-Bibliography, 1724-1979*. 399pp. Metuchen, N.J.: The Scarecrow Press, 1985.

Smith, Curtis C., ed. *Twentieth-Century Science-Fiction Writers*. 2nd ed. 933pp. Chicago: St. James Press, 1986.

Smith, Murphy D. *Historical American Sketches: An Illustrated Guide to Sketches in the Manuscript Collections of the American Philosophical Society*. 275pp. Boston: G.K. Hall, 1984.

Smithsonian Institution. *Guide to the Smithsonian Archives*. 431pp. Archives and Special Collections of the Smithsonian Institution, 4. Washington, D.C.: Smithsonian Institution Press, 1983.

Sokal, Michael M., and Patrice A. Rafail. *A Guide to Manuscript Collections in the History of Psychology and Related Areas*. 212pp. Millwood, N.Y.: Kraus International Publications, 1982.

Stannard, Jerry. "Early American Botany and Its Sources." In *Bibliography & Natural History: Essays Presented at a Conference Convened in June 1964*, pp.73-102. Lawrence: University of Kansas Libraries, 1966.

Stanton, William. *American Scientific Explorations, 1803-1860: Manuscripts in Four Philadelphia Libraries*. 140pp. APS Library Publication, 15. Philadelphia: American Philosophical Society, 1991.

Stapleton, Darwin H. *Accounts of European Science, Technology and Medicine Written by American Travellers Abroad, 1735-1860, in the Collections of the American Philosophical Society*. 48pp. APS Library Publication no. 9. Philadelphia: American Philosophical Society, 1985.

Stapleton, Darwin H. *The History of Civil Engineering Since 1600: An Annotated Bibliography*. 232pp. Bibliographies of the History of Science and Technology, 14. New York: Garland Publishing, 1986.

Stroud, Richard H., ed. *National Leaders of American Conservation*. 432pp. Washington, D.C.: Smithsonian Institution Press, 1985.

Sturchio, Jeffrey L., ed. *Corporate History and the Chemical Industries: A Resource Guide*. 53pp. Center for History of Chemistry Publication, 4. Philadelphia: Center for History of Chemistry, 1985.

[Technology and Culture] "Current Bibliography in the History of Technology, 1962-." *Technology and Culture*, vol. 5-, 1964-.

Teich, Albert H., and Jill H. Pace, with others. *Science and Technology in the USA*. 408pp. Longman Guide to World Science and Technology, 5. Harlow, Essex: Longman, 1986.

Thompson, Susan J. *A Chronology of Geological Thinking from Antiquity to 1899*. 320pp. Metuchen, N.J.: Scarecrow Press, 1988.

Travers, Bridget, ed. *World of Invention*. 770pp. Detroit: Gale Research, 1994.

Travers, Bridget, ed. *World of Scientific Discovery*. 776pp. Detroit: Gale Research, 1994.

Tselos, George D., and Colleen Wickey, comps. *A Guide to Archives and Manuscript Collections in the History of Chemistry and*

Chemical Technology. 198pp. Center for History of Chemistry Publication, 7. Philadelphia: Center for History of Chemistry, 1987.

Tucher, Andrea J., comp. *Agriculture in America, 1622-1860: Printed Works in the Collections of the American Philosophical Society, the Historical Society of Pennsylvania, the Library Company of Philadelphia*. 212pp. Americana to 1860, 2. New York: Garland Publishing, 1984.

Tucher, Andrea J., comp. *Natural History in America, 1609-1860: Printed Works in the Collections of the American Philosophical Society, the Historical Society of Pennsylvania, the Library Company of Philadelphia*. 287pp. Americana to 1860, 4. New York: Garland Publishing, 1985.

U.S. Congress. House. Committee on Science and Technology. *Bibliography of Studies and Reports on Science Policy and Related Topics, 1945-1985: Report*. Prepared for the Task Force on Science Policy, Committee on Science and Technology, House of Representatives, Ninety-ninth Congress, Second Session. 221pp. Science Policy Study, Background Report, no. 2 part A. Superintendent of Documents no. Y4.Sci2.99/HH. Washington, D.C.: Government Printing Office, 1987.

U.S. National Academy of Sciences. *Biographical Memoirs*, vol. 1-. Washington, D.C.: The Academy, 1877-.

U.S. National Aeronautics and Space Administration. *Aeronautics and Astronautics: An American Chronology of Science and Technology in the Exploration of Space, 1915-1960*. By Eugene M. Emme. 240pp. Washington, D.C., 1961. Continued in annual or multi-year volumes, covering the period to 1985 (last published 1990), under varying titles, chiefly, *Astronautics and Aeronautics*, in NASA History Series.

University of Toronto. Institute for the History and Philosophy of Science and Technology. *Science in Society: An Annotated Guide to Resources*. Project directed by David McGee, with contributors. 82pp. Toronto: Wall & Thompson, 1989.

Van Keuren, David K. *"The Proper Study of Mankind:" An Annotated Bibliography of Manuscript Sources on Anthropology and Archeology in the Library of the American Philosophical Society.* 79pp. APS Library Publication, 10. Philadelphia, Penn.: American Philosophical Society, 1986.

Viney, Wayne, Michael Wertheimer, and Marilyn Lou Wertheimer. *History of Psychology: A Guide to Information Sources.* 502pp. Gale Information Guide Library. Detroit: Gale Research, 1979.

Warnow, Joan N. *A Selection of Manuscript Collections at American Repositories.* 73pp. AIP National Catalog of Sources for History of Physics, Report 1. New York: American Institute of Physics, 1969.

Warnow, Joan N., and others. *Guidelines for Records Appraisal at Major Research Facilities: Selection of Permanent Records of DOE Laboratories: Institutional Management and Policy, and Physics Research.* 20 [+11]pp. AIP Publication R-324. New York: Center for History of Physics, American Institute of Physics, 1985.

Warnow-Blewett, Joan. "Documenting Recent Science: Progress and Needs." *Osiris* 2nd ser. 7 (1992): 267-198.

Warnow-Blewett, Joan, and Jurgen Teichmann. *Catalog of Sources for History of Solid State Physics.* 156pp. AIP International Catalog of Sources for History of Physics and Allied Sciences, Report 6. New York: American Institute of Physics, 1992.

Wasson, Tyler, ed. *Nobel Prize Winners: An H.W. Wilson Biographical Dictionary.* 1165pp. New York: H.W. Wilson, 1987.

Watson, Robert Irving, ed. *Eminent Contributors to Psychology.* 2 vols. New York: Springer Publishing Co., 1974-1976.

Watson, Robert Irving. *The History of Psychology and the Behavioral Sciences: A Bibliographic Guide.* 241pp. New York: Springer, 1978.

Weisbard, Phyllis Holman, with Rima Apple, eds. *The History of Women and Science, Health, and Technology: A Bibliographic*

Guide to the Professions and the Disciplines. 2nd ed. 100pp. Madison, Wis.: University of Wisconsin System Women's Studies Librarian, Memorial Library, 1993.

Wellcome Institute for the History of Medicine, London. *Subject Catalogue of the History of Medicine and Related Sciences.* 18 vols. Munchen: Kraus International, 1980.

Wheaton, Bruce R. *Inventory of Sources for History of Twentieth-Century Physics (ISHTCP): Report and Microfiche Index to 700,000 Letters.* 61 microfiches and 294pp. report. Stuttgart: Verlag fur die Geschichte der Naturwissenschaften und der Technik, 1992.

Wheaton, Bruce R., and John L. Heilbron. *An Inventory of Published Letters to and from Physicists, 1900-1950.* 102pp. and 4 microfiches. Berkeley Papers in History of Science, 6. Berkeley: Office for History of Science and Technology, University of California, 1982.

Who Was Who in American History: Science and Technology. 688pp. Chicago: Marquis's Who's Who, 1976.

Williams, Trevor I., ed. *Collins Biographical Dictionary of Scientists.* 4th ed. 602pp. Glasgow: HarperCollins, 1994. Previous editions as *Biographical Dictionary of Scientists.*

World Who's Who in Science: A Biographical Dictionary of Notable Scientists from Antiquity to the Present. Edited by Allen G. Debus. 1855pp. Chicago: Marquis-Who's Who, 1968.

Zusne, Leonard. *Biographical Dictionary of Psychology.* 563pp. Westport, Conn.: Greenwood Press, 1984. Previous edition as *Names in the History of Psychology.* 1975.

Index:
Chronology and Scientist Cohorts

The index contains references to personal and institutional names and to names of devices (e.g., spacecraft) and projects, the Nobel Prize, journal titles (when the subject of an entry), titles of legislative acts, and the like. Organizational names are given in normal word order (e.g., University of Chicago). Most federal agencies are filed under "U.S." For guidance in searching by subject, see the topical headings that accompany each entry in the Chronology. The Research Guide and Bibliography are not included in the index.

Abbe, Cleveland, 96, 102, 103, 124, 136, 193, 371
Abbot, Charles Greeley, 388
Abbot, Henry L., 93
Abbot, John, 37
Abel, John Jacob, 164, 174, 184, 213, 380
Abelson, Philip Hauge, 256, 429
Academy of Natural Sciences of Philadelphia, 47, 50, 75
Academy of Science of St. Louis, 89
Academy of Sciences of Cracow, 151
Acosta, Jose, 4
Acoustical Society of America, 226, 233, 298
Acta Crystallographica, 276
Adams, Edward F., 222
Adams, John, 28
Adams, John Quincy, 56, 59, 62, 71, 154
Adams, Leason Heberling, 211, 410
Adams, Roger, 192, 410
Adams, Walter S., 188, 212, 399
Adrain, Robert, 42, 45, 56, 358

Advisory Committee on X-Ray and Radium Protection, 226
Aerospace Corporation, 306
Agassiz Association, 110
Agassiz, Alexander, 107, 371
Agassiz, Louis, 76, 79, 82, 89, 91, 92, 95, 96, 97, 100, 106, 108, 109, 364
Agramonte y Simoni, Aristides, 152
Ahlquist, Jon E., 342
Aiken, Howard Hathaway, 266, 420
Air Quality Act, 319
Aitken, Robert G., 195, 234, 380
Akeley, Carl Ethan, 380
Albert, Abraham Adrian, 254, 429
Albright, Fuller, 277, 420
Aldrin, Edwin (Buzz), 323
Alexanderson, Ernst F.W., 167, 399
Allee, Warder Clyde, 251, 280, 410
Allen, Edgar, 213, 410
Allen, Eugene Thomas, 380
Allen, Joel Asaph, 106, 124, 371
Allison, Samuel King, 244, 420
Allison Commission, 126
Allport, Gordon Willard, 248, 420

Alpher, Ralph, 278, 436
Alter, David, 88
Altman, Sidney, 344
Aluminum Company of America, 129
Alvarez, Luis W., 255, 320, 337, 344, 429
Alvarez, Walter, 337
American Academy of Arts and Sciences, 28, 29, 30, 34, 35, 36, 40, 67, 79
American Anthropological Association, 157
American Anthropologist, 133
American Antiquarian Society, 54
American Association for Cancer Research, 168
American Association for the Advancement of Science, 69, 77, 92, 107, 109, 122, 124, 143, 153, 186, 204, 216, 251, 252
American Association of Anatomists, 132, 213
American Association of Economic Entomologists, 135, 289
American Association of Engineering Societies, 204, 336
American Association of Neurological Surgeons, 236
American Association of Physical Anthropologists, 224
American Association of Physics Teachers, 231
American Association of Scientific Workers, 251
American Association of Variable Star Observers, 180
American Association of Zoological Parks and Aquariums, 262
American Astronautical Society, 291
American Astronomical Society, 150, 239, 270
American Botanical Club, 124
American Botanist, 155
American Breeders Association, 185
American Bryological Society, 149
American Cancer Society, 185
American Chemical Journal, 118
American Chemical Society, 112, 118, 130, 147, 154, 239

American Council on Education, 243
American Crystallographic Association, 282
American Dental Association, 91
American Engineering Council, 204
American Entomological Society, 91
American Ethnological Society, 71
American Eugenics Society, 217
American Farmer, 53
American Federation of Mineralogical Societies, 275
American Federation of the Mathematical and the Natural Sciences, 168
American Forestry Association, 122
American Genetic Association, 185
American Geographical Society, 190
American Geographical Society of New York, 83
American Geological Institute, 277
American Geological Society, 52
American Geologist, 133
American Geophysical Union, 200
American Historical Association. History of science section, 200
American Institute for the History of Pharmacy, 260
American Institute of Aeronautics and Astronautics, 312
American Institute of Biological Sciences, 277
American Institute of Chemical Engineers, 170
American Institute of Electrical Engineers, 126, 312
American Institute of Mining and Metallurgical Engineers, 106
American Institute of Nutrition, 224
American Institute of Physics, 233, 296
American Journal of Anatomy, 155
American Journal of Mathematics, 116
American Journal of Physical Anthropology, 224
American Journal of Physiology, 149
American Journal of Psychology, 130
American Journal of Science and Arts, 52
American Mathematical Monthly, 143

Index

American Mathematical Society, 134, 139, 143, 153, 196
American Medical and Philosophical Register, 46
American Medical Association, 77, 171
American Men of Science, 165
American Meteorological Society, 200
American Metrological Society, 108
American Microscopical Society, 116
American Midland Naturalist, 173
American Mineralogical Journal, 46
American Mineralogical Society, 38
American Morphological Society, 137
American Museum Journal, 153
American Museum of Natural History (New York), 102, 122, 153, 178
American Mycological Society, 160, 166
American Naturalist, 100
American Nature Study Society, 171
American Nuclear Society, 291
American Ornithologists' Union, 124, 129
American Pharmaceutical Association, 85
American Philosophical Society, 17, 24, 25, 31, 34, 43, 49, 66
American Physical Society, 142, 150, 185, 233, 303
American Physiological Society, 130
American Phytopathological Society, 170
American Psychological Association, 141, 244, 245
American Public Health Association, 106, 119, 165
American Rocket Society, 231, 312
American Scientific Affiliation, 260
American Social Science Association, 97
American Society for Biochemistry and Molecular Biology, 166
American Society for Control of Cancer, 185
American Society for Experimental Pathology, 185
American Society for Horticultural Science, 160
American Society for Microbiology, 150
American Society for Pharmacology and Experimental Therapeutics, 171
American Society for Promoting and Propagating Useful Knowledge 24
American Society for Psychical Research, 127
American Society for the Dissemination of Science, 200
American Society for X-ray and Electron Diffraction, 282
American Society Held at Philadelphia for Promoting Useful Knowledge, 24
American Society of Agronomy, 168
American Society of Biological Chemists, 166
American Society of Civil Engineers, 84, 100
American Society of Human Genetics, 277
American Society of Limnology and Oceanography, 277
American Society of Mammalogists, 201
American Society of Mechanical Engineers, 121
American Society of Medical History, 212
American Society of Naturalists, 124
American Society of Plant Physiologists, 212
American Society of Professional Geographers, 161
American Society of Protozoologists, 275
American Society of Zoologists, 137
American Sociological Association, 164
American Statistical Association, 67, 133
American Telephone and Telegraph Company, 114, 216, 311
American Union Academy of Literature, Science and Art, 102
American University, 198
Ames, Joseph Sweetman, 380
Ames, Nathaniel, 12
Analyst, or Mathematical Companion, 45
Anders, William A., 321
Anderson School of Natural History, 108

Anderson, Carl David, 236, 244, 248, 429
Anderson, Edgar, 286, 420
Anderson, Herbert L., 263
Anderson, Philip Warren, 333, 436
Andrews, Roy Chapman, 399
Anfinsen, Christian Boehmer, 305, 326, 436
Angell, James Rowland, 388
Annals of Mathematics Studies, 258
Annual of Scientific Discovery, 82
Apollo (spacecraft), 309, 318, 321, 323, 331
Argonne National Laboratory, 271
Armstrong, Edwin H., 196, 239, 410
Armstrong, Neil, 323
Arnold, Harold De Forest, 216, 399
Arthur D. Little Company, 128, 179
Asaro, Frank, 337
Aserinsky, Eugene, 289
Asilomar conference, 330
Associated Universities, Inc., 272, 299
Association for Maintaining the American Women's Table at the Zoological Station (at Naples), 149
Association for Women in Science, 325
Association of Academies of Science, 217
Association of American Agricultural Colleges and Experiment Stations, 130
Association of American Geographers, 161
Association of American Geologists and Naturalists, 69, 70, 77
Association of Official Agricultural Chemists, 126
Association to Aid Scientific Research by Women, 149
Astronomical and Astrophysical Society of America, 150
Astronomical Diary and Almanac, 12
Astronomical Journal, 81
Astronomical Society of the Pacific, 135
Astronomy and Astro-Physics, 139
Astrophysical Journal, 144
Atanasoff, John V., 262
Atoms for Peace program, 290
Atwater, Wilbur Olin, 110, 371

Atwood, Wallace Walter, 203, 388
Audubon, John James, 52, 58, 84, 360
Audubon Magazine, 150
Austin, Louis Winslow, 170, 388
Avery, Oswald T., 265, 295, 399
Axelrod, Julius, 323, 429
Ayers, S. Henry, 175

Baade, Walter, 286, 307, 410
Babcock, Ernest Brown, 196, 273, 399
Babcock, Harold D., 222, 286
Babcock, Horace W., 286
Babcock, Stephen Moulton, 137, 371
Bache, Alexander Dallas, 64, 72, 82, 85, 95, 105, 364
Backus, John, 296
Badger, Richard McLean, 237, 420
Baekeland, Leo H., 169, 380
Bailey, Liberty Hyde, Jr., 132, 160, 380
Bailey, Loring Woart, 371
Bailey, Solon Irving, 139, 376
Baird, Spencer Fullerton, 90, 105, 131, 367
Balchen, Bernt, 228
Baldwin, James Mark, 143
Ballard, Harlan H., 110
Baltimore, David, 324, 331, 443
Bancroft, Wilder Dwight, 147, 239, 388
Bandy, Orville Lee, 436
Banister, John, 7, 9, 353
Barbour, Henry Gray, 183, 410
Barbour, Thomas, 195
Bardeen, John, 275, 296, 299, 326, 429
Barghoorn, Elso S., 320
Barnard, Edward Emerson, 140, 198, 210, 380
Baron, George, 42
Barrell, Joseph, 388
Barringer, Daniel, 164
Bartlett, Paul D., 235
Bartol Research Foundation, 212
Barton, Benjamin Smith, 37, 38, 41, 43, 44, 357
Barton, Otis, 241
Bartram, John, 14, 15, 17, 18, 19, 23, 24, 354
Bartram, William, 23, 26, 35, 41, 355
Baruch-Compton-Conant Commission (rubber supply), 263
Barus, Carl, 380

Index

Bassler, Raymond Smith, 399
Bateman, Alan Mara, 201, 410
Bateman, Harry, 399
Bates, Marston, 281, 294, 429
Bates, Robert W., 237
Bateson, Gregory, 420
Battelle Memorial Institute, 226
Bauer, Louis Agricola, 147, 161, 388
Beach, Alfred, 75
Beadle, George Wells, 259, 301, 420
Beal, Foster E.L., 121
Beams, Jesse W., 246
Bearden, Joyce Alvin, 216
Beatley, Ralph, 261
Beaumont, William, 56, 360
Becker, George Ferdinand, 376
Beckwith, Jonathan, 321
Beebe, (Charles) William, 241, 399
Begle, Edward Griffith, 429
Bekesy, Georg von, 308, 420
Bell, Alexander Graham, 113, 125, 190, 207, 376
Bell, Eric Temple, 248, 399
Bell Telephone Laboratories, 216, 220, 221, 234, 240, 250, 255, 275, 291, 293
Bellevue Hospital medical school (New York), 117
Benacerraf, Baruj, 337, 436
Benedict, Francis Gano, 168, 212, 388
Benedict, Ruth, 240, 410
Bennett, Floyd, 218
Bennett, Hugh H., 243, 399
Bensley, Robert Russell, 240, 388
Benson, Andrew A., 279
Berg, Ernst J., 148
Berg, Paul, 296, 329, 337, 440
Bergman, Stefan, 322, 420
Bergmann, Max, 410
Berkner, Lloyd Viel, 301, 429
Berkshire Agricultural Society, 46
Berman, Harry M., 205
Bernstein, Felix, 400
Berry, Clifford E., 262
Bessey, Charles Edwin, 120, 142, 376
Bethe, Hans Albrecht, 242, 246, 251, 264, 278, 318, 429
Bible-Science Association, 314
Bigelow, Jacob, 48, 59
Billings, John Shaw, 118, 119

Binney, William G., 60
Biological Bulletin, 150
Biologics Corporation, 342
Biophysical Society, 299
Bird Lore, 150
Birge, Edward Asahel, 177, 376
Birge, Raymond Thayer, 410
Birkhoff, George David, 179, 183, 184, 261, 400
Bishop, John Michael, 345
Bishop, Katharine Scott, 206
Bjerknes, Jacob Aall Bonnevie, 420
Black, Harold Stephen, 221, 421
Blake, William P., 82
Blalock, Alfred, 266, 421
Bliss, Gilbert Ames, 213, 271, 400
Bliss, Nathaniel, 22
Bloch, Felix, 255, 272, 286, 293, 341, 429
Bloch, Konrad E., 313, 429
Bloembergen, Nicolaas, 339, 436
Blumberg, Baruch Samuel, 332, 440
Boas, Franz, 140, 149, 177, 180, 194, 222, 249, 380
Bocher, Maxime, 388
Bochner, Salomon, 421
Bohr, Niels, 255
Bohune, Lawrence, 4
Boltwood, Bertram Borden, 160, 168, 190, 221, 388
Bolza, Oskar, 380
Bonaparte, Charles Lucien, 57, 362
Bond, George Phillips, 78, 81, 93, 369
Bond, Thomas, 17
Bond, William Cranch, 67, 78, 81, 91, 360
Booth, James Curtis, 65
Bordley, John Beale, 29, 31
Boring, Edwin Garrigues, 411
Borlaug, Norman Ernest, 323, 430
Borman, Frank, 321
Boss, Benjamin, 246
Boss, Lewis, 166, 376
Boston Nutrition Laboratory, 168
Boston Society of Civil Engineers, 79
Boston Society of Natural History, 48, 60, 92
Botanical Gazette, 110
Botanical Society, 143, 166
Botanical Society of America, 160, 166

Boulder Dam, 245
Bowditch, Henry Pickering, 105, 107, 127, 371
Bowditch, Nathaniel, 40, 44, 45, 59, 357
Bowdoin, James, 28
Bowen, Harold G., 271
Bowen, Ira Sprague, 219, 421
Bowen, Norman Levi, 206, 223, 411
Bowen, Rufus, 444
Bowie, William, 220, 388
Bowman, Isaiah, 178, 190, 400
Boyer, Herbert W., 328
Boylston, Zabdiel, 12, 13
Brace, DeWitt Bristol, 162, 381
Bradford, Joshua Taylor, 367
Brashear, John A., 122, 371
Brattain, Walter Houser, 275, 296, 421
Brattle, Thomas, 8
Brauer, Richard Dagobert, 421
Bray, William C., 219
Breit, Gregory, 216
Brewster, William, 124
Bridges, Calvin B., 189, 411
Bridgman, Percy W., 221, 233, 270, 309, 400
Briggs, Lyman James, 165, 254, 389
Brillouin, Leon Nicolas, 411
Britton, Elizabeth Knight, 138, 149, 157
Britton, Nathaniel Lord, 138, 381
Brode, Wallace Reed, 421
Brodel, Max, 389
Bronk, Detlev Wulf, 282, 411
Brookhaven National Laboratory, 272, 330
Brookings Institution, 220
Brooklyn Bridge, 125
Brooks, Alfred Hulse, 136, 389
Brooks, William Keith, 151, 376
Brooks, William Robert, 371
Brouwer, Dirk, 284, 421
Brown, Amos Peaslee, 172
Brown, Ernest W., 198, 208, 389
Brown, George Harold, 246, 430
Brown, Herbert Charles, 335, 430
Brown, Michael Stuart, 328, 342, 443
Brown, Samuel, 53
Brown, William, 27
Brown-Sequard, Charles-Edouard, 367
Bruce, Archibald, 46

Bruner, Jerome Seymour, 436
Bryan, Kirk, 206, 411
Bucher, Walter Herman, 238, 411
Buckingham, Edgar, 389
Buddington, Arthur Francis, 411
Buffon, Georges-Louis Leclerc, Comte de, 30
Bull, William, 15
Bulletin of the Atomic Scientists, 268
Bumpus, Hermon Carey, 151, 381
Burbank, Luther, 218
Burbidge, E. Margaret, 298
Burbidge, Geoffrey R., 298
Burke, Bernard F., 293
Burnham, Sherburne Wesley, 151, 371
Burrill, Thomas J., 115
Burroughs Corporation, 299
Burton, William Merriam, 181, 389
Bush, Vannevar, 232, 253, 258, 261, 267, 331, 411
Bushnell, David, 26
Byrd, Richard Evelyn, 218, 223, 228, 300
Byrd, William, II, 9

Cabell, James L., 119
California Academy of Sciences, 86
California Institute of Technology, 202, 224, 226, 245
California State Earthquake Investigation Commission, 171
Calkins, Gary Nathan, 154, 389
Calvin, Melvin, 279, 308, 430
Cambridge Miscellany of Mathematics, Physics and Astronomy, 72
Cameron, Angus Ewan, 278, 430
Campbell, Douglas Houghton, 381
Campbell, Ian, 421
Campbell, William Wallace, 145, 176, 381
Cannon, Annie Jump, 135, 145, 198, 214, 261, 381
Cannon, Walter Bradford, 171, 228, 237, 389
Cape Canaveral rocket-testing facility, 280
Carleton College, 142
Carlson, Anton Julius, 400
Carlson, Chester, 252, 430
Carmichael, Leonard, 259

Index

Carmichael, Robert D., 179, 220
Carnap, Rudolf, 242, 283, 411
Carnegie Corporation, 179, 193, 197
Carnegie Foundation, 174
Carnegie Institution of Washington, 157, 159, 161, 163, 165, 166, 168, 176, 188, 189, 194, 206, 224, 233, 245, 253, 260
Carnegie Mellon University, 184
Carnes, Peter, 29
Carothers, Wallace H., 235, 249, 255, 421
Carrel, Alexis, 163, 181, 232, 389
Carroll, James, 152, 376
Carson, Rachel Louise, 310, 430
Carus, Paul, 133, 376
Carver, George Washington, 264
Castle, William Ernest, 172, 185, 187, 293, 389
Catesby, Mark, 11, 12, 14, 353
Catherwood, Frederick, 70
Cattell, James McKeen, 132, 143, 144, 153, 165, 186, 190, 205, 209, 381
Cech, Thomas Robert, 344
Centennial Exhibition (Philadelphia), 112
Central High School, Philadelphia, 66
Challenger (spacecraft), 341, 343
Chalmers, Lionel, 27
Chamberlain, Alexander F., 140
Chamberlain, Charles Joseph, 381
Chamberlain, Owen, 294, 304, 436
Chamberlin, Thomas Chrowder, 123, 142, 163, 225, 371
Champlain, Samuel de, 4
Chandler, Abiel, 84
Chandler, Charles F., 112
Chandler, Seth Carlo, 122, 138, 376
Chandrasekhar, Subrahmanyan, 241, 252, 341, 430
Chaney, Ralph Works, 194, 411
Chang, Min Chueh, 292
Chapman, Alvan Wentworth, 92, 364
Chapman, Frank Michler, 150, 178, 381
Chapman, Nathaniel, 54
Chapman, Sydney, 411
Charleston Botanic Garden and Society, 43
Chase, Agnes, 241
Chase, Martha Cowles, 286

Chemical Abstracts, 144
Chemical Foundation, 197
Chemical Society of Philadelphia, 35, 39
Chesapeake Zoological Laboratory, 117
Chicago Academy of Sciences, 89
Chicago Century of Progress International Exposition, 232
Chicago Museum of Science and Industry, 232
Child, Charles Manning, 180, 224, 389
Chittenden, Russell Henry, 111, 158, 230, 381
Chovet, Abraham, 26
Christofilos, Nicholas C., 436
Christy, James W., 335
Church, Benjamin, 27
Cincinnati Astronomical Society, 71
Cincinnati Observatory, 102
Clap, Thomas, 17
Clapp, Cornelia M., 132
Clark, Alvan, 85, 129, 362
Clark, Alvan G., 148, 369
Clark, George Bassett, 369
Clark, Kenneth Bancroft, 430
Clark, William, 43
Clark, William Mansfield, 202, 400
Clark University, 140, 203
Clarke, Frank Wigglesworth, 170, 376
Claude, Albert, 329
Clausen, Jens Christen, 284, 412
Clausen, Roy E., 196
Clayton, John (1657-1725), 9
Clayton, John (1694-1773), 16, 26
Clean Waters Restoration Act, 317
Cleaveland, Parker, 49, 53, 358
Cleland, Ralph Erskine, 326, 412
Clemence, Gerald Maurice, 284
Clements, Frederic Edward, 172, 191, 253, 389
Clinton, DeWitt, 49
Cobb, Doris M., 246
Coblentz, William Weber, 389
Cochran, William Gemmell, 283, 312, 430
Coffin, James H., 86
Coghill, George Ellett, 227, 390
Cohen, E. Richard, 299
Cohen, Morris Raphael, 400
Cohen, Stanley, 342
Cohen, Stanley Norman, 328

Cohn, Edwin Joseph, 412
Colbert, Edwin H., 315, 430
Cold Spring Harbor Biological Laboratory, 260, 268, 294
Cold Spring Harbor Laboratory, 136
Colden, Cadwallader, 13, 17, 18, 19, 23, 353
Colden, Jane, 23
Cole, Frank Nelson, 381
Cole, Leon Jacob, 155, 175, 400
College of William and Mary, 10, 21
Collinson, Peter, 14, 15, 18, 19, 23, 24
Colt, Samuel, 65
Columbia (spacecraft), 339
Columbia University, 39, 108, 134, 150, 205, 238-39, 254, 281, 316
  School of Mines, 97
Columbian Chemical Society, 35
Columbian Institute for the Promotion of Arts and Sciences, 49, 69
Committee of 100 on Research, 186
Commoner, Barry, 436
Compton, Arthur Holly, 210, 216, 219, 232, 244, 256, 262, 269, 311, 412
Compton, Karl T., 182, 239, 412
Comstock, Anna Botsford, 134, 145, 179
Comstock, George Cary, 381
Comstock, John Henry, 134, 145, 196
Conant, James B., 235, 254, 267, 282, 412
Condon, Edward U., 227, 244, 322, 421
Conklin, Edwin Grant, 381
Connecticut Academy of Arts and Sciences, 38
Connecticut agricultural experiment station, 110, 134
Conrad, Frank, 202
Conrad, Timothy A., 60, 63, 65, 362
Contemporary Psychology, 297
Cook County Hospital (Chicago), 248
Cooke, Josiah Parsons, Jr., 87, 97, 100, 369
Cooley, Denton Arthur, 322
Coolidge, Julian Lowell, 215, 390
Coolidge, William David, 176, 185, 390
Coon, Carleton Stevens, 421
Cooper, Curtice, 141
Cooper, Leon N., 299, 326, 440

Cooper, Thomas, 35, 46, 47, 49, 68, 356
Cope, Edward Drinker, 98, 101, 104, 110, 119, 124, 146, 148, 372
Copernicus, Nicolaus, 8
Corey, Elias James, 345
Corey, Robert B., 284
Cori, Carl F., 210, 273, 421
Cori, Gerty Radnitz, 210, 273, 422
Cormack, Allan MacLeod, 335
Cornell University, 128, 132, 142, 147
Cornuti, Jacques Philippe, 5
Cottrell, Frederick G., 182, 400
Coues, Elliott, 107, 124, 372
Coulter, John Merle, 110
Council for a Livable World, 310
Councilman, William Thomas, 376
Courant, Richard, 412
Cournand, Andre F., 295, 422
Cowan, Clyde Lorrain, 297
Cowles, Henry Chandler, 390
Cox, Gertrude M., 283
Coxe, John R., 47
Cram, Donald James, 343
Crampton, Henry Edward, 195, 400
Creation Research Society, 312
Crew, Henry, 224, 232, 382
Crile, George Washington, 164
Croghan, George, 23-24
Cronin, James Watson, 314, 337, 440
Cross, Charles Whitman, 159, 176, 377
Crowe, Kenneth M., 299
Crystallographic Society of America, 282
Curie, Marie, 205
Curtis, Heber Doust, 197, 204, 262, 390
Curtiss, Ralph Hamilton, 400
Cushing, Harvey Williams, 182, 236, 390
Cushman, Joseph Augustine, 187, 223, 400
Cutler, Manasseh, 30

Dakin, Henry Drysdale, 401
Dale, David Owen, 84
Dall, William Healey, 103, 137, 205, 377
Dalton, John Call, 82, 91, 111, 369
Daly, Reginald Aldworth, 390

Index

Dana, James Dwight, 66, 94, 95, 115, 364
Dana, James Freeman, 51
Dana, Samuel L., 51, 63
Danforth, Charles Haskell, 401
Danforth, Samuel, 11
Daniels, Farrington, 412
Dantzig, George Bernard, 430
Darlington, William, 57, 358
Dartmouth College, 84
Darwin, Charles, 89, 92, 111, 114, 121
Davenport, Charles Benedict, 149, 161, 176, 178, 390
Davis, Charles Henry, 80, 95
Davis, D. Dwight, 314, 431
Davis, Edwin H., 79
Davis, Hallowell, 251
Davis, Harold T., 220
Davis, John, 48
Davis, William Morris, 135, 142, 377
Davisson, Clinton Joseph, 220, 247, 401
Day, David Talbot, 206, 382
Dean, Bashford, 193, 390
Deane, Samuel, 34
Debye, Peter Joseph William, 401
Deep Sea Drilling Project, 321
DeForest, Lee, 169, 390
DeGolyer, Everette Lee, 196, 412
Dehmelt, Hans Georg, 345
Dehn, Max, 401
DeKruif, Paul, 217
Delafield, Francis, 127
Delbruck, Max, 264, 321, 339, 431
Demerec, Milislav, 257, 260, 422
Dennison, David Mathias, 217, 221, 422
Desor, Pierre Jean Edouard, 364
DeVries, William C., 340
Dexter, Aaron, 33
Dick, George, 209
Dick, Gladys, 209
Dickinson, Roscoe Gilkey, 202, 412
Dickson, Leonard Eugene, 146, 201, 390
Dietz, Robert S., 307
Dirac, Paul A.M., 325, 422
Dixon, Jeremiah, 22, 355
Doan, Richard L., 216
Dobzhansky, Theodosius, 215, 222, 247, 311, 332, 422
Dochez, Alphonse R., 209

Doering, William Von Eggers, 266
Doisy, Edward A., 213, 253, 263, 412
Donaldson, Henry Herbert, 139, 167, 382
Doolittle, James H., 228
Douglas, Jesse, 222, 422
Douglass, Andrew Ellicott, 390
Douglass, William, 13, 16
Dow, Herbert Henry, 148
Dow Chemical Company, 148, 266
Drake, Daniel, 52, 82
Drake, Edwin, 91
Drake, Frank Donald, 304, 440
Draper, Henry, 96, 106, 121, 372
Draper, John William, 68, 70, 74, 81, 83, 100, 102, 110, 112, 364
Drinker, Cecil, 184
Drinker, Philip, 222
Dryden, Hugh Latimer, 422
Duane, William, 194, 390
Dubins, Lester E., 316
Dubos, Rene Jules, 254, 422
DuBridge, Lee, 259
Dudley, Paul, 11, 12
Dudley Observatory, 85
Dugan, Raymond Smith, 219, 401
Duggar, Benjamin Minge, 172, 277, 391
Dulbecco, Renato, 312, 331, 431
DuMond, Jesse W.M., 299, 412
DuMont, Allen Balcom, 250
Dunbar, Carl Owen, 279, 413
Dunbar, William, 42
Duncan, John Charles, 204, 217, 401
Dunglison, Robley, 56, 62, 362
Dunham, Edward, 333
Dunn, Leslie Clarence, 215, 413
DuPont chemical company, 41, 160, 220, 232, 250, 255
Duportail, Louis Lebeque de Presle, 27
Durant, William Crapo, 170
Duryea, Charles E., 141
Duryea, J. Frank, 141
Dutton, Clarence Edward, 121, 122, 372
DuVigneaud, Vincent, 288, 293, 335, 422
Dykshorn, Simon W., 237

Early Bird (satellite), 316
East, Edward Murray, 174, 199, 401
Eastman, George, 118

Eastman Kodak Company, 118
Eaton, Amos, 50, 51, 55, 358
Echo (communication satellite), 307
Eckart, Carl Henry, 422
Eckert, John Presper, Jr., 270, 281, 285, 436
Eckert, Wallace John, 238, 259, 278, 284, 422
Ecological Society of America, 190, 272
Ecologists' Union, 272
Econometric Society, 231
Economic Geology, 133, 164, 201
Eddy, Mary Baker, 111
Edelman, Gerald Maurice, 326, 440
Edgerton, Harold Eugene, 233, 423
Edinger, Johanna Gabrielle Ottilie (Tilly), 278, 423
Edison, Thomas Alva, 112, 117, 118, 123, 125, 132, 190, 234, 377
Edlefsen, Niels, 228
Eggers, David F., Jr., 278
Eigenmann, Carl H., 195, 382
Einstein, Albert, 205, 238, 254, 295, 401
Eisenhart, Luther Pfahler, 401
Electron Microscope Society of America, 262
Elgin Botanic Garden, 39
Eli Lily and Company, 340
Elion, Gertrude Belle, 343
Eliot, Charles W., 145
Elkin, William Lewis, 382
Elliot, James L., 333
Elliott, Stephen, 48, 50, 53, 86
Elliott Society of Natural History, 86
Ellis, John, 20
Ellsworth, Henry L., 64
Elsasser, Walter, 240
Elvehjem, Conrad Arnold, 423
Emergency Committee in Aid of Displaced German Scholars, 238
Emerson, Alfred E., 280
Emerson, Benjamin Kendall, 372
Emerson, Robert, 423
Emmons, Ebenezer, 58, 65, 69, 70, 362
Emmons, Samuel Franklin, 372
Emporium of Arts and Sciences, 47
Enders, John Franklin, 187, 280, 290, 302, 423

Engelmann, George, 68, 90, 91, 364
Engineering Foundation, 187
Engineers Joint Council, 204, 336
Enterprise (spacecraft), 334
Entomological Society of America, 166, 289
Epstein, Paul Sophus, 248, 401
Erie Canal, 51
Erlanger, Joseph, 265, 316, 391
Esau, Katherine, 423
Espy, James Pollard, 70, 73, 360
ESSA (weather satellite), 317
Eugenics Records Office (Cold Spring Harbor, N.Y.), 176
Eugenics Research Association, 196
European Commission for Nuclear Research, 293
Evans, Alexander William, 391
Evans, Griffith Conrad, 196, 230, 413
Evans, Herbert McLean, 206
Evans, Lewis, 21, 354
Evans, Oliver, 31, 36, 53
Ewald, Paul Peter, 413
Ewing, James, 200, 391
Ewing, (William) Maurice, 281, 296, 316, 331, 431
Explorer (satellite), 303, 305
Eyring, Henry, 241, 423

Fahlberg, Constantin, 117
Fairchild, David Grandison, 250, 391
Fairchild, Herman LeRoy, 115
Fairchild Tropical Gardens, 250
Fajans, Kasimir, 413
Fankuchen, Isidor, 276, 282, 431
Faraday, Michael, 61
Farrar, John, 52, 358
Featherstonhaugh, George W., 61, 63, 358
Federal Insecticide Act, 175
Federated American Engineering Societies, 204
Federation of American Scientists, 268
Federation of American Societies for Experimental Biology, 185
Federation of Atomic Scientists, 268
Feller, William, 255, 284, 431
Fenn, Wallace Osgood, 246, 413
Fenneman, Nevin Melancthon, 191, 391
Fenner, Clarence Norman, 391

Index

Fermi, Enrico, 254, 263, 268, 269, 293, 313, 423
Fermi National Accelerator Laboratory, 320, 341
Fernald, Merritt Lyndon, 214, 282, 391
Fernow, Bernhard E., 129
Ferrel, William, 92, 367
Fessenden, Reginald Aubrey, 167, 391
Feynman, Richard Phillips, 315, 344, 437
Field, Cyrus W., 90
Fieser, Louis, 241, 253, 265, 319, 423
Fieser, Mary Peters, 265, 319
Fine, Henry Burchard, 382
Finlay, Carlos, 152
Firmin, Giles, 6
Fischer, Hermann Otto Lorenz, 413
Fiske, John, 109
Fitch, Asa, 87, 364
Fitch, John, 33
Fitch, Val Logsdon, 314, 337, 437
Fleischmann, Martin, 345
Fleming, Williamina P.S., 135, 382
Fletcher, Alice C., 177
Fletcher, Harvey, 240, 401
Flexner, Abraham, 174
Flexner, Simon, 382
Florida State University, 325
Flory, Paul J., 288, 329, 431
Foerste, August Frederick, 382
Folin, Otto K.O., 152, 199, 391
Food and Drug Act, 166
Food, Drug and Cosmetic Act, 252
Forbes, Alexander, 203, 207, 402
Forbes, Stephen Alfred, 121, 131, 372
Ford, Henry, 171, 184
Ford Foundation, 245
Forest Reserve Act, 139
Forrester, Jay Wright, 279
Foshag, William Frederick, 264, 413
Fothergill, John, 15, 20, 26
Fourcroy, Antoine F. de, 33
Fowler, William Alfred, 298, 341, 431
Francis, John W., 46
Franck, James, 269, 402
Frank, Philipp G., 402
Franklin, Benjamin, 13, 17, 18, 19, 20, 22, 24, 27, 29, 31, 35, 354
Franklin, Kenneth L., 293

Franklin Institute (Philadelphia), 55, 64, 111, 212
Freeman, Frank N., 249
Fremont, John, 72
Friedman, Jerome Isaac, 345
Frost, Edwin Brant, 144, 391
Fuchs, Klaus, 264
Fuller, R. Buckminster, 275, 341
Fulton, John Farquhar, 228, 423
Fulton, Robert, 44, 49
Funk, Casimir, 402

Gabb, William More, 372
Gaines, Walter Lee, 402
Gajdusek, Daniel Carleton, 332, 437
Galileo, 8
Gamow, George, 245, 258, 278, 295, 321, 423
Garcia, Celso-Ramon, 292
Garden, Alexander, 19
Gas Light Company of Baltimore, 50
Gasser, Herbert Spencer, 265, 413
Gay, Frederick Parker, 244, 391
Gell-Mann, Murray, 289, 308, 314, 321, 327, 440
Gemini (spacecraft), 316
Genentech, 332
General Education Board, 157, 211
General Electric Company, 117, 152, 176, 184, 235, 338
General Foods Company, 225
General Motors Company, 170, 179
General Motors Research Corporation, 206
Genetics, 192
Genth, Frederick Augustus, 78, 83, 88, 367
Geochemical Society, 294
Geological Society of America, 133, 233
Georgi, Howard M., 330
Georgia Institute of Technology, 127
Germer, Lester H., 220, 247
Gesell, Arnold Lucius, 212, 402
Giacconi, Riccardo, 309
Giaever, Ivar, 328, 440
Giauque, William Francis, 211, 279, 340, 423
Gibbon, John H., Jr., 289
Gibbs, George, 46, 52

Gibbs, Josiah Willard, 95, 113, 154, 158, 160, 372
Gibbs, Oliver Wolcott, 82, 88, 146, 171, 367
Gilbert, Grove Karl, 128, 372
Gilbert, Walter, 337, 440
Gill, Theodore Nicholas, 372
Gilliss, James M., 60, 81
Gilluly, James, 280, 424
Gilmor, Robert, Jr., 53
Gingrich, Newell Shiffer, 216
Glaser, Donald A., 287, 306, 440
Glashow, Sheldon Lee, 314, 319, 330, 336, 440
Glenn, John H., Jr., 311
Glomar Challenger, 321
Glover, Townend, 87
Goddard, Robert Hutchings, 201, 218, 269, 402
Godel, Kurt Friedrich, 257, 335, 431
Godfrey, Thomas, 14, 354
Godman, John D., 57
Goeddel, David V., 336
Gold, Thomas, 437
Goldberger, Joseph, 192, 391
Goldring, Winifred, 253, 413
Goldsborough, L.M. (Lieutenant), 60
Goldschmidt, Richard Benedict, 402
Goldstein, Joseph Leonard, 328, 342, 443
Gomberg, Moses, 152, 392
Good, Robert Alan, 437
Goodman, Nelson, 295
Goodpasture, Ernest William, 413
Goodspeed, Thomas Harper, 290, 413
Goodyear, Charles, 68
Gorgas, William, 161
Gorham, John, 47, 53
Goudsmit, Samuel Abraham, 266, 303, 424
Gould, Augustus Addison, 68, 70, 79, 364
Gould, Benjamin Apthorp, 78, 81, 82, 85, 93, 95, 104, 367
Gould, Stephen Jay, 443
Grabau, Amadeus William, 150, 392
Granger, Walter Willis, 392
Grant, Madison, 194
Gray, Asa, 64, 67, 68, 74, 78, 89, 92, 111, 114, 134, 282, 364

Gray, Francis Calley, 91
Green, Jacob, 62, 360
Greenwood, Isaac, 13, 14, 16, 354
Gregory, William King, 207, 285, 402
Griffin, Roger B., 128
Grinnell, Henry, 86
Grinnell, Joseph, 402
Griscom, John, 39
Gronovius, Johann Friedrich, 15, 16
Grote, Augustus Radcliffe, 123, 372
Groves, Leslie (General), 261
Grube, Emil H., 146
Guillemin, Roger C.L., 333, 437
Gulick, John Thomas, 134, 163, 369
Gursky, Herbert, 309
Gutenberg, Beno, 242, 259, 413
Guthrie, Edwin Ray, Jr., 414
Guthrie, Samuel, 61
Gutte, Bernd, 321
Guyer, Michael Frederic, 392
Guyot, Arnold Henri, 80, 93, 126, 364

Hadley, John, 14
Haeckel, Ernst, 98
Hague, Arnold, 372
Hakluyt, Richard, 3
Hale, George Ellery, 138, 139, 144, 148, 150, 161, 170, 193, 214, 224, 252, 392
Hale Observatories, 224
Hale Solar Laboratory, 286
Hall, Asaph, 114, 369
Hall, Charles Martin, 128, 382
Hall, Edwin Herbert, 118, 128, 382
Hall, Granville Stanley, 124, 125, 130, 141, 162, 207, 212, 377
Hall, James, Jr., 65, 78, 133, 365
Halsted, George Bruce, 377
Halsted, William Stewart, 377
Hamilton, Howard L., 171
Hansen, William Webster, 431
Hare, Robert, 39, 52, 53, 54, 58, 67, 90, 358
Hariot, Thomas, 3
Harkins, William Draper, 392
Harkness, William, 373
Harlan, Richard, 57, 362
Harper, Robert Almer, 382
Harper, Roland McMillan, 402

Index

Harriet Newell Lowell Society for Dental Research, 182
Harriman, Mary Averell (Mrs. E.H.), 176
Harriman Alaska Expedition, 150
Harris, Thaddeus W., 87
Harrison, Ross Granville, 162, 167, 392
Hart, Edward, 130
Hart, Edwin Bret, 392
Hartline, Haldan Keffer, 318, 341, 424
Harvard University, 8, 13, 14, 16, 18, 22, 28, 29, 31, 33, 40, 48, 52, 76, 84, 96, 128, 144, 154, 157, 183, 188, 189, 194, 211, 263, 266
   Lawrence Scientific School, 77, 83
   Medical School, 105, 321, 344
   Museum of Comparative Zoology, 91, 95
   Observatory, 67, 77, 78, 122, 135, 139, 145, 159, 169, 183, 198, 213
Harvey, Edmund Newton, 192, 287, 414
Harvey Cushing Society, 236
Hassall, Albert, 157
Hassler, Ferdinand R., 44, 62, 357
Hauptman, Herbert Aaron, 342, 437
Hawaii Volcano Observatory, 180
Hayden, Ferdinand V., 99, 119, 369
Hayford, John Fillmore, 173, 392
Hazeltine, (Louis) Alan, 201
Heaviside, Oliver, 158
Hedrick, Henry B., 208
Heezen, Bruce C., 296, 301, 325, 437
Heidelberger, Michael, 199
Hektoen, Ludvig, 162, 382
Hellinger, Ernest, 402
Helly, Eduard, 403
Hench, Philip Showalter, 281
Henderson, Lawrence J., 186, 195, 403
Henderson, Yandell, 392
Henry, Joseph, 58, 61, 66, 74, 75, 79, 80, 82, 85, 95, 117, 362
Henson, Matthew A., 174
Hentz, Nicholas M., 71
Herndon, William Lewis, 83
Heroult, Paul-Louis-Toussaint, 128
Herrick, Charles Judson, 297, 392
Herrick, Clarence L., 139, 383
Herschbach, Dudley Robert, 342

Hershey, Alfred Day, 286, 321, 431
Herter, Christian A., 164
Hess, Harry Hammond, 302, 304, 307, 323, 431
Hess, Victor Franz, 244, 403
Hewitt, Abram S., 115
Hewitt, Peter C., 155
High Voltage Engineering Corporation, 272
Hilgard, Eugene Woldemar, 165
Hill, George William, 116, 373
Hill, Julian W., 235
Hill, Lester Sanders, 414
Hiss, Philip H., Jr., 175
History of Science Society, 212
Hitchcock, Albert S., 230, 241, 392
Hitchcock, Edward, 63, 69, 83, 360
Hitchings, George Herbert, 343
Hoagland, Dennis Robert, 403
Hoagland, Hudson, 267
Hobday, John, 26
Hober, Rudolf Otto Anselm, 392
Hodge, Frederick Webb, 169
Hoffmann, Roald, 315, 338, 443
Hofstadter, Robert, 308, 437
Holbrook, John Edwards, 72, 360
Holden, Edward Singleton, 109, 135, 377
Hollerith, Herman, 137, 147, 180
Holley, Alexander, 111
Holley, Robert W., 316, 320, 437
Hollis, Thomas, 13
Hollister, Charles D., 325
Holmes, Oliver Wendell, 73
Holmes, William Henry, 198
Holzinger, Karl J., 249
Hooton, Earnest Albert, 414
Hoover, Herbert, 204, 214, 215
Hoover Dam, 245
Hopkins, Albert, 66
Hopkinson, Thomas, 17
Hopper, Grace Murray, 305
Horn, George Henry, 125, 373
Horner, William E., 59
Horsford, Eben Norton, 73, 367
Horton, Arthur W., Jr., 293
Hosack, David, 39, 46, 357
Hough, George Washington, 373
Houghton, Douglass, 365
Houghton, Roland, 15

Howard, Leland Ossian, 143, 383
Howe, Elias, 75
Howe, James Lewis, 383
Howell, William Henry, 383
Hrdlicka, Ales, 219, 224, 393
Hubbard, Gardiner G., 125, 132-33
Hubble, Edwin Powell, 210, 213, 225, 289, 414
Hubble Space Telescope, 346
Hubbs, Carl Leavitt, 330, 414
Hubbs, Laura C., 330
Hubel, David Hunter, 339, 440
Hudson, Claude Silbert, 403
Huggins, Charles Brenton, 317
Hull, Albert Wallace, 205, 403
Hull, Clark Leonard, 264, 403
Human Biology, 227
Human Genome Project, 344
Humane Society of Massachusetts, 31
Humphreys, Andrew A., 93
Hunt, Frederick Vinton, 292, 432
Hunt, Reid, 393
Hunt, T. Sterry, 76, 369
Huntington, Edward V., 193, 223, 393
Huntington, Ellsworth, 403
Huntington, George S., 155
Hussey, William Joseph, 383
Hutchison, Paul C., 290
Huxley, Thomas Henry, 111
Hyatt, Alpheus, 96, 98, 100, 104, 120, 142, 373
Hyatt, John Wesley, 101
Hyman, Libbie Henrietta, 258, 414

Ichbiah, Jean D., 336
Iddings, Joseph Paxson, 159, 383
Imperial University of Tokyo, 118
Index Medicus, 118
Indiana Geological Survey, 65
Indiana University, 274
Infrared Astronomy Satellite, 340
Institute for Advanced Study, 230, 238, 257, 259, 275
Institute for Defense Analyses, 302
Institute for Sex Research, 274
Institute of Aeronautical Sciences, 236
Institute of Aerospace Science, 312
Institute of Electrical and Electronics Engineers, 312
Institute of Environmental Sciences, 299

Institute of Mathematical Statistics, 244
Institute of Radio Engineers, 182, 312
International Astronomical Union, 234
International Atomic Energy Agency, 290, 300
International Business Machines Company, 147, 179, 239, 266, 278, 296, 314, 324, 340
International Conference on the Peaceful Uses of Atomic Energy, 295
International Congress of Arts and Sciences (St. Louis), 161
International Council of Scientific Unions, 200
International Critical Tables of Numerical Data, 207
International Education Board, 209, 211, 224
International Encyclopedia of Unified Science, 251
International Geodetic Association, 135
International Geophysical Year, 295, 301
International Journal of American Linguistics, 194
International Research Council, 200
International Solar Union, 183
Iowa State University, 102
Iron and Steel Magazine, 149
Irving, John Duer, 201
Isaacs, Charles Edward, 89, 365
Iselin, Columbus O'Donnell, 424
Ives, Herbert E., 220, 403
Ives, Joseph C., 93

Jackson, Charles Thomas, 65, 75, 76, 121, 365
Jackson, James, 47
Jackson Laboratory, 226
Jacobs, Walter Abraham, 199, 403
Jaggar, Thomas Augustus, Jr., 175, 180, 274, 393
James, Edwin, 53
James, Finley, 39
James, Thomas Potts, 125
James, William, 104, 137, 168, 177, 373
Jansky, Karl Guthe, 234, 432
Jefferson, Thomas, 30, 31, 34, 38, 40, 42, 44, 58, 355

Index

Jeffrey, Edward Charles, 214, 393
Jeffries, John, 30
Jeffries, Zay, 267, 414
Jenkins, Louise F., 229
Jennings, Herbert Spencer, 167, 393
Jensen, Arthur R., 322
Jepsen, Glenn Lowell, 273, 424
Jepson, Willis Linn, 393
Jesup North Pacific Expedition, 149
Jet Propulsion Laboratory, 266, 310
Jewett, Frank Baldwin, 216, 255, 403
Johannsen, Albert, 234, 393
Johanson, Donald Carl, 329, 443
Johns Hopkins University, 110, 111, 112, 116, 117, 118, 124, 156
 Medical School, 142, 194
Johnson, Douglas Wilson, 233, 403
Johnson, Robert U., 137
Johnson, Samuel W., 110, 111
Johnson, Virginia Eshelman, 317
Johnson, Willard Drake, 383
Jones, Donald F., 199
Jones, Harry Clary, 393
Jones, Hugh (?-1701/02), 9
Jones, Hugh (ca.1692-1760), 10
Jones, Walter Jennings, 186, 393
Jordan, David Starr, 113, 377
Jordan, Edwin Oakes, 393
Josselyn, John, 7
Journal of Analytical and Applied Chemistry, 130
Journal of Animal Behavior, 180
Journal of Applied Psychology, 194
Journal of Biological Chemistry, 164
Journal of Biophysical and Biochemical Cytology, 294
Journal of Cell Biology, 294
Journal of Chemical Physics, 239
Journal of Comparative Neurology, 139
Journal of Experimental Medicine, 147
Journal of Experimental Zoology, 162
Journal of General Physiology, 196
Journal of Geology, 142
Journal of Geophysical Research, 147
Journal of Heredity, 185
Journal of Infectious Diseases, 162
Journal of Morphology, 130
Journal of Pharmacology and Experimental Therapeutics, 174
Journal of Physical Chemistry, 147
Journal of Sedimentary Petrology, 233
Journal of Wildlife Management, 248
Joyner, Mary C., 259
Juday, Chancey, 177, 243, 393
Julian, Percy Lavon, 242, 291, 424
Junto (Philadelphia), 13
Just, Ernest Everett, 253

Kahlenberg, Louis Albrecht, 393
Kalm, Pehr, 18
Kane, Elisha Kent, 86
Karle, Jerome, 342, 437
Karman, Theodore von, 403
Karpinski, Louis Charles, 179, 257, 404
Katz, Samuel L., 302
Kaufmann, Berwind P., 257
Kay, George Frederick, 394
Kay, Marshall, 285, 315, 318, 424
Keeler, James Edward, 144, 151, 383
Keen, William W., Jr., 106
Keibel, Franz, 177
Keith, William, 11
Kellogg, Albert, 365
Kellogg, Vernon Lyman, 170, 394
Kelly, William, 85
Kelser, Raymond Alexander, 220, 414
Kemble, Edwin C., 218
Kemeny, John G., 315
Kendall, Edward Calvin, 186, 241, 281, 414
Kendall, Henry Way, 345
Kennelly, Arthur Edwin, 158, 383
Kentucky geological survey, 87
Kerst, Donald William, 255, 432
Kettering, Charles F., 180, 206, 404
Kharasch, Morris Selig, 424
Khorana, Har Gobind, 320, 324, 437
Kilborne, Fred Lucius, 141
Killian, James R., Jr., 301
King, Clarence R., 99, 119, 373
King, Helen Dean, 199, 394
Kingsley, James L., 44
Kinnersley, Ebenezer, 17, 21, 354
Kinsey, Alfred Charles, 274, 278, 289, 414
Kirkwood, Daniel, 80, 97, 365
Kirkwood, John Gamble, 432
Kitt Peak National Observatory, 302
Kleitman, Nathaniel, 289

Kline, Nathan Schellenberg, 437
Knight, Samuel Howell, 414
Knipling, Edward Fred, 432
Knopf, Adolph, 226, 404
Knudsen, Vern Oliver, 236, 298, 415
Kofoid, Charles Atwood, 394
Kohler, Wolfgang, 227, 242, 415
Kolff, Willem Johan, 340
Kornberg, Arthur, 296, 304, 308, 438
Kornei, Otto, 252
Kowal, Charles T., 331
Kraus, Charles August, 404
Krieg, David, 9
Kuhn, Adam, 22, 23
Kuhn, Thomas S., 310, 438
Kuiper, Gerard Peter, 276, 432
Kunitz, Moses, 415
Kurtz, Thomas E., 315
Kusch, Polykarp, 293, 432

Ladd-Franklin, Christine, 141
Ladenburg, Rudolf Walther, 404
Lamb, Willis Eugene, Jr., 275, 293, 432
Lancefield, Rebecca Craighill, 334, 424
Lancefield Society, 334
Land, Edwin Herbert, 248, 274
Landsat (satellite), 327
Landsteiner, Karl, 229, 246, 258, 394
Lane, Alfred Church, 211, 226, 383
Lane, Jonathan Homer, 101, 367
Langley, Samuel Pierpont, 120, 129, 136, 138, 147, 167, 369
Langmuir, Irving, 184, 199, 235, 300, 404
Laplace, Pierre-Simon, 59
Lark-Horovitz, Karl, 415
Larsen, Esper Signius, Jr., 205, 404
Lashley, Karl Spencer, 227, 415
Latimer, Wendell Mitchell, 249, 415
Laura Spelman Rockefeller Memorial, 196, 211
Lavoisier, Antoine-Laurent, 33, 35
Lawrence, Ernest O., 228, 236, 252, 260, 269, 303, 424
Lawrence Berkeley Laboratory, 292
Lawrence Radiation Laboratory, 294
Lawson, Andrew Cowper, 172, 383
Lawson, John, 10
Lazear, Jesse W., 152
Lazzaroni, 82, 85

Lea, Isaac, 65, 360
Leavitt, Henrietta Swan, 169, 206, 394
LeConte, John, 90, 367
LeConte, John L., 51, 125
LeConte, Joseph, 83, 115, 367
Leder, Philip, 344
Lederberg, Joshua, 274, 287, 301, 441
Lederman, Leon Max, 343
Lee, Tsung-Dao, 297, 298, 441
Lee, Yuan Tseh, 342
LeFevre, ?, 10
Lefschetz, Solomon, 230, 262, 404
Leidy, Joseph, 75, 77, 79, 98, 102, 118, 368
Leopold, Aldo, 239, 243, 415
Lesley, J. Peter, 88, 368
Lesquereux, Leo, 87, 120, 125, 365
Lesueur, Charles, 51, 56, 358
Levene, Phoebus Aaron, 172, 225, 394
Leverett, Frank, 383
Levi-Montalcini, Rita, 342
Levine, Philip, 258
Levinson, Norman, 432
Lewis, Gilbert Newton, 174, 191, 208, 237, 273, 404
Lewis, Meriwether, 43
Lewis, Sinclair, 214
Lewis and Clark expedition, 31, 42-43, 45, 48, 64
Libby, Willard Frank, 273, 306, 338, 432
Library Company of Philadelphia, 15, 24
Library of Congress, 226
Lick, James, 109
Lick Observatory, 109, 129, 140, 145, 151, 176, 186, 234
Liebig, Justus von, 69, 73
Lilienthal, David, 271
Lillie, Frank Rattray, 171, 195, 230, 394
Limnological Society of America, 243
Lindbergh, Charles A., 232
Lindgren, Waldemar, 383
Lining, John, 27
Linnaean Society of New England, 48
Linnaean Society of Philadelphia, 44
Linnaeus, Carolus, 17, 20, 21, 22
Liotta, Domingo, 322
Lipmann, Fritz A., 288

Index

Lipscomb, William Nunn, Jr., 332, 438
Literary and Philosophical Society of Charleston, S.C., 48
Literary and Philosophical Society of New York, 49
Little, Arthur D., 128
Little, Clarence Cook, 188, 226, 415
Livingston, Burton Edward, 404
Livingston, Milton Stanley, 228
Lloyd, Curtis G., 127
Lloyd, John Uri, 127, 377
Locke, John, 79
Loeb, Jacques, 151, 167, 183, 196, 212, 384
Loeb, Leo, 394
Loewi, Otto, 394
Loewner, Charles, 415
Logan, James, 10, 14, 15, 353
London, Fritz, 284, 424
Long, Crawford Williamson, 72
Long, Cyril Norman Hugh, 424
Long, Stephen H., 53
Long Island Biological Association, 260
Loomis, Elias, 73, 92, 365
Loomis, Mahlon, 98
Lotka, Alfred James, 216, 404
Lovejoy, Arthur Oncken, 245, 394
Lovell, James A., Jr., 321
Lovell, John Harvey, 384
Lowell, Francis Cabot, 48
Lowell, Percival, 143, 163, 165, 193, 229, 384
Lowell Institute (Boston), 67, 76
Luria, Salvador Edward, 264, 289, 321, 432
Lusk, Graham, 165, 394
Lyceum of Natural History of New York, 50
Lyman, Benjamin Smith, 98, 373
Lyman, Chester Smith, 98, 365
Lyman, Theodore, 188, 395

Maanen, Andriaan Van, 404
MacArthur, Robert Helmer, 318, 326, 441
Maclean, John, 36, 37, 357
MacLeod, Colin M., 265, 432
Maclure, William, 45, 47, 53, 56, 69, 356

Macmillan, William Duncan, 220, 222, 395
Magnus-Levy, Adolf, 395
Maiman, Theodore Harold, 307, 441
Mall, Franklin Paine, 155, 177, 384
Manhattan Project, 254, 261, 263, 264, 267, 269, 271
Mansfield, Jared, 39
Marcou, Jules, 85, 90, 368
Marine Biological Laboratory (Woods Hole, Mass.), 132, 150
Mariner (spacecraft), 310, 316, 325
Markoe, Francis, Jr., 74
Marsh, George Perkins, 96
Marsh, Othniel Charles, 98, 109, 114, 119, 121, 369
Marshall, Humphry, 29
Martin, Henry Newell, 122, 130, 377
Martin, Robert G., 310
Marvel, Carl Shipp, 344, 415
Mason, Charles, 22, 355
Mason, Max, 405
Massachusetts General Hospital, 76
Massachusetts geological and natural history survey, 65, 70
Massachusetts Institute of Technology, 93, 100, 109, 116, 180, 232, 259, 272
Massachusetts Society for Promoting Agriculture, 35
Massachusetts state board of health, 101
Mast, Samuel Ottmar, 178, 395
Masters, William Howell, 317
Mathematical Association of America, 143, 190, 209, 213
Mathematical Correspondent, 42
Mathematical Diary, 56
Mathematical Reviews, 255
Mather, Cotton, 10, 11, 12, 13, 14
Mather, Increase, 8
Mather, Kirtley Fletcher, 415
Mather, Stephen Tyng, 191
Mather, William Williams, 65, 66, 362
Mauchly, John William, 270, 281, 285, 338, 432
Maury, Antonia, 145, 395
Maury, Matthew Fontaine, 60, 71, 77, 81, 85, 88, 108, 365
Mayer, Alfred Marshall, 116, 373

Mayer, Maria Goeppert, 278, 295, 311, 327, 433
Mayr, Ernst, 273, 425
McAtee, Waldo Lee, 237, 248, 405
McCarthy, John, 296
McCarty, Maclyn, 265
McClintock, Barbara, 285, 341, 425
McClung, Clarence Erwin, 156, 395
McClurg, James, 26
McCollum, Elmer Verner, 183, 191, 206, 405
McCormick, Cyrus, 61
McDonald Observatory, 256
McDowell, Ephraim, 46
McGraw-Hill publishing, 227
McLean Hospital for the Insane, 152
McMahon, Bernard, 43
McMillan, Edwin Mattison, 256, 273, 284, 433
Mead, Margaret, 222, 229, 335, 425
Medical Repository, 37
Medical society (Boston), 16
Medical Society of New Haven County, 33
Meek, Fielding Bradford, 89, 368
Meggers, William Frederick, 415
Mehl, Robert Franklin, 425
Meinzer, Oscar Edward, 184, 208, 405
Mellon Institute, 184
Melsheimer, Frederick V., 44
Meltzer, Samuel James, 377
Mendel, Lafayette Benedict, 163, 183, 395
Menkin, Miriam F., 267
Mercury (spacecraft), 309, 311
Mergenthaler, Ottmar, 126
Merica, Paul Dyer, 416
Merriam, Clinton Hart, 128, 136, 150, 384
Merriam, John Campbell, 395
Merrifield, Robert Bruce, 304, 321, 341, 438
Merrill, Elmer Drew, 207, 298, 405
Merrill, George Perkins, 61, 138, 148, 378
Merton, Robert K., 250, 328
Meselson, Matthew, 298
Metallographist, 149
Metaphysical Club, 104
Meteorological Society of Mannheim, 29

Meyerhof, Otto, 405
Michael, Arthur, 117, 378
Michaelis, Leonor, 405
Michaux, Andre, 30, 41
Michaux, Francois A., 41, 72
Michel, Helen V., 337
Michelson, Albert Abraham, 116, 130, 160, 167, 202, 218, 221, 234, 378
Midgley, Thomas, Jr., 206, 231, 416
Miller, Dayton Clarence, 159, 171, 216, 261, 395
Miller, George Abram, 151, 384
Miller, Robert Rush, 330
Miller, Samuel, 41
Miller, Stanley Lloyd, 284, 441
Millikan, Robert A., 176, 183, 192, 193, 208, 216, 290, 395
Mills, Robert L., 291
Mineralogical Society of America, 201
Mink, Douglas J., 333
Minot, Charles Sedgwick, 125, 129, 141, 155, 378
Minot, George Richards, 218, 240, 416
Mirsky, Alfred Ezra, 425
Mises, Richard Von, 405
Missouri Botanical Garden, 91
Mitchel, Ormsby MacKnight, 71, 76
Mitchell, Elisha, 360
Mitchell, John, 17, 18, 21
Mitchell, Maria, 78, 79, 368
Mitchell, Silas Weir, 106, 369
Mitchill, Samuel Latham, 36, 37, 38, 50
Moessbauer, Rudolf Ludwig, 308
Molina, Mario J., 330
Monist, 133
Monsanto corporation, 155
Montgomery, Edmund Duncan, 160, 373
Montgomery, Thomas Harrison, Jr., 154, 395
Monthly American Journal of Geology and Natural Science, 61
Moore, Charlotte E., 222
Moore, Eliakim Hastings, 384
Moore, Joseph Haines, 235, 405
Moore, Raymond Cecil, 209, 233, 277, 416
Moore, Robert Lee, 235, 405
Moore, Stanford, 306, 326, 433

Index

Mordell, Louis Joel, 416
Morehouse, George R., 106
Morgan, Agnes Fay, 188
Morgan, Herbert Rollo, 286, 405
Morgan, John, 23, 27
Morgan, Lewis Henry, 114
Morgan, Thomas Hunt, 174, 178, 189, 199, 217, 221, 222, 235, 237, 270, 395
Morgan, W. Jason, 318
Morgenstern, Oskar, 266
Morley, Edward W., 130, 159, 210, 373
Morrill Act (federal land-grant college bill), 94
Morris, Henry M., 308
Morse, Edward Sylvester, 96, 100, 115, 373
Morse, Jedidiah, 29, 33, 356
Morse, Philip M., 227
Morse, Samuel F.B., 66, 71, 74
Morton, Samuel George, 64, 68, 362
Morton, Thomas, 5
Morton, William Thomas Green, 76
Moss, William Lorenzo, 405
Mottelson, Ben Roy, 331, 441
Moulton, Forest Ray, 164, 252, 395
Mount Wilson Observatory, 161, 170, 186, 188, 194, 210, 224
Moyes, Henry, 31
Muhlenberg, Gotthilf Henry Ernest, 47, 49
Muir, John, 137
Muller, Hermann J., 189, 219, 270, 319, 416
Mulliken, Robert Sanderson, 317, 425
Mumford, Lewis, 319
Munn, Orson, 75
Murphy, James Bumgardner, 405
Murphy, William P., 218, 240
Murray, Joseph E., 345
Myer, Albert J., 103

Nantucket Maria Mitchell Association, 157
Nathans, Daniel, 323, 334, 441
National Academy of Engineering, 314
National Academy of Sciences, 95, 102, 104, 105, 115, 119, 127, 136, 146, 171, 190, 193, 197, 200, 204, 213, 214, 239, 243, 255, 260, 261, 262, 282, 293, 302, 314, 329, 344
National Advisory Cancer Council, 247
National Association of Biology Teachers, 231
National Association of Inventors, 75
National Association of Science Writers, 240
National Audubon Society, 153
National Board of Health, 119
National Committee on Radiation Protection, 226
National Council of Teachers of Mathematics, 203
National Defense Education Act, 304
National Electric Signalling Company, 167
National Geographic Magazine, 133
National Geographic Society, 132
National Institute for the Promotion of Science, 69, 74
National Inventors Council, 257
National Radio Astronomy Observatory, 299, 305, 337
National Research Council, 193, 197, 211, 218, 272
National Research Fund, 214
National Resources Board, 243
National Resources Committee, 247
National Roster of Scientific and Specialized Personnel, 259
National Science Teachers Association, 144
National Wildlife Federation, 245
National Zoological Park, 137
Natural History, 153
Natural History Survey of New York, 65
Nature Conservancy, 272
Nautical Almanac, 80
Nautilus (nuclear submarine), 303
Naval Consulting Board, 190, 193
Nef, John Ulric, 384
New England Journal of Medicine and Surgery, 47
New Orleans Academy of Sciences, 86
New York Academy of Sciences, 50
New York Botanical Garden, 138
New York Mathematical Society, 134, 139, 143

New York University, 100
New York World's Fair, 256
Newberry, John Strong, 93, 368
Newcomb, Simon, 104, 123, 144, 165, 174, 373
Newman, Horatio Hackett, 195, 249, 406
Newton, Hubert Anson, 96, 100, 138, 370
Newton, Isaac, 8, 10, 12, 18
Newton, Isaac (1800-1867), 94
Neyman, Jerzy, 250, 416
Nicholas, John Spangler, 425
Nichols, Edward L., 142
Nichols, Ernest Fox, 143, 396
Nicholson, Seth Barnes, 186, 416
Nieuwland, Julius Arthur, 173, 232, 406
Nirenberg, Marshall Warren, 308, 320, 441
Nobel Prize:
    chemistry, 186, 232, 235, 240, 241, 270, 279, 284, 288, 290, 293, 306, 308, 315, 317, 320, 326, 329, 332, 335, 337, 338, 340, 341, 342, 343, 344, 345
    medicine or physiology, 173, 181, 229, 237, 240, 248, 263, 265, 270, 273, 280, 281, 286, 288, 290, 295, 301, 304, 308, 309, 313, 317, 318, 320, 321, 323, 325, 326, 329, 331, 332, 333, 334, 335, 337, 339, 341, 342, 343, 345
    peace, 309, 323
    physics, 167, 208, 219, 244, 247, 252, 254, 259, 263, 265, 270, 273, 275, 286, 293, 294, 296, 298, 300, 304, 306, 308, 311, 313, 314, 315, 318, 319, 320, 321, 326, 328, 330, 331, 332, 333, 334, 336, 337, 339, 340, 341, 343, 345
Noble, Gladwyn Kingsley, 233, 416
Noether, Amalie Emmy, 238
Noguchi, (Seisaku) Hideyo, 406
North Carolina geological survey, 55

North Pacific Exploring and Surveying Expedition, 86
Northrop, John Howard, 229, 270, 416
Norton, John Pitkin, 76, 81, 368
Nott, Josiah, 74
Novy, Frederick George, 132, 384
Noyes, Arthur Amos, 131, 140, 144, 157, 199, 219, 396
Noyes, William Albert, 144, 384
Nuttall, Thomas, 51, 62, 63, 72, 360

Ochoa, Severo, 304, 433
Office of Science and Technology, 311, 328
Office of Science and Technology Policy, 332
Ogburn, William F., 247
Ohio geological survey, 66
Ohio University, 339
Olmsted, Denison, 55
Onsager, Lars, 232, 320, 425
Operations Research Society of America, 287
Oppenheimer, J. Robert, 262, 264, 269, 271, 275, 279, 289, 313, 318, 425
Optical Society of America, 192, 233
Ord, George, 43, 45
Orton, James, 99, 112, 370
Osborn, Henry Fairfield, 194, 384
Osborne, Thomas Burr, 134, 163, 180, 183, 384
Osgood, William Fogg, 168, 384
Osler, William, 141, 168
Osterhout, Winthrop J.V., 196, 396
Ostwald, Wilhelm, 131
Ott, Isaac, 131, 378
Oviedo y Valdes, Gonzalo Ferdandez de, 3
Owen, David Dale, 65, 87, 365
Owen, Richard, 84
Owen, Robert Dale, 56

Packard, Alpheus Spring, Jr., 96, 100, 113, 133, 145, 374
Page, Charles G., 80
Painter, Theophilus Shickel, 225, 238, 416
Palade, George Emil, 329, 433
Paleontological Society, 171

Index

Palmer, Edward, 105, 370
Palomar Observatory, 224, 276, 286, 305
Paolini, Francis R., 309
Papanicolaou, George Nicholas, 223, 406
Park, Orlando, 280
Park, Thomas, 280
Park, William Hallock, 142, 152, 385
Parkhurst, John Adelbert, 181, 385
Patterson, John Thomas, 287, 396
Patterson, Robert, 45
Pauli, Wolfgang, 259
Pauling, Linus Carl, 253, 284, 290, 309, 324, 425
Payne, Fernandus, 406
Payne, William W., 139
Payne-Gaposchkin, Cecilia H., 213, 336
Peabody Academy of Science, 99
Peale, Charles Willson, 31, 34, 39, 43, 79, 355
Peale, Rembrandt, 39, 358
Peale, Rubens, 31
Peale, Titian R., 51, 53, 57, 362
Pearl, Raymond, 202, 217, 227, 406
Peary, Robert E., 174
Pease, Francis Gladhelm, 406
Pedersen, Charles John, 343
Pegram, George Braxton, 256, 406
Peirce, Benjamin, 72, 76, 82, 84, 85, 88, 95, 99, 103, 365
Peirce, Benjamin Osgood, 378
Peirce, Charles Sanders, 103, 104, 115, 374
Penington, John, 35
Pennsylvania Horticultural Society, 58
Pennsylvania Hospital (Philadelphia), 20
Penrose, Richard F., Jr., 233
Penzias, Arno, 313, 334, 441
Perrine, Charles Dillon, 162, 396
Peters, Christian Heinrich Friedrich, 366
Peters, John P., 234
Peterson, Roger Tory, 241, 433
Petiver, James, 9, 10, 11
Petrologists' Club, 176
Pharmacopoeial Convention, 54
Phelps, Almira Hart Lincoln, 59, 63
Philadelphia Humane Society, 28
Philadelphia Journal of the Medical and Physical Sciences, 54
Philadelphia Medical and Physical Journal, 43
Philadelphia Society for Promoting Agriculture, 31, 57
Phillips, John C., 172, 187
Philosophical Society (Boston), 8
Philosophy of Science Association, 237
Physical Review, 142, 185
Physical Review Letters, 303
Physiological Zoology, 224
Pickering, Edward Charles, 109, 125, 139, 159, 169, 201, 378
Pickering, William Hayward, 433
Pickering, William Henry, 139, 149, 385
Pierce, George Washington, 176, 201, 396
Pierce, John Robinson, 293
Pike, Nicholas, 33
Pikl, Josef, 242
Pincus, Gregory Goodwin, 246, 267, 292, 316, 319, 425
Pioneer (spacecraft), 304, 327
Pirsson, Louis V., 159, 189
Planned Parenthood Federation of America, 317
Poinsett, Joel R., 69, 74
Polaroid Corporation, 248, 274
Pons, B. Stanley, 345
Poor, Charles Lane, 396
Pope, Joseph, 33
Popular Astronomy, 142
Popular Science Monthly, 107
Porter, Rufus, 75
Post, Emil Leon, 425
Pound, Robert Vivian, 306
Pourtales, Louis Francois de, 79, 103, 105, 106, 368
Powell, John Wesley, 101, 103, 110, 114, 115, 119, 131, 158, 169, 209, 370
Power, Frederick Belding, 378
Pratt, Frederick Haven, 396
President's Science Advisory Committee, 301, 310, 311, 328
Press, Frank, 281
Priestley, Joseph, 33, 35, 36, 37, 42, 355
Prince, Thomas, 20

Princeton Mathematical Series, 256
Princeton University, 36, 108, 179, 182, 273
Proctor, Richard Anthony, 374
Project ALSOS, 266
Project Hindsight, 307
Project Mohole, 304
Prudden, Theophil Mitchell, 127, 378
Psychological Corporation, 205
Psychological Review, 143
Psychometric Society, 244
Psychometrika, 244
Pumpelly, Raphael, 108, 374
Pupin, Michael I., 146, 244, 385
Purcell, Edward Mills, 272, 286, 433
Pursh, Frederick, 39, 43, 48, 357
Putnam, Frederic Ward, 96, 100, 374

Quarterly Review of Biology, 217
Queeny, John Francisco, 155

Rabi, Isidor Isaac, 265, 425
Rademacher, Hans, 416
Radiation Laboratory, 258
Radiation Research Society, 287
Radio Corporation of America, 246
Rado, Tibor, 426
Rafinesque, Constantine S., 43, 54, 59, 358
Rainwater, (Leo) James, 331, 438
Raleigh, Walter, 3
Ramsdell, Lewis Stephen, 426
Ramsey, Norman Foster, Jr., 345
Rand Corporation, 277, 299
Ray, John, 8
Raymond, Percy Edward, 203, 255, 406
Reber, Grote, 246, 262, 433
Rebka, Glen A., 306
Redfield, William C., 61, 77, 360
Reed, Walter, 152, 378
Reichenbach, Hans, 297, 417
Reichert, Edward Tyson, 172
Reid, Harry Fielding, 178, 385
Reifenstein, Edward C., Jr., 277
Reines, Frederick, 297
Remington Rand, 285
Remsen, Ira, 112, 115, 117, 118, 156, 221, 378
Rensselaer Polytechnic Institute, 55, 64
Research Corporation, 182

Revelle, Roger, 433
Rhine, Joseph Banks, 240, 248, 275, 338, 426
Richards, Alfred Newton, 184, 406
Richards, Dickinson Woodruff, 295
Richards, Theodore William, 154, 186, 188, 225, 396
Richardson, Owen W., 182
Richter, Burton, 330, 332, 441
Richter, Charles Francis, 242, 259, 426
Richtmyer, Floyd Karker, 224, 406
Ricketts, Howard Taylor, 166, 192, 396
Rickover, Hyman G., 292
Riddle, Oscar, 231, 237, 406
Ridgway, Robert, 130, 156, 378
Riley, Charles V., 114
Ringgold, Cadwalader, 86
Ritchey, George Willis, 385
Ritchie, John, 122
Ritt, Joseph Fels, 417
Rittenhouse, David, 25, 32, 35, 36, 355
Ritter, William Emerson, 182, 385
Riverbank Laboratory, 200
Robbins, Frederick Chapman, 280, 290, 438
Robie, Thomas, 11
Robinson, Abraham, 438
Robinson, Denis M., 272
Rocha-Lima, Henrique da, 192
Rock, John, 267, 292
Rockefeller, John D., 103
Rockefeller Foundation, 184, 193, 197, 230, 232-33, 231, 238, 239, 250, 260
Rockefeller Institute for Medical Research, 155, 171, 213, 265
Rockefeller Sanitary Commission for the Eradication of Hookworm Disease, 173
Rockefeller University, 155
Rodgers, John, 86
Roetter, Paul, 90
Rogers, Henry Darwin, 71, 89, 90, 366
Rogers, William Barton, 71, 92, 93, 123, 362
Rolfe, John, 4
Romans, Bernard, 26
Romer, Alfred Sherwood, 297, 417
Rood, Ogden Nicholas, 370
Roosevelt, Theodore, 154, 159

Index

537

Rose, Wickliffe, 211
Ross, Frank E., 259, 396
Rossby, Carl-Gustaf A., 236, 256, 426
Rossi, Bruno B., 309
Rous, (Francis) Peyton, 173, 317, 406
Rowe, Allan Winter, 237, 407
Rowland, Frank Sherwood, 330
Rowland, Henry Augustus, 94, 110, 123, 124, 135, 150, 156, 378
Royal Astronomical Society, 93
Royal Society of Arts, 20, 21
Royal Society of London, 6, 7, 9, 10, 11, 20, 23, 25, 28, 36
Royal Swedish Academy of Sciences, 18
Ruark, Arthur E., 230
Rudenberg, Reinhold, 407
Ruedemann, Rudolf, 274, 385
Rush, Benjamin, 23, 25, 32, 35, 47, 48, 356
Russell, Henry Norris, 181, 183, 219, 223, 270, 300, 407
Russell, Richard Joel, 426
Rutgers University, 255
Rutherfurd, Lewis Morris, 94, 368

Sabin, Albert Bruce, 299, 306, 433
Sabin, Florence Rena, 155, 194, 213, 396
Sabine, Paul Earls, 200, 236, 407
Sabine, Wallace C.W., 144, 153, 200, 298, 396
Sagan, Carl Edward, 337, 441
Sager, Ruth, 288, 438
St. John, Charles Edward, 222, 385
Salisbury, Rollin Daniel, 385
Salk, Jonas Edward, 291, 299, 312, 433
Salk Institute for Biological Studies, 312
Salmon, Daniel E., 126
Sandage, Allan Rex, 305
Sanderson, Ezra Dwight, 175, 407
Sandia National Laboratories, 280
Sane, National Committee for a Sane Nuclear Policy, 299
Sanger, Margaret, 191, 317
Sargent, Charles Sprague, 120, 146
Sarton, George A.L., 189, 222, 298, 407
Saunders, Frederick Albert, 407
Sauveur, Albert, 145, 149, 181, 385
Savage, Leonard Jimmie, 292, 316, 438

Savage, Thomas S., 77
Save-the-Redwoods League, 194
Sax, Karl, 249, 294, 417
Saxton, Joseph, 74, 83, 362
Say, Thomas, 29, 43, 50, 51, 53, 56, 60, 360
Scatchard, George, 332, 417
Schaeberle, John Martin, 379
Schairer, John Frank, 426
Schally, Andrew Victor, 333, 441
Schawlow, Arthur Leonard, 294, 302, 339, 438
Scheffe, Henry, 305, 433
Schiff, Leonard Isaac, 280, 438
Schlesinger, Frank, 229, 397
Schmidt, Karl Patterson, 280-81, 417
Schmidt, Maarten, 311, 441
Schoenheimer, Rudolf, 244, 426
Schoolcraft, Henry R., 62, 84, 361
Schopf, Johann David, 32
Schott, Charles Anthony, 370
Schrader, Franz, 417, 265
Schrieffer, John R., 300, 326, 442
Schuchert, Charles, 148, 176, 189, 242, 279, 385
Schwartz, Melvin, 343
Schwinger, Julian Seymour, 315, 438
Science, 125, 144, 153
Science for the People, 322
Science Press, 209
Science Service, 204
Scientific American, 75
Scientific Monthly, 190
Scientific Research Society of America, 274
Scientists and Engineers for Johnson, 314
Scopes, John Thomas, 214, 319
Scott, William Berryman, 385
Scripps, Edward W., 204
Scripps Institution of Oceanography, 182
Scudder, Samuel H., 96, 117, 134, 374
Seaborg, Glenn Theodore, 257, 284, 288, 434
Seares, Frederick Hanley, 234, 259, 397
Seasat (satellite), 335
Sedgwick, William T., 128
See, Thomas J.J., 397
Segre, Emilio Gino, 294, 304, 434
Seismological Society of America, 166

Semple, Ellen Churchill, 178
SESPA (Scientists and Engineers for Social and Political Action 322
Setchell, William Albert, 202, 386
Shaler, Nathaniel Southgate, 138, 374
Shannon, Claude Elwood, 249, 279, 439
Shapley, Harlow, 181, 198, 204, 229, 283, 306, 328, 417
Shaw, Henry, 91
Shaw, Louis A., 222
Shelford, Victor Ernest, 185, 253, 312, 407
Shepard, Alan Bartlett, Jr., 309
Shepard, Charles U., 62
Shepard, Francis Parker, 277, 426
Sherard, William, 12
Sherrill, Miles, 157
Shewhart, Walter Andrew, 211, 253, 417
Shippen, William, Jr., 23
Shockley, William Bradford, 275, 294, 296, 434
Shortley, George H., 244
Shull, Aaron Franklin, 407
Sibley, Charles G., 342
Sidereal Messenger, 76
Sierra Club, 140
Sigma Xi, 128, 274
Silicon Valley, 294
Silliman, Benjamin, 40, 44, 46, 52, 60, 67, 69, 76, 97, 358
Silliman, Benjamin, Jr., 76, 88, 91, 368
Simpson, George Gaylord, 226, 273, 426
Sinnott, Edmund Ware, 215
Skinner, Burrhus Frederic (B.F.), 251, 276, 426
Skinner, John Stuart, 53
Skylab (spacecraft), 329
Slater, John Clarke, 426
Slater, Samuel, 34
Slichter, Louis Byrne, 427
Slipher, Earl C., 309, 407
Slipher, Vesto Melvin, 172, 181, 225, 323, 407
Sloan-Kettering Institute for Cancer Research, 268
Sloane, Hans, 12, 20
Slosson, Edwin E., 204

Small, William, 21
Smith, David Eugene, 179
Smith, Edgar Fahs, 187, 379
Smith, Erwin Frink, 165, 202, 379
Smith, Hamilton Othanel, 323, 334, 442
Smith, Homer William, 249, 285, 427
Smith, James Edward, 37
Smith, John (Captain), 4, 5
Smith, John Lawrence, 73
Smith, Philip Edward, 203, 231, 407
Smith, Samuel S., 32
Smith, Sidney I., 108, 374
Smith, Theobald, 141, 386
Smithson, James L.M., 60, 357
Smithsonian Institution, 60, 69, 75, 79, 80, 90, 94, 99, 103, 109, 112, 117, 157, 189
  Astrophysical Observatory, 136
  Bureau of American Ethnology, 119, 169
Smyth, Henry DeWolf, 269
Snell, George Davis, 337, 427
Social Science Research Council, 209, 243
Society for Experimental Biology and Medicine, 159
Society for Industrial and Applied Mathematics, 287
Society for Nuclear Medicine, 291
Society for Plant Morphology and Physiology, 148, 166
Society for Social Responsibility in Science, 280
Society for the Promotion of Agriculture, Arts, and Manufactures, 37
Society for the Psychological Study of Social Issues, 245
Society for the Study of Evolution, 272
Society for the Study of Speciation, 272
Society of American Bacteriologists, 150
Society of Economic Geologists, 203
Society of Experimental Psychologists, 162
Society of General Physiologists, 272
Society of Professional Zoologists, 268
Society of Rheology, 226, 233
Society of Systematic Zoology, 277
Society of Vertebrate Paleontology, 257

Index

Soil Conservation Society of America, 260
Solar Maximum Mission (satellite), 338
Sonneborn, Tracy Morton, 247, 434
Spencer, Archibald, 17
Sperry, Elmer Ambrose, 179, 386
Sperry, Roger Wolcott, 339, 434
Sperry Rand Corporation, 285
Spock, Benjamin, 271
Springfield Armory, 36
Sputnik, 300
Squier, Ephraim G., 79, 114
Stahl, Franklin W., 298
Stallo, Johann Bernhard, 123, 368
Standard Oil Company, 103, 181
Stanford Linear Accelerator, 310
Stanford Positron-Electron Accelerating Ring, 330
Stanford Research Institute, 272
Stanford University, 252, 272
Stanley, Wendell Meredith, 241, 270, 325, 427
Starkey, George, 6, 353
Stebbins, George Ledyard, 265, 282, 434
Steenbock, Harry, 215, 417
Stein, William Howard, 306, 326, 434
Steinberger, Jack, 343
Steinmetz, Charles Proteus, 148, 210, 397
Stejneger, Leonhard Hess, 195, 379
Stephens, John Lloyd, 70
Stern, Curt, 279, 427
Stern, Otto, 263, 323, 417
Stevens, John, 41
Stevens, Nettie Maria, 163, 386
Stevens, Stanley Smith, 251, 285, 300, 434
Stevens Institute of Technology, 105
Stewart, George Neil, 145, 386
Stewart, John Quincy, 219
Stewart, Timothy A., 344
Stiles, Charles Wardell, 157, 173, 397
Stimpson, William, 370
Stine, Charles M.A., 220, 407
Stock, Chester, 215, 417
Stommel, Henry Melson, 274, 439
Stone, Wilson S., 287
Strategic Defense Initiative, 341
Stratton, Samuel Wesley, 156, 209, 386

Streeter, George Linius, 197, 203, 397
Stromberg, Gustaf Benjamin, 408
Struve, Otto, 256, 305, 427
Sturtevant, Alfred Henry, 178, 189, 418
Sullivant, William S., 78, 125
Sullivant Moss Society, 149
Sumner, Francis Bertody, 397
Sumner, James Batcheller, 217, 270, 418
Surveyor (spacecraft), 317
Sutherland, Earl Wilbur, Jr., 325, 439
Sutton, Walter Stanborough, 158, 408
Swann, William Francis Gray, 212, 408
Swarthmore College, 212, 242
Swasey, Ambrose, 187
Sylvester, James Joseph, 70, 112, 116, 366
Synthetic Fuels Corporation, 338
System Development Corporation, 299
Szent-Gyorgyi, Albert von, 418
Szilard, Leo, 251, 254, 268, 305, 310, 315, 427

Takamine, Jokichi, 153, 207, 379
Tarski, Alfred, 427
Tatum, Edward Lawrie, 259, 274, 301, 434
Taube, Henry, 340, 439
Taussig, Helen, 266
Taylor, Charles Vincent, 252, 418
Taylor, Frank Bursley, 175, 386
Taylor, Frederick Winslow, 386
Taylor, Hugh Stott, 418
Taylor, Richard E., 345
Television Engineers Institute of America, 257
Teller, Edward, 245, 254, 283, 288, 434
Telstar (communication satellite), 311
Temin, Howard Martin, 324, 331, 442
Tennent, David Hilt, 397
Tennessee Valley Authority, 239
Terman, Frederick E., 263
Terman, Lewis Madison, 192, 216
Terrestrial Magnetism and Atmospheric Electricity, 147
Tesla, Nikola, 126, 132, 139, 386
Thacher, Catharine, 203
Thacher, Thomas, 7
Tharp, Marie, 296
Thaxter, Roland, 386

Thayer, William Sydney, 386
Theiler, Max, 248, 427
Thomas, Edward Donnall, 345
Thompson, Benjamin (Count Rumford), 36, 38, 49, 67, 356
Thompson, Elizabeth, 107, 125
Thomson, Elihu, 379
Thorndike, Edward Lee, 221
Three Mile Island nuclear power plant, 336
Thurston, Robert Henry, 121, 374
Ting, Samuel C.C., 330, 332, 443
TIROS (weather satellite), 306
Titchener, Edward Bradford, 156, 162, 177
Tolman, Edward C., 236
Tolman, Richard Chace, 174, 194, 251, 408
Tombaugh, Clyde William, 229, 434
Tonegawa, Susumu, 343
Torrey, John, 53, 55, 67, 72, 92, 363
Torrey Botanical Club, 100, 138
Townes, Charles Hard, 291, 294, 302, 313, 439
Transit of Venus Commission, 104
Traut, Herbert F., 223
Trelease, William, 386
Troost, Gerard, 47, 56, 57, 359
Trowbridge, John, 374
Trump, John G., 272
Trumpler, Robert Julius, 229, 418
Tucker, Albert William, 256, 258
Tuve, Merle A., 216, 264, 427
Twenhofel, William H., 217, 408
Tyler, Samuel, 74
Tyndall, John, 107

U.S. Advanced Research Projects Agency, 303
U.S. Advisory Committee on Uranium, 254
U.S. Air Force, 306, 322
U.S. and Mexican Boundary Survey, 90
U.S. Army Engineers, 62, 93, 95
U.S. Army Signal Service, 103, 109, 126
U.S. Atomic Energy Commission, 271, 272, 283, 285, 289, 292, 313, 320, 330, 333
U.S. Bureau of Animal Industry, 141

U.S. Bureau of Mines, 175, 215
U.S. Centers for Disease Control, 329, 339
U.S. Chemical Warfare Service, 198
U.S. Coast Survey, 44, 62, 72, 79, 84, 99, 106, 116, 126
U.S. Committee on Medical Research, 261
U.S. Congress. Office of Technology Assessment, 327
U.S. Department of Agriculture, 87, 94, 114, 126, 128, 131, 135, 143, 165, 166, 175, 243, 262, 342
    Bureau of Biological Survey, 146, 257
    Division of Forestry, 129
U.S. Department of Commerce, 215
U.S. Department of Energy, 333
U.S. Department of the Interior, 80
U.S. Entomological Commission, 114
U.S. Environmental Protection Agency, 324, 327
U.S. Environmental Science Services Administration, 136, 317
U.S. Exploring Expedition (Wilkes), 66, 96
U.S. Fish and Wildlife Service, 105, 146, 257
U.S. Fish Commission, 105
U.S. Fixed Nitrogen Research Laboratory, 199
U.S. Food and Drug Administration, 166, 340, 343
U.S. Forest Service, 164
U.S. Geographical and Geological Survey of the Rocky Mountain Region, 103
U.S. Geographical Surveys West of the 100th Meridian, 101
U.S. Geological and Geographical Survey of the Territories, 99
U.S. Geological Exploration of the Fortieth Parallel, 99
U.S. Geological Survey, 116, 119, 123-24, 126, 136, 170, 184, 191, 209
U.S. House Committee on Science and Astronautics (Science and Technology), 305

Index

U.S. Joint Committee on Atomic Energy (Congress), 271
U.S. Marine Hospital Service, 120, 129, 154, 158
U.S. Military Academy (West Point), 40, 50, 69
U.S. National Advisory Committee for Aeronautics, 189, 254, 302
U.S. National Aeronautics and Space Administration, 189, 254, 302, 305, 307, 311, 316, 327, 333, 335, 338
U.S. National Bureau of Standards, 155, 170, 209
U.S. National Cancer Institute, 247
U.S. National Defense Research Committee, 254, 258, 258-59, 261, 264, 267
U.S. National Institute of Standards and Technology, 156
U.S. National Institutes of Health, 230, 243, 247, 344
U.S. National Oceanic and Atmospheric Administration, 324
U.S. National Park Service, 191
U.S. National Science Foundation, 267, 282, 304
U.S. Nautical Almanac Office, 259
U.S. Naval Academy, 75
U.S. Naval Observatory, 60, 71, 77, 80, 81, 98, 123
U.S. Naval Research Laboratory, 170, 208
U.S. Navy, 95, 303
U.S. Nuclear Regulatory Commission, 330
U.S. Office of Naval Research, 271
U.S. Office of Scientific Research and Development, 261, 263
U.S. Patent Office, 64, 67, 90, 94, 215, 343, 344
U.S. Public Health Service, 38, 158, 230, 247, 268, 329
U.S. Science Advisory Board, 239
U.S. Surgeon General, 315
U.S. Weather Bureau, 136, 258
Ulam, Stanislaw, 283, 434
Ulrich, Edward Oscar, 386
Union of American Biological Societies, 205

Union of Concerned Scientists, 322
University of California, 129, 182, 188, 228, 236, 250, 257, 273, 292, 297
University of Chicago, 142, 146, 256, 263, 271
University of Illinois, 192
University of Michigan, 225
University of Pennsylvania, 22, 23, 25, 132, 270
University of Texas, 256
University of Utah, 345
University of Virginia, 56, 70
University of Wisconsin, 175, 324
Urey, Harold Clayton, 230, 235, 239, 240, 284, 339, 418

Van Allen, James Alfred, 303, 434
Van de Graaff, Robert Jemison, 227, 272, 318, 427
Van Hise, Charles Edward, 387
Van Slyke, Donald Dexter, 234, 408
Van Vleck, John H., 218, 236, 333, 338, 427
Vandiver, Harry Schultz, 408
Vanguard I (satellite), 303
Vanuxem, Lardner, 58, 65, 361
Varmus, Harold Elliot, 345
Vaughan, Henry Earle, 293
Vaughan, Victor Clarence, 132
Veblen, Oswald, 177, 207, 408
Vernon, William, 9
Verrill, Addison E., 96, 108, 374
Very Large Array radio telescope, 337
Vickery, Hubert Bradford, 418
Viking (spacecraft), 333
Vinograd, Jerome Ruben, 298, 312, 435
Virginia Society for Promoting Useful Knowledge, 26
Vishniac, Roman, 428
Vogt, Marguerite, 312
Volney, Constantin-F. Chasseboeuf, Count de, 41
Von Braun, Werner, 303, 334, 435
Von Karman, Theodore, 226
Von Neumann, John, 266, 267, 283, 287, 300, 428
Voyager (spacecraft), 334

Wagner, Thomas E., 339
Waksman, Selman Abraham, 221, 255, 266, 286, 329, 418
Walcott, Charles Doolittle, 149, 173, 182, 189, 379
Wald, Abraham, 274, 428
Wald, George, 318, 435
Walker, Sears C., 66
Wallace, Alfred Russel, 89
Walter, Thomas, 33
Wang, Hsien Chung, 439
Ware, Louise M., 222
Warner and Swasey Company, 129
Warren, John Collins, 47, 54, 76
Washburn, Edward Wight, 207, 408
Washington, Henry Stephens, 159, 161, 397
Wasserburg, Gerald Joseph, 442
Waterhouse, Benjamin, 31, 39, 47
Watson, Elkanah, 46
Watson, James Dewey, 288, 309, 320, 344, 442
Watson, John Broadus, 185, 201, 224, 304, 408
Watson, Sereno, 113, 370
Watson, Thomas A., 190
Watson, Thomas J., 179, 239
Weaver, Warren, 231, 250, 260, 279
Webster, Arthur Gordon, 150, 162, 387
Webster, David Locke, 197, 418
Webster, John White, 69, 81
Wedderburn, Joseph H.M., 408
Weil, Roger, 312
Weinberg, Steven, 319, 327, 336, 442
Weisskopf, Victor Frederick, 435
Welch, W.H., 165
Welch, William Henry, 117, 193, 379
Weller, Thomas Huckle, 280, 290, 439
Welles, (George) Orson, 249
Wellesley College, 116
Wells, David A., 82
Wells, Harry Gideon, 169, 180, 408
Wells, Horace, 73
Wells, William Charles, 51, 356
Wesleyan University, 110
Western Academy of Natural Sciences (Cincinnati), 64
Western Electric Company, 211, 218
Western Museum (Cincinnati), 52

Western Quarterly Reporter of Medical, Surgical and Natural Science, 55
Westinghouse, George, 102, 132
Westinghouse Electric Company, 132, 202
Wetherill, Charles M., 94
Wheeler, George M., 101, 119
Wheeler, Helen-Mar, 290
Wheeler, John Archibald, 255, 435
Wheeler, William Morton, 177, 397
Whipple, Fred Lawrence, 283
Whipple, George Hoyt, 213, 240
Whitcomb, John C., Jr., 308
White, Andrew Dickson, 147
White, (Charles) David, 189, 387
White, Israel Charles, 127, 379
White, John, 3
Whitehead, Alfred North, 211, 227, 387
Whitfield, Robert Parr, 122, 370
Whiting, Sarah Frances, 116
Whitman, Charles Otis, 118, 130, 132, 374
Whitney, Eli, 35, 56
Whitney, Josiah Dwight, 87, 368
Whitney, Willis R., 193
Wiener, Alexander S., 258
Wiener, Norbert, 276, 315, 418
Wiesel, Torsten N., 339
Wigner, Eugene Paul, 254, 311, 428
Wild Flower Preservation Society of America, 157
Wilderness Society, 243
Wiley, Harvey Washington, 166, 375
Wilhelm, Richard Herman, 435
Wilkes, Charles, 60, 66
Wilks, Samuel Stanley, 435
Willard, Joseph, 28
Williams, Hannah, 10
Williams, Henry Shaler, 128, 379
Williams, Robert Runnels, 187, 245, 419
Williams, Samuel, 28, 29
Williams College, 66, 99
Willis, Bailey, 141, 387
Williston, Samuel Wendell, 114, 133, 180, 379
Wilson, Alexander, 43, 45, 357
Wilson, Edmund Beecher, 128, 147, 163, 256, 387

Index

Wilson, Edward O., 318, 325, 331, 442
Wilson, Edwin Bidwell, 154, 182, 243, 409
Wilson, Harold Albert, 397
Wilson, Henry (Senator), 95
Wilson, Kenneth Geddes, 340, 443
Wilson, Robert Woodrow, 313, 334, 443
Wilson, Thomas B., 75
Winchell, Alexander, 133, 368
Winchell, Alexander Newton, 173, 397
Winchell, Horace Vaughn, 397
Winchell, Newton Horace, 133, 173, 375
Winlock, Joseph, 370
Winstein, Saul, 435
Winthrop, John (1714-1779), 16, 18, 20, 21, 22, 23, 24, 25, 28, 354
Winthrop, John, Jr. (1606-1676), 5, 6, 7, 8, 353
Wintner, Aurel, 260, 428
Wisconsin Alumni Research Foundation, 215
Wistar, Caspar, 38, 47, 356
Wistar Institute of Anatomy and Biology, 140, 167
Wolcott, Alexander Simon, 68
Wolfowitz, Jacob, 435
Wolfrom, Melville Lawrence, 428
Wood, Alphonso, 74
Wood, Horatio Charles, 375
Wood, Robert Williams, 164, 398
Wood, William, 5
Woodhouse, James, 35, 37
Woods Hole Oceanographic Institution, 230
Woodward, Robert B., 266, 276, 284, 296, 306, 315, 325, 336, 439
Woodward, Robert Simpson, 379
Worcester Foundation for Experimental Biology, 267
Wright, Frederick Eugene, 179, 409
Wright, George Frederick, 111, 140, 375

Wright, Orville, 160, 398
Wright, Sewall, 321, 344, 419
Wright, Wilbur, 160, 398
Wright, William Hammond, 398
Wu, Chien-Shiung, 297, 435
Wu, Hsien, 199, 419
Wyman, Jeffries, 77, 366

Xerox Corporation, 252

Yale University, 40, 46, 53, 59, 95, 98, 215, 228, 230
   Sheffield Scientific School, 76, 93, 111
Yalow, Rosalyn Sussman, 333, 439
Yang, Chen Ning, 291, 297, 298, 439
Yanofsky, Charles, 313
Yeager, Charles, 276
Yellow Fever Board, 152
Yellowstone National Park, 106
Yerkes, Ada Watterson, 228
Yerkes, Robert M., 139, 180, 195, 204, 228, 297, 409
Yerkes Observatory (University of Chicago), 140, 148, 150, 181
Yerkes Regional Primate Research Center, 228
Yosemite National Park, 137
Youden, William John, 428
Youmans, Edward Livingston, 107
Young, Charles Augustus, 102, 111, 131, 156, 219, 370
Young, John Richardson, 42, 359
Young, John Wesley, 177, 209, 409
Young, William, 23

Zachariasen, (Fredrik) William Houlder, 268, 435
Zinn, Walter, 251-52, 263
Zinsser, Hans, 175, 187, 243, 409
Zweig, George, 314
Zwicky, Fritz, 428
Zworykin, Vladimir, 208, 225